CW01163104

SMART INNOVATIONS AND TECHNOLOGICAL ADVANCEMENTS IN CIVIL AND MECHANICAL ENGINEERING

SMART INNOVATIONS AND TECHNOLOGICAL ADVANCEMENTS IN CIVIL AND MECHANICAL ENGINEERING

Edited by
Satish Chinchanikar, PhD
Ashok Mache, PhD
Shardul G. Joshi, PhD
Preeti Kulkarni, PhD

AAP | APPLE ACADEMIC PRESS

First edition published 2025

Apple Academic Press Inc.
1265 Goldenrod Circle, NE,
Palm Bay, FL 32905 USA

760 Laurentian Drive, Unit 19,
Burlington, ON L7N 0A4, CANADA

CRC Press
2385 NW Executive Center Drive,
Suite 320, Boca Raton FL 33431

4 Park Square, Milton Park,
Abingdon, Oxon, OX14 4RN UK

© 2025 by Apple Academic Press, Inc.

Apple Academic Press exclusively co-publishes with CRC Press, an imprint of Taylor & Francis Group, LLC

Reasonable efforts have been made to publish reliable data and information, but the authors, editors, and publisher cannot assume responsibility for the validity of all materials or the consequences of their use. The authors are solely responsible for all the chapter content, figures, tables, data etc. provided by them. The authors, editors, and publishers have attempted to trace the copyright holders of all material reproduced in this publication and apologize to copyright holders if permission to publish in this form has not been obtained. If any copyright material has not been acknowledged, please write and let us know so we may rectify in any future reprint.

Except as permitted under U.S. Copyright Law, no part of this book may be reprinted, reproduced, transmitted, or utilized in any form by any electronic, mechanical, or other means, now known or hereafter invented, including photocopying, microfilming, and recording, or in any information storage or retrieval system, without written permission from the publishers.

For permission to photocopy or use material electronically from this work, access www.copyright.com or contact the Copyright Clearance Center, Inc. (CCC), 222 Rosewood Drive, Danvers, MA 01923, 978-750-8400. For works that are not available on CCC please contact mpkbookspermissions@tandf.co.uk

Trademark notice: Product or corporate names may be trademarks or registered trademarks and are used only for identification and explanation without intent to infringe.

Library and Archives Canada Cataloguing in Publication

CIP data on file with Canada Library and Archives

Library of Congress Cataloging-in-Publication Data

CIP data on file with US Library of Congress

ISBN: 978-1-77491-412-0 (hbk)
ISBN: 978-1-77491-413-7 (pbk)
ISBN: 978-1-00353-958-2 (ebk)

About the Editors

Satish Chinchanikar, PhD
Professor, Vishwakarma Institute of Information Technology, India

Satish Chinchanikar, PhD, is currently working as a Professor at the Vishwakarma Institute of Information Technology, India. His main research interest is in advanced manufacturing processes and machining of hard alloys using coated tools. He has over 25 years of teaching and industry experience and has published over 100 papers in international journals and conferences. He has authored a book chapter on finish machining of hardened steels published by Elsevier and a textbook on advanced manufacturing processes. He has been awarded an excellent paper certificate at the International Conference on Key Engineering Materials in Malaysia. He is working as a reviewer for many peer-reviewed international journals. His work has received over 2000+ citations, and he has an h-index 21 for his work. Dr. Chinchanikar has published five patents and holds three copyrights to date. He strongly believes that teaching and research should go hand in hand. He received his PhD from the Indian Institute of Technology Kanpur and received a master's from Pune University, India.

Ashok Mache, PhD
Associate Professor, Vishwakarma Institute of Information Technology, India

Ashok Mache, PhD, is currently working as Associate Professor at Vishwakarma Institute of Information Technology, India. His main research interest is in composite materials, advanced finite element modelling, and vehicle dynamics. He has over 18 years of teaching and industry experience and has published more than 20 papers in international journals and conferences. He has authored two book chapters in *Bio-composites for Automotive Applications*. He is working as a reviewer for peer-reviewed international journals. He has received 400+ citations and has an h-index 11 for his work and holds one copyright. He received his PhD from the Indian Institute of Science, Bangalore, India.

Shardul G. Joshi, PhD

Professor, Department of Civil Engineering at Vishwakarma Institute of Information Technology, Pune, India

Shardul G. Joshi, PhD, is working as Professor in the Department of Civil Engineering at the Vishwakarma Institute of Information Technology, Pune, India. He has over 20 years of teaching experience. He has published eight research papers in conferences and journals and holds one patent. He has completed his doctoral studies at Thapar University Patiala, Punjab, India.

Preeti Kulkarni, PhD

Associate Professor at Vishwakarma Institute of Information Technology, India

Preeti Kulkarni, PhD, is currently working as Associate Professor at the Vishwakarma Institute of Information Technology, India. Her main research interests are construction management, concrete technology-recycled materials, and soft computing applications in civil engineering. She has more than 18 years of teaching and industry experience and has published over 20 papers in international journals and over 25 in international conferences. She is working as a reviewer for peer-reviewed international journals. She has also authored one book, titled *Infrastructure Engineering and Construction Economics*. She received her PhD in Concrete Technology from Shivaji University, India.

Contents

Contributors .... *ix*

Preface .... *xiii*

PART I: Materials and Manufacturing Engineering 1

1. **A Nanoclay-Reinforced Jute-Polyester Nanocomposite for Enhanced Mechanical Performance** 3
   Anindya Deb, Ashok Mache, and Gude Venkatesh

2. **Effect of Epoxidized Soyabean Oil on Mechanical and Structural Properties of Sepiolite-Filled Polypropylene/Polyolefin Elastomer Composites** 25
   M. B. Kulkarni, R. Kumbhakarna, D. S. Bhutada, Bhushan Hazare, S. Thorat, S. Radhakrishnan, and Y. S. Munde

3. **Process Parameter Optimization by Ant Lion Algorithm of Austenitic Stainless Steel (SS 304) for Cutting Force in Turning using PVD-Coated Tools Deposited with TiAlN/TiSiN Coating Materials** 39
   Christoph Schiffers, Manish Adwani, Omkar Kulkarni, Atul Kulkarni, and Ganesh Kakandikar

4. **Electrical Discharge Machining Process Optimization During Machining of EN19 Alloy Steel Using Desirability Concept** 57
   Vijay Kumar S. Jatti, Vinay Kumar S. Jatti, Savita V. Jatti, Pawandeep Dhall, and Satish Chinchanikar

5. **Some Studies on Hole Feature Recommendation for Additive Manufacturing Processes** 77
   Dama Y. B., Bhagwan F. Jogi, and R. S. Pawade

PART II: Design Engineering, Automation, and Electric Vehicle Technology 93

6. **Comparative Evaluation of Modal and Harmonic Analysis of Stepped Horn for Ultrasonic Vibration Assisted Turning** 95
   Govind S. Ghule, Sudarshan Sanap, Satish Chinchanikar, and Avez Shaikh

7. **Design and Analysis of Wing for Unmanned Aerial Vehicle** 115
   Pranit Dhole, Ratnakar Ghorpade, and Chetan Patil

8. Estimation of Modal Loss Factor of Viscoelastic Material Using Modal Strain Energy Method 133
Gorakh Pawar, Pravin Hujare, Anil Sahasrabudhe, and Deepak Hujare

9. Crop Yield Prediction and Leaf Disease Detection Using Machine Learning 147
Sachin S. Sawant, Dyuti Bobby, Atharva Dusane, and Gaurav Durge

PART III: Structural Engineering 165

10. Buckling Analysis of Thin-Walled Cylinder under Axial Compression and Internal Hydrostatic Pressure Using Finite Element Model 167
Saurabh Patil, Ashok Mache, and Shardul Joshi

11. Seismic Behavior of Buckling-Restrained Brace-Installed Steel Buildings 181
Prajakta Shete, Shweta Sajjanshetty, and Suhasini Madhekar

12. Non-Linear Dynamic Analysis of Multi-Story Reinforced Cement Concrete (RCC) Building Having Different Geometry 195
Mahesh Pathare and Ramchandra Apte

13. Application of an Average Response Spectrum for Analysis of Structures 209
Praveen Ashok Patil, Shradul Joshi, and Rahul Joshi

14. Fluid Dynamics Analysis of Liquid Sloshing in a Rectangular Container under Lateral Excitation 219
Saurabh Patil and Shardul Joshi

15. Study of Mechanical and Micro Structural Properties of Fly Ash and GGBS-Based Geopolymer Concrete 227
Shivdatta B. Bhosale, S. G. Joshi, R. A. Joshi and J. P. Watve

PART IV: Environmental and Water Resources Engineering 241

16. Application of Solar Dryer for Drying of Agricultural Products 243
Tanvi Shah and Krishnakedar Gumaste

17. Flow Estimation and Flood Forecasting over Narmada River Using Three Data-Driven Techniques 259
Rohit Gaikwad, Ramchandra Kavanekar, Pradnya Dixit, and Preeti Kulkarni

18. Correlating Stage Measurement Stations Using Three Data-Driven Techniques: A Comparative Assessment 273
Aalisha Lanjewar, Vishakha Kondhalkar, Pradnya Dixit, and Preeti Kulkarni

Index *291*

Contributors

Manish Adwani
India and South East Asia, CemeCon AG, India

Ramchandra Apte
Civil Engineering Department, Vishwakarma Institute of Information Technology, Slavitribai Phule Pune University, Pune, Maharashtra, India

Shivdatta B. Bhosale
Department of Civil Engineering, Vishwakarma Institute of Information Technology, Pune, Maharashtra, India

D. S. Bhutada
School of Petroleum, Polymer and Chemical Engineering, Dr. Vishwanath Karad MIT World Peace University, Pune, Maharashtra, India

Dyuti Bobby
Department of Engineering, Sciences and Humanities (DESH), Vishwakarma Institute of Technology, Pune, Maharashtra, India

Satish Chinchanikar
Department of Mechanical Engineering, Vishwakarma Institute of Information Technology, Pune, Maharashtra, India

Y. B. Dama
DBATU, HCL, USA

Anindya Deb
Center for Product Design and Manufacturing, Indian Institute of Science, Bangalore, Karnataka, India

Pawandeep Dhall
DY Patil college of Engineering, Akurdi, Pune, Maharashtra, India

Pranit Dhole
School of Mechanical Engineering, MIT World Peace University, Pune, Maharashtra, India

Pradnya Dixit
Department of Civil Engineering, Vishwakarma Institute of Information Technology, Pune, Maharashtra, India

Gaurav Durge
Department of Engineering, Sciences and Humanities (DESH), Vishwakarma Institute of Technology, Pune, Maharashtra, India

Atharva Dusane
Department of Engineering, Sciences and Humanities (DESH), Vishwakarma Institute of Technology, Pune, Maharashtra, India

Rohit Gaikwad
MTECH-Water Resources and Environmental Engineering, Department of Civil Engineering, Vishwakarma Institute of Information Technology, Pune, Maharashtra, India

Ratnakar Ghorpade
Faculty, School of Mechanical Engineering, MIT World Peace University, Pune, Maharashtra, India

Govind S. Ghule
Reseach Scholar, Department of Mechanical Engineering, MIT-SOE, MIT-ADTU, Pune, Maharashtra, India

Krishnakedar Gumaste
Faculty of Civil-Environmental Engineering, Walchand College of Engineering, Sangli, Maharashtra, India

Bhushan Hazare
School of Petroleum, Polymer and Chemical Engineering, Dr. Vishwanath Karad MIT World Peace University, Pune, Maharashtra, India

Deepak Hujare
MIT World Peace University, Pune, Maharashtra, India

Pravin Hujare
Vishwakarma Institute of Information Technology, India

Savita V Jatti
DY Patil college of Engineering, Akurdi, Pune, Maharashtra, India

Vijaykumar S Jatti
DY Patil college of Engineering, Akurdi, Pune, Maharashtra, India
Symbiosis Institute of Technology, Pune

Bhagwan F. Jogi
Department of Mechanical Engineering, Dr. Babasaheb Ambedkar Technological University, Lonere-Raigad, Maharashtra, India

Rahul Joshi
Department of Civil Engineering, Vishwakarma Institute of Information Technology, India

Shardul Joshi
Department of Civil Engineering, Vishwakarma Institute of Information Technology, Pune, Maharashtra, India

Ganesh Kakandikar
Department of Mechanical Engineering, MIT World Peace University, Pune, Maharashtra, India

Ramchandra Kavanekar
MTECH-Water Resources and Environmental Engineering, Department of Civil Engineering, Vishwakarma Institute of Information Technology, Pune, Maharashtra, India

Vishakha Kondhalkar
MTECH-Water Resources and Environmental Engineering, Department of Civil Engineering, Vishwakarma Institute of Information Technology, Pune, Maharashtra, India

Atul Kulkarni
Department of Mechanical Engineering, Vishwakarma Institute of Information Technology, Pune, Maharashtra, India

M. B. Kulkarni
School of Petroleum, Polymer and Chemical Engineering, Dr. Vishwanath Karad MIT World Peace University, Pune, Maharashtra, India

Omkar Kulkarni
Department of Mechanical Engineering, MIT World Peace University, Pune, Maharashtra, India

Contributors

Preeti Kulkarni
Department of Civil Engineering, Vishwakarma Institute of Information Technology, Pune, Maharashtra, India

R. Kumbhakarna
School of Petroleum, Polymer and Chemical Engineering, Dr. Vishwanath Karad MIT World Peace University, Pune, Maharashtra, India

Aalisha Lanjewar
MTECH-Water Resources and Environmental Engineering, Department of Civil Engineering, Vishwakarma Institute of Information Technology, Pune, Maharashtra, India

Ashok Mache
Department of Mechanical Engineering, Vishwakarma Institute of Information Technology, Pune, Maharashtra, India

Suhasini Madhekar
Department of Civil Engineering, College of Engineering Pune, Pune, Maharashtra, India

Y. S. Munde
MKSSS's Cummins College of Engineering for Women, Karvenagar, Pune, Maharashtra, India

Mahesh Pathare
Civil Engineering Department, Vishwakarma Institute of Information Technology, Savitribai Phule Pune University, Pune, Maharashtra, India

Chetan Patil
Faculty, School of Mechanical Engineering, MIT World Peace University, Pune, Maharashtra, India

Praveen Ashok Patil
Department of Civil Engineering, Vishwakarma Institute of Information Technology, Pune, Maharashtra, India

Saurabh Patil
Mechanical Engineering, Vishwakarma Institute of Information Technology, Pune, Maharashtra, India

R. S. Pawade
Department of Mechanical Engineering, Dr. Babasaheb Ambedkar Technological University, Lonere-Raigad, Maharashtra, India

Gorakh Pawar
University of Utah, Salt Lake City, UT, USA

S. Radhakrishnan
School of Petroleum, Polymer and Chemical Engineering, Dr. Vishwanath Karad MIT World Peace University, Pune, Maharashtra, India

Anil Sahasrabudhe
All India Council for Technical Education (AICTE), New Delhi, India

Shweta Sajjanshetty
M. Tech (Structural Engineering), College of Engineering Pune, Pune, Maharashtra, India

Sudarshan Sanap
Mechanical Engineering, MIT-SOE, MIT-ADTU, Pune, Maharashtra, India

Sachin S. Sawant
Department of Engineering, Sciences and Humanities (DESH, Vishwakarma Institute of Technology, Pune, Maharashtra, India

Christoph Schiffers
Product Manager, CemeCon AG, Germany

Tanvi Shah
Department of Civil-Environmental Engineering, Walchand College of Engineering, Sangli, Maharashtra, India

Avez Shaikh
Department of Mechanical Engineering, VIIT, Pune, Maharashtra, India

Prajakta Shete
Research Scholar, Department of Civil Engineering, College of Engineering, Pune, Maharashtra, India

S. Thorat
School of Petroleum, Polymer and Chemical Engineering, Dr. Vishwanath Karad MIT World Peace University, Pune, Maharashtra, India

Gude Venkatesh
Visvesvaraya Technological University, Bangalore, Karnataka, India

J. P. Watve
Department of Mechanical Engineering, Vishwakarma Institute of Information Technology, Pune, Maharashtra, India

Preface

Innovation in the mechanical and civil engineering industries emerges continuously to increase efficiency, quality, and sustainability of a product and building construction. Artificial intelligence, cloud-based collaboration, nanomaterials, additive manufacturing, design engineering, and automation all contribute to improving the efficiency of the mechanical and civil engineering industries.

This book, *Smart Innovation and Technological Advancements in Mechanical and Civil Engineering,* encompasses recent topics of knowledge, intelligence, and innovation in materials and manufacturing engineering, design engineering, automation and electric vehicle technology, structural engineering, and environmental and water resources engineering. This book offers multidisciplinary research on the above-mentioned topics in a readily accessible form. This book covers the latest research in core engineering that employs knowledge and artificial intelligence to provide solutions to real-life problems in industry, the environment, and the community. The topics have been discussed in light of technological advancements for economic development and infrastructural stability by the adequate and pragmatic deployment of science and technology in preparation for the future.

This new volume, *Smart Innovations and Technological Advancements in Civil and Mechanical Engineering,* offers knowledge, innovation, and sustainability topics. It discusses converging technologies and innovations that are the primary drivers of Industry 4.0. The emphasis is on advanced manufacturing, composite materials, viscoelastic materials, solar power satellite, solar dryers, structural engineering, water resource engineering, multiscale problems, and simulation methods. The book effectively offers many aspects of multidisciplinary research in the civil and mechanical engineering fields. The authors have included recent research and technological advancements in materials, machining, analysis of engineering structures, advanced optimization techniques, viscoelastic materials, and machine learning applications to civil and mechanical engineering fields.

The book is divided into three sections.

The first section on advanced materials and machining processes examines studies on advanced composite materials, ultrasonic vibration-assisted turning, additive manufacturing, and process optimization using advanced techniques.

The authors address smart innovations and technological advancements in advanced materials and machining processes. The chapters investigate process optimization under the given cutting conditions and with work material properties to attain the desired characteristics within the required ranges, which ultimately helps to provide quality products to end-customers. The optimization techniques and key results are discussed in detail.

The first chapter examines the effect of nanoclay addition on the mechanical properties of a jute–polyester composite. The addition of nanoclay increases the strength and modulus of the composite laminate as compared with its virgin counterpart. This research provides valuable information where significant increase in stiffness is possible with the addition of a small amount of nanoclay, which can be useful for applications where stiffness is important, such as windmill blades.

The second chapter investigates the effects of different sepiolite contents on various properties of PP/POE/SEP/ESBO composites. Mechanical and thermal properties, as well as morphological characteristics, melting and crystallization properties, are all investigated. The interaction of PP and SEP, as well as sepiolite distribution, were studied. According to the mechanical results, 10% sepiolite loading to the PP/POE/ESBO composite balances stiffness and toughness.

The third chapter discusses process parameter optimization of austenitic stainless steel (SS 304) for cutting force in turning using PVD-coated tools deposited with TiAlN/TiSiN coating materials. An ant lion optimizer is used, which can predict a much better result. It has been seen that feed has a significant influence on cutting force compared with the depth of cut and cutting speed.

The fourth chapter discusses the design aspects as well as the modal and harmonic analysis of a stepped horn, which is used to transfer high-frequency vibrations from the source to cutting tools made of various materials. The titanium alloy stepped horn has the lowest deformation under the specified conditions, followed by aluminum alloy, mild steel, and stainless steel stepped horns.

The fifth chapter focuses on determining the best input process parameters to maximize material removal rate while minimizing tool wear rate and surface roughness. This study provides shopfloor operators with guidelines for setting input parameters based on the required response values.

The second section discusses modeling and analysis of engineering structures that include studies on the analysis of wing for unmanned aerial vehicle (UAV), viscoelastic materials, and crop yield prediction and leaf disease detection using machine learning. In today's world, additive manufacturing (AM) is quickly becoming the dominant manufacturing technology. SLS, FDM, DMLS, SLA, PolyJet, LTP, FDM, LENS, binder jet printing, and other manufacturing processes are developed using additive manufacturing techniques.

The sixth chapter presents ideas based on a 3D-printing study of hole features for various additive manufacturing processes. The research presented in this chapter is extremely useful for researchers, academicians, and those working in the 3D printing-based manufacturing domain.

The seventh chapter describes the process of designing and analyzing a wing for an unmanned aerial vehicle. Recent advances in computer simulations, composite materials, and fabrication technology have enabled the design of unmanned aerial vehicles with improved performance and structural stability while minimizing weight. The work in this chapter focuses on a comparative study of various performance factors, such as lift force, drag force, and lift to drag ratio for NACA 4412 and Clark Y airfoils.

The eighth chapter presents a study on estimation of modal loss factor of viscoelastic material using modal strain energy method. The governing equations representing viscoelastic damping analysis are solved using finite element method based on modal frequency response analysis solver.

The ninth chapter discusses the use of machine learning for crop yield prediction and leaf disease detection. In the growing season, a farmer always must deal with two prominent issues: the first one is to make the correct choice of crops to grow, and the second being the detection of infestations. Indeed, crop yield is drastically affected by the above-mentioned factors. The work presented in this chapter provides solution to address this issue using machine learning techniques.

The third section discusses smart innovations and machine learning applications in civil and mechanical engineering *using advanced simulation tools*. The authors address smart innovations in the fields of solar power satellites and solar dryers and machine learning applications to predict crop yield, detect leaf disease, and estimate water flow and flood forecasting.

The tenth chapter presents a finite element analysis of the buckling of thin-walled cylinders under axial compression and internal hydrostatic pressure. The numerical results of the study demonstrated that the internal hydrostatic pressure had a significant strengthening effect on the cylindrical shell.

Structures are vulnerable to large inelastic displacements or collapse during seismic events and thus require special attention to limit deformations and forces. A fundamental requirement is the design of earthquake-resistant structures with adequate ductility and stiffness. The eleventh chapter in this section demonstrates the seismic behavior of steel moment resisting frame (SMRF) buildings with and without buckling restrained braces (BRBs). When compared with other types of dampers, the use of BRBs for seismic response control proves to be very beneficial because BRBs are a cost-effective option that can be manufactured locally.

Chapter 12 presents nonlinear dynamic analysis on a multi-story reinforced cement concrete (RCC) building with variable geometry. ETABS, a finite element-based software, is used to do the analysis. The authors' investigations claimed that square RCC geometry is preferable for dynamic analysis than rectangular and circular geometry.

The response spectrum method is discussed in Chapter 13 for seismic analysis. The average response spectra for the northeast India region was obtained from the recent history data of seven sites in Assam. Seismo-Signal software is used to generate response spectra. A 10-story RCC building is modeled in SAP 2000, and a comparative study is conducted for this structure, taking into account the average response spectra and response spectra defined in IS 1893 2016.

Sloshing is a common phenomenon observed in liquid containers with a free surface under external excitations. The efforts to describe sloshing behavior using the analytical method are limited due to its highly nonlinear nature. In the past studies, numerical methods have proved to provide accurate solutions to sloshing problems. The study in Chapter 14 provides fundamental information for computational fluid dynamics analysis of sloshing phenomena using ready to use CFD codes.

Chapter 15 discusses the investigation of the mechanical and microstructural properties of alkali activated fly ash/GGBS (ground granulated blast furnace slag) concrete with varying proportions of fly ash to GGBS, as well as the effect of alkaline activators on compressive strength.

Part IV mainly covers the research work in the field of environmental and water resources engineering. The construction of a solar dryer with natural circulation is presented in Chapter 16. Experiments were conducted on turmeric, onions, and grapes to reduce their moisture content. Traditionally, open sun drying has been used to dry perishable crops because it is the most cost-effective and environment-friendly method. However, there are numerous drawbacks to this method, including the possibility of contaminating agricultural products due to the presence of insects, birds, and animals, as well as color loss and nutrient loss due to uncontrolled exposure to sunlight. The solar dryer proposed here is found to be the most efficient technique.

Designing a system that can be used for early flood detection is important for controlling and minimizing flood-related losses. The study presented in Chapter 17 attempts to predict daily, 1-day, and 2-day flow values at the Mandleshwar station on the Narmada River in India. For discharge prediction, three data-driven techniques are used: model tree (MT), artificial neural networks (ANN), and support vector regression (SVR).

Chapter 18 discusses three data-driven techniques, namely, model tree (MT), artificial neural networks (ANN), and a support vector regression (SVR) which are employed for daily water level correlation estimation. Modeling the stage in river flow is more important in controlling floods, planning sustainable development, managing water resources and economic development, and sustaining the ecosystem. Based on the findings, it is possible to conclude that data-driven techniques such as MT and SVR can be used for forecasting in various areas with various model combinations.

This volume will keep readers abreast of smart innovations and technological advancements in the competitive area of civil and mechanical engineering fields.

PART I
Materials and Manufacturing Engineering

CHAPTER 1

A Nanoclay-Reinforced Jute-Polyester Nanocomposite for Enhanced Mechanical Performance

ANINDYA DEB[1], ASHOK MACHE[2], and GUDE VENKATESH[3]

[1]Department of Design and Manufacturing, Indian Institute of Science, Bangalore, India

[2]Department of Mechanical Engineering, Vishwakarma Institute of Information Technology, Pune, India

[3]Visvesvaraya Technological University, Bangalore, India

ABSTRACT

Natural fiber-reinforced composites are gaining popularity due to various advantages over commonly used composites such as glass fiber-reinforced composites in terms of environmental friendliness and not posing health hazards. Additionally, such composites can be economic with competitive mechanical properties when compared to synthetic fiber-reinforced composites. Among the many natural fiber-based materials, jute appears to be particularly attractive as it is easily available commercially in the form of woven mats which are convenient for composite manufacturing. The present study is aimed at gaining quantitative information on use of polyester in jute-polyester composites with various proportions of nanoclay (in the range of 3–9% by weight); in particular, attention is paid to the mechanical properties such as elastic modulus and failure strength, which are determined experimentally. Specimens of nanoclay (montmorillonite)-reinforced jute–polyester composite for testing are obtained through the hand-layup technique supplemented with compression molding. To ensure homogeneous dispersion of nanoclay clusters in polymer matrix, sonication was performed on liquid polyester with a given proportion of modified Indian nanoclay using high-frequency ultrasonic waves. Mechanical properties were determined for jute–polyester composite laminates with various proportions of nanoclay and compared with the baseline case in which no nanoclay was added. The study shows that the addition of nanoclay to polyester results in substantial increase in tensile modulus and strength of jute–polyester composites. Further, improved mechanical properties were obtained by incorporating steel wire mesh (SWM) as reinforcement along with jute fiber in nanoclay-loaded (6 wt%) polyester resin to get jute–polyester–steel wire mesh-nanoclay hybrid composite.

1.1 INTRODUCTION

Natural fibers have been used as reinforcement in various materials for over 3000 years. Several types of natural fibers have been investigated for their use in composites such as kenaf, flax, jute, hemp, wood fiber, jowar, sisal, bamboo, vakka, banana, etc. [1-3] Among the many natural fiber-based materials, jute appears to be particularly attractive as it is readily available commercially in the form of woven mats, which are convenient for composite manufacturing. Jute fiber-based composites have been investigated for their mechanical properties such as tensile strength, flexural, compression, and shear strength under static loading, damping properties and also for their energy absorption under impact loading.[4-11] Ray et al.[12,13] reported alkali-treated jute fiber reinforcement with matrix as vinyl ester. In these studies, the authors compared the mechanical, dynamic, thermal, and impact fatigue behavior compared to that of untreated jute fiber-reinforced vinyl ester composites. They claimed that alkali treatment resulted in superior dynamic, mechanical, thermal, and impact properties. Mohanty et al.[14] studied the effects of surface modification on the mechanical properties and biodegradability of jute-Biopol and jute-Poly Amide composites. Enhancements in tensile, bending, and impact strength were observed to be more than 50, 30, and 90% respectively when compared to pure Biopol sheets. For the past few years, numerous studies have been performed on jute fiber-reinforced polyester composites. Fraga et al.[15] studied the relationship between moisture absorption and dielectric behavior, Sabeel et al.[16] reported the elastic properties, notched strength, and fracture criteria, Santulli[17] performed impact damage characterization using acoustic emission, Dash et al.[18] studied thermal and weathering behavior, and Sever et al.[19] examined the effect of silane treatment.

Nanoparticle additives, like nanoclay, are being used as reinforcement for composites manufacturing to improve their properties. Well-dispersed nanoparticles in the matrix give better mechanical and thermo-mechanical properties for the nanocomposites. Using atomic microscope and micromechanical simulation studies,[20–22] Venkatesh et al.[23, 24] showed that well-dispersed montmorillonite/polypropylene nanocomposites gave better mechanical and thermo-mechanical properties than the composites with nanoclay agglomeration. Mechanical, morphological, and thermal properties of nanoclay/polymer composites have been studied by using mechanical stirring and sonication processes. Subramaniyan and Sun[25] claimed that sonication process was observed to be an effective mixing method to separate nanoclay stacks. However, homogeneity was not achieved as their TEM images indicated mixing of resin in the gallery spaces of nanoclay and some areas of exfoliated nanoclay with random orientation and they attributed it to inadequate mixing procedures. Bensadoun et al.[26] reported a study on bio-based composite materials reinforced with nanoclay particles. A soy-based unsaturated polyester resin was used as a matrix, and glass and flax fibers were used as reinforcement. They found that the addition of 3% of nanoclays by weight improved the flammability by up to 30% as compared to conventional composites, further the combination of nanoclay and bioresin doubled this value. Behra et al.[27] reported the fabrication of jute fiber reinforced biodegradable nano-biocomposite using soymilk as matrix and Cloisite 30B nanoclay (0, 1, 3, 5, 7, and 10 wt%) as additive and characterized for mechanical and dynamic mechanical properties, and moisture sensitivity. Their experimental result showed that, with 5% clay loaded (by weight) composite, the tensile strength, flexural strength, and storage modulus were 1.55, 1.51, and 2.62 times greater than that of jute–soy composites without nanoclay respectively. Even though there is an increase in tensile strength and modulus, maximum tensile strength and modulus are limited to 51.5 MPa and 1358 MPa, respectively. Dewan et al.[28] studied thermo-mechanical and flexural properties of a nanocomposite based on montmorillonite K10 nanoclay dispersed into B-440 premium polyester resin to fabricate jute fabric reinforced composite. They found 50% increase in interlaminar shear strength, 40% increase in flexural strength, and 34% increase in compressive strength with the addition of 1 wt% of nanoclay in polyester resin.

A through study was carried out to investigate the effect of nanoclay on mechanical properties of jute fiber reinforced polyester composites fabricated by simple hand-layup technique supplemented with compression molding. To ensure homogeneous dispersion of nanoclay clusters in polymer matrix, sonication was performed on liquid polyester with a given proportion of modified Indian nanoclay using high-frequency ultrasonic waves. To the best knowledge of the authors, alkali modified Crysnano1010, montmorillonite nanoclay has been used for the first time in the current study as reinforcement in a jute–polyester composite resulting in significant increases in tensile modulus and strength (more than 100% and 75% respectively) compared to the baseline jute–polyester composite without nanoclay.

Further, improved mechanical properties were obtained by incorporating steel wire mesh (SWM) as reinforcement along with jute fiber in nanoclay-loaded (6 wt%) polyester resin to get jute–polyester–steel wire mesh-nanoclay hybrid composite (6J4SPN6). This hybrid composite showed significant increase in tensile and flexural properties compared to the baseline jute–polyester–steel wire mesh hybrid composite and pure jute–polyester composite without nanoclay for the same volume fraction of polyester.

1.2 EXPERIMENTAL METHODS AND MATERIALS

1.2.1 MATERIALS

Commercially available bidirectional woven jute mat having a count of 17 by 15 (15 yarns in weft and 17 yarns in warp direction per square inch) was used as a reinforcement for jute–polyester composite. The woven jute mat with 220 GSM (grams per square meter) was procured from Champdany Industries Ltd., Kolkata, West Bengal, India through the supplier Jute Cottage, Bangalore. Isothalic polyester resin, cobalt naphthenate accelerator, and MEKP catalyst were procured from Swathi Chemicals, Bangalore. Table 1.1 shows the detailed specifications of jute fabric and steel wire mesh. Steel wire mesh and woven jute mats were used in the preparation of plain jute-based composites and hybrid composites. Alkali modified Crysnano1010, a montmorillonite nanoclay, was procured from Crystal Nanoclay Pvt. Ltd., Pune, India.

TABLE 1.1 Specifications of Jute Fabric and Steel Wire Mesh.

Material	Number of fibers/wires per square inch		Diameter of fiber/wire (mm)	GSM (gram per square meter)
	warp	Weft		
Jute	17	15	0.4	220
Steel wire mesh	15	15	0.2	292

1.2.2 FABRICATION OF LAMINATES

The jute–polyester composite laminates were initially fabricated by hand layup technique and then subjected to sustained pressure in a compression molding machine at room temperature for two hours. Mylar sheets were used at the bottom and top of a laminate at the time of compression resulting in a glassy finish of the external faces. The resin system is prepared from commercially available general purpose polyester resin, catalyst, and accelerator in the weight ratio of 1:0.020:0.015, respectively. Each layer of jute fabric is pre-impregnated with resin and placed one over the other, taking care to maintain practically achievable tolerances on fabric alignment. Jute–polyester composite laminate consists of 10-ply of jute fabric.

For the fabrication of nanoclay (montmorillonite)-reinforced jute–polyester composites, Crysnano1010, montmorillonite nanoclay is used as reinforcement in the range of 3–9% by weight. Dispersion of nanoclay particles into the resin is the key to obtain the desired properties of the composites. In this work, proper dispersion of the nanoparticles in the unsaturated polyester resin is carried out by hand mixing followed by sonication. A high-frequency ultrasonication bath was used for the dispersion of nanoparticles at 25 kHz and 100 W for 30 min (Figure 1.1). Predetermined quantities of nanoclay are dispersed with polyester resin to get the jute–polyester nanoclay hybrid nanocomposites in the range of 3 to 12 wt%. Jute–polyester nanoclay hybrid nanocomposites are labeled as JPNX (X being the concentration of nanoclay by weight varying from 0 to 12%) in the present study.

Both types (i.e., plain and hybrid) jute–polyester composite laminates were fabricated by simple hand-layup technique supplemented with compression molding. The jute fiber volume fraction (V_f) in a given composite was determined using the following relation:

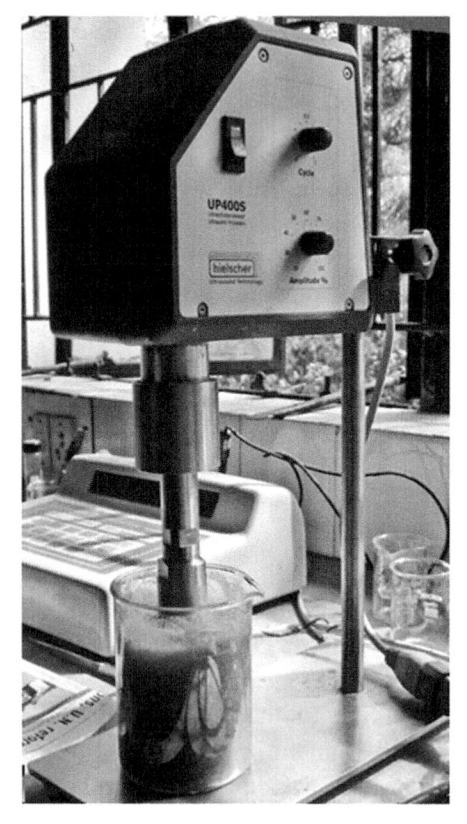

$$V_j = \frac{\dfrac{w_j}{\rho_j}}{\dfrac{w_j}{\rho_j} + \dfrac{w_m}{\rho_m} + \dfrac{w_n}{\rho_n}} \tag{1.1}$$

FIGURE 1.1 Ultrasonic sonicator.

Here w_j (ρ_j), w_m (ρ_m), and w_n (ρ_n) are weights (densities) of jute, matrix (i.e., polyester resin), and nanoclay respectively. Volume fraction of jute fiber in jute–polyester and jute–polyester–nanoclay hybrid composites is maintained approximately at 40 ± 1% (Table 1.2).

Additionally, with a view to obtaining superior properties, one more hybrid composite laminate was fabricated by a similar technique using jute and steel wire mesh as reinforcement in nanoclay-loaded (6 wt%) polyester resin. In this case, jute fiber volume fraction (V_j) in a given composite was determined using the following relation:

$$V_j = \frac{\dfrac{w_j}{\rho_j}}{\dfrac{w_j}{\rho_j} + \dfrac{w_m}{\rho_m} + \dfrac{w_n}{\rho_n} + \dfrac{w_s}{\rho_s}} \qquad (1.2)$$

where w_s and ρ_s are weight and density of steel wire mesh.

TABLE 1.2 Volume Fraction of Composite Laminates.

Composite laminate configuration	Thickness (mm)	Jute fiber VF (%)	Steel wire mesh VF (%)	Polyester VF (%)	Nanoclay VF (%)
10JP60	5.0 (±0.1)	40.00(±1)	N/A	60.00(±1)	N/A
10JPN3	5.0 (±0.1)	39.34(±1)	N/A	59.40(±1)	1.26
10JPN6	5.0 (±0.1)	40.12(±1)	N/A	57.38(±1)	2.50
10JPN9	5.0 (±0.1)	39.45(±1)	N/A	56.71(±1)	3.84
10JPN12	5.0 (±0.1)	39.59(±1)	N/A	55.25(±1)	5.16
10JP66	7.0 (±0.2)	34.00(±1)	N/A	66.00(±1)	N/A
6J4SP	3.8 (±0.1)	30.10(±1)	4.22	65.69(±1)	N/A
6J4SPN6	4.6 (±0.1)	30.93(±1)	2.82	63.53(±1)	2.73

A jute–steel wire mesh–polyester–nanoclay hybrid composite laminate consists of 6 plies of jute and 4 plies of SWM. It may be noted that the jute (J) and steel (S) wire mesh plies are placed in the sequence J/S/J/S/J/J/S/J/S/J. Jute–steel wire mesh–polyester–nanoclay hybrid composite are labeled as 6J4SPN6.

1.2.3 SPECIMEN PREPARATION

Specimens of required dimension were cut from the composite laminates as per ASTM standards. Specimens were cut each in weft and warp direction from the same laminate for tensile and flexural tests. Diamond saw was used to cut specimens from the laminates with sufficient allowance for finishing. Final dimensions were arrived by finishing the specimens using medium grade emery paper.

1.3 MECHANICAL TESTING

In order to study the effects of reinforcing nanoclay in a jute–polyester composites with various proportions, a number of quasi-static tensile and flexural tests were performed in a 100 kN UTM of make BiSS, Bangalore.

1.3.1 TENSILE TESTS

Tensile specimens of composite were prepared as per ASTM-D3039 standard. Tests on specimens of jute–polyester composite and jute–polyester–nanoclay hybrid composite were carried out on Universal Testing Machine at a crosshead speed of 1 mm/min to determine tensile strength and modulus of elasticity.

1.3.2 FLEXURAL TESTS

Flexural tests were performed on specimens using a three-point bend fixture as shown in Figure 1.2. Tests were performed as per ASTM D790-10 standard. The flexural test was initiated by applying the load perpendicular to the fiber direction. The flexural test was conducted on JP composite and JP composite hybrid laminates to get apparent flexural strength given by the equation:

$$\sigma_f = \frac{3PL}{2bd^2} \quad (1.3)$$

where:
σ_f = stress in outer fiber at midpoint (MPa)
P = load at a given point on the load–deflection curve (N)
L = support span (mm)
b = width of beam tested (mm); d = depth of beam tested (mm)

As per ASTM D790, flexural strain is given by the equation:

$$\varepsilon_f = \frac{6Dd}{L^2} \quad (1.4)$$

where:
ε_f = strain in the outer surface (mm/mm)
D = maximum deflection of the center of the beam (mm)
L = support span (mm), and d = depth of beam tested (mm)

Tensile flexural strength (σ_{tf}) and compressive flexural strength (σ_{cf}) are determined as follows:

$$\frac{PL}{4} = \frac{1}{3} \frac{\sigma_{tf} \sigma_{cf}}{\sigma_{tf} + \sigma_{cf}} bd^2 \quad (1.5)$$

By knowing the ratio of uniaxial tensile strength and compressive strength

$$\frac{\sigma_t}{\sigma_c} = \frac{A}{B}$$

tensile flexural strength and compressive flexural strength can be related as

$$\frac{\sigma_{tf}}{\sigma_{cf}} = \frac{A}{B} \qquad (1.6)$$

From eqs 1.3 and 1.4, we get

$$\sigma_{tf} = \frac{3PL}{4bd^2}\frac{A}{(1+A)} \text{ and } \sigma_{cf} = \frac{3PL}{4bd^2}\frac{B}{(1+B)} \qquad (1.7)$$

FIGURE 1.2 Flexural test using three-point bending fixture.

1.4 RESULTS AND DISCUSSION

1.4.1 SCANNING ELECTRON MICROSCOPE (SEM)

Addition of a small amount of nanoclay as filler particles in matrix can enhance the mechanical and thermal properties of polymer-based composites significantly.[26–34] Optimal loading of nanoclay in the resin system with its uniform distribution provides a better interaction between the nanoparticles, matrix, and fiber due to the high aspect ratio and large specific surface area of the nanoclay. Several scientists used montmorillonite nanoclay as a filler in the polymer matrix composites. The aspect ratio of montmorillonite layer in a well-dispersed state can go as high as thousand.[26] The mechanism of reinforcement and bonding characteristic between the nanoclay and its surrounding matrix has been thoroughly investigated by many researchers[27-29] and also widely discussed in the literature.[26-29] Chan et al.[27] studied the mechanism of reinforcement of nanoclay–polymer composite. Based on the images obtained from SEM and the results from TEM, they proved the existence of interlocking and bridging effects in the composites. They clarified that nanoclay clusters with the diameter of 10 nm could enhance the mechanical interlocking inside the composites, thus breaking up the crack propagation. As stated by the author, a bridging effect because of the good bonding between matrix and nanoclay clusters could resist the crack which is located near the cluster and opening in the sample. The clusters of nanoclay can interlock the polymer chains and eventually form strong barriers to stop crack propagation. Chan et al.[28] analyzed the chemical composition between epoxy matrix and nanocomposite by conducting Fourier transform infrared spectroscopy (FTIR) and X-ray photoelectron spectroscopy (XPS). These experiments showed that a chemical bonding exists at an interface between the nanoclay and matrix of the composites. Thus, such a bonding can increase the thermal and mechanical properties of resultant polymer matrix composites. Dewan et al.[26] reported in their studies that the addition of 1 wt.% of nanoclay in the resin system resulted in a better interaction between the matrix and fiber which in turn leads to an effective stress transfer from the matrix to the dispersed fiber reinforcement by means of interlocking of nanoclay with fibers. Greenland[35] in his investigation found that the attachment between the polymer and clay (montmorillonite) was probably being through the oxygens of the clay surface and hydrogen bonds between the hydroxyl groups of the polymer.

In the present study, commercially available nanoclay called Crysnano 1010P were loaded in various wt.% in polyester resin. The SEM images were obtained by examining the dispersion condition of nanoclay inside the polyester matrix. The samples with 3 wt.%, 6 wt.%, 9 wt.%, and 12 wt.% of nanoclay are shown in Figures 1.3(a)–(d) respectively. The particles as marked with circles are nanoclay clusters. All nano-sized nanoclay clusters were found to be uniformly dispersed. Similar findings as discussed earlier could be observed from the SEM images which supported better interaction between the fiber and the matrix. This was confirmed from the experimental test results showing improved mechanical properties of jute–polyester nanocomposite for optimal loading of nanoclay as compared to plane jute–polyester (i.e., without the addition of nanoclay) composite.

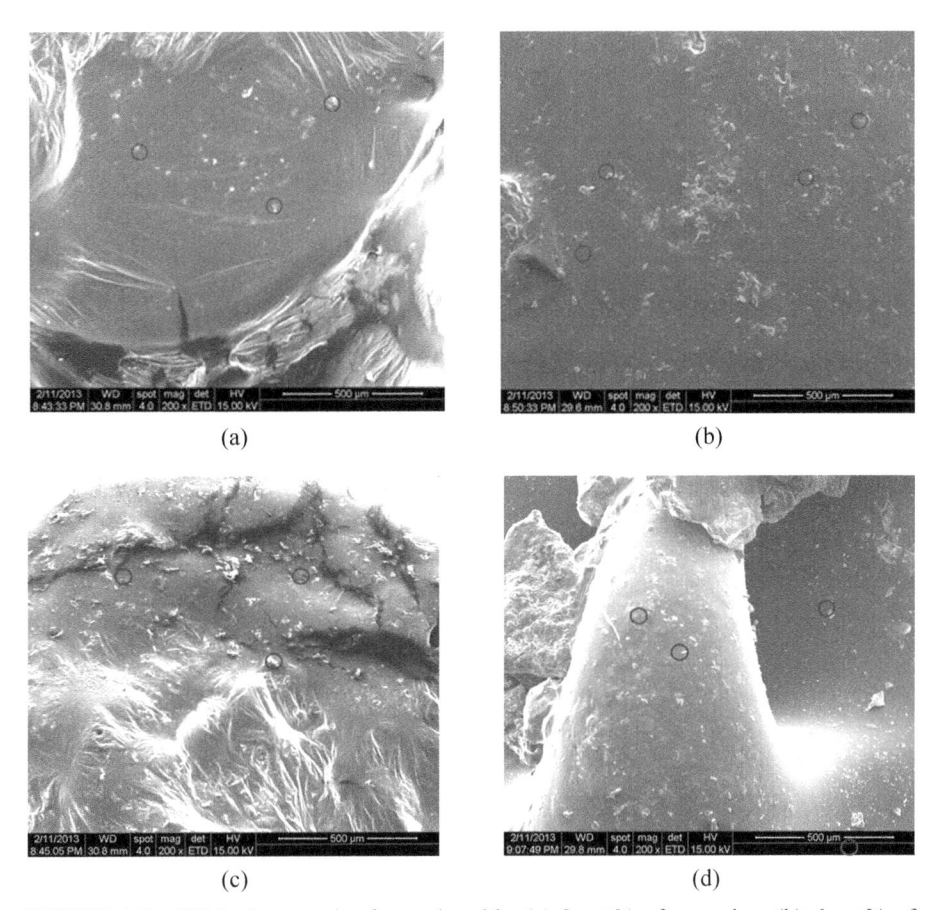

FIGURE 1.3 SEM photograph of sample with: (a) 3 wt.% of nanoclay, (b) 6 wt.% of nanoclay, (c) 9 wt.% of nanoclay, and (d) 12 wt.% of nanoclay.

1.4.2 TENSILE TEST RESULTS

A range of mechanical properties were evaluated in the tests conducted with jute–polyester composites and the jute–polyester–nanoclay hybrid composites with different percentages of nanoclay content. The stress–strain curves for jute–polyester and jute–polyester–nanoclay hybrid composites are shown in Figure 1.4. The main objective for these tests has been to ascertain the expected increase in mechanical properties such as tensile strength and modulus and flexural strength.

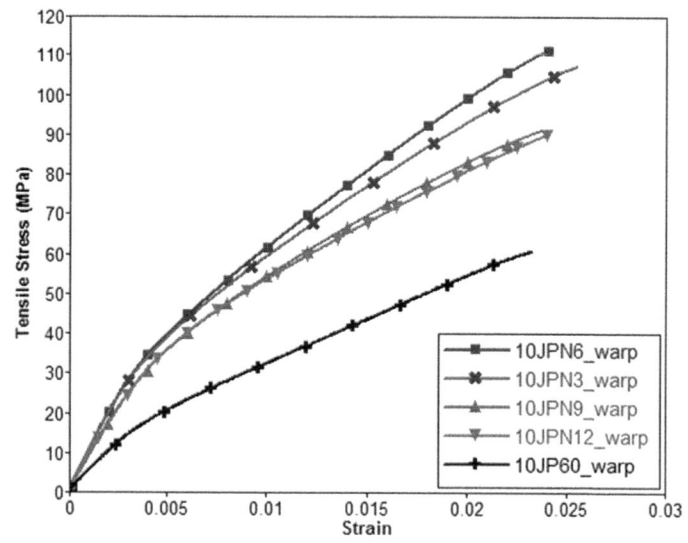

FIGURE 1.4(a) Tensile stress–strain curves of jute–polyester (JP) and jute–polyester–nanoclay (JPN) hybrid composites in warp direction.

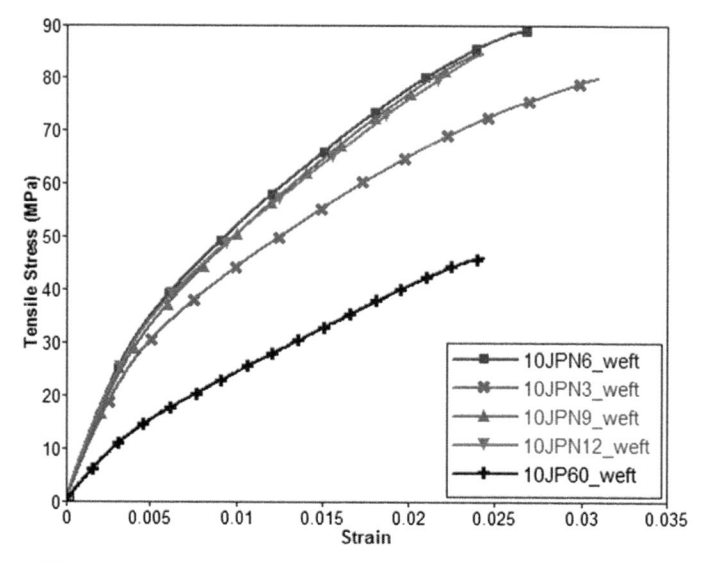

FIGURE 1.4(b) Tensile stress–strain curves of jute–polyester (JP) and jute–polyester–nanoclay (JPN) hybrid composites in weft direction.

The mean tensile modulus obtained from tests for each composition of nanocomposite designated earlier as JPNX (X being the concentration of nanoclay by weight varying from 0–12%) is presented in Table 1.3 for values of nanoclay concentration (i.e., X being equal to 0%, 3%, 6%, 9%, and 12%).

TABLE 1.3 Tensile Strength and Modulus of Jute–Polyester, Jute–Polyester–Nanoclay, and Jute–Polyester–Steel Wire Mesh and Jute–Polyester–Steel Wire Mesh-Nanoclay Hybrid Composites.

Composite laminate configuration	Tensile strength (MPa) warp	Tensile strength (MPa) weft	Tensile modulus (GPa) warp	Tensile modulus (GPa) weft
10JP60	60.30 (3.60)	46.50 (1.10)	4.45 (0.10)	3.45 (0.10)
10JPN3	107.01 (1.43)	78.07 (2.08)	9.15 (0.33)	7.25 (0.10)
10JPN6	112.67 (3.72)	88.67 (2.46)	9.16 (0.38)	7.83 (0.72)
10JPN9	91.87 (0.59)	84.40 (0.71)	8.66 (0.23)	7.38 (0.06)
10JPN12	90.12 (0.78)	84.17 (0.78)	8.40 (0.19)	7.30 (0.94)
10JP66	36.36 (1.04)	29.80 (0.74)	2.52 (0.20)	2.48 (0.14)
6J4SP	62.9 (1.8)	46.3 (1.1)	6.2 (0.1)	5.5 (0.1)
6J4SPN6	114.00 (1.95)	82.37 (1.95)	13.10 (0.40)	11.27 (0.14)

Woven jute mat used in jute polyester composites and jute–polyester nanoclay hybrid composites consists of 16 yarns in warp and 15 yarns in weft direction per square inch. Owing to the bidirectional nature of the mats used with different numbers of yarns used in the warp and weft directions, tensile modulus in the warp and weft direction are different for jute–polyester composites (JPN0) and are found to be 4.45 GPa and 3.45 GPa respectively. From the experimental test results, it was found that the tensile modulus and tensile strength of the composites increased with increasing the nanoclay content. In jute–polyester–nanoclay hybrid composites, tensile modulus increased as nanoclay content increased up to 6% nanoclay loading. Tensile modulus was found to be 9.15 GPa and 7.25 GPa for 3% nanoclay loading (JPN3) and it was 9.17 GPa and 7.57 GPa for 6% nanoclay loading (JPN6) in warp and weft directions respectively. For 3% nanoclay loading (JPN3) increase in tensile modulus in hybrid nano composites compared to plain jute–polyester composites was found to be 112.7% and 113.2% in the warp and weft directions, respectively. Similarly, for 6% nanoclay loading (JPN6) increase in tensile modulus in hybrid nano composites compared to plain jute–polyester composites was found to be 113.0% and 122.6% in the warp and weft directions, respectively. There was a marginal increase in the tensile modulus between 3% and 6% nanoclay loading. As the nanoclay content increased beyond 6%, tensile modulus found to be decreased. The optimal amount of nanoclay should not exceed 6 wt.%. Since as nanoclay loading increases to 9 wt.%, nanoclays start to agglomerate thereby decreasing modulus. Tensile modulus for JPN9 was found to be 8.40 GPa and 7.38 GPa (Table 1.3) in the warp and weft directions respectively which was lower than JPN6. Similarly, tensile modulus for JPN12 was found to be 8.26 GPa and 7.30 GPa (Table 1.3) in the warp and weft directions, respectively.

The trend in tensile strength is no different from tensile modulus for JPN0 and JPNX. Tensile strength in the warp and weft direction was different for jute–polyester composites (JPN0) and was found to be 60.30 MPa and 46.50 MPa, respectively. Tensile strength of 107.01 MPa and 78.07 MPa was observed for 3 wt.% nanoclay loading (JPN3) and it was 112.67 MPa and 88.67 MPa for 6 wt.% nanoclay loading (JPN6) in warp and weft directions, respectively. With 6% nanoclay loading (JPN6) in polyester resin resulted about 93.3% and 94.5% increase in tensile strength in the warp and weft directions respectively as compared to plain jute–polyester composites. As the nanoclay content increased beyond 6 wt.%, tensile strength was found to be decreased. Tensile strength for JPN9 was found to be 91.87 MPa and 84.40 MPa (Table 1.3) in the warp and weft directions respectively.

Thus, it can be concluded that the optimal loading of 6 wt.% of nanoclay resulted in maximum tensile strength and tensile modulus of jute–polyester nanocomposite. Thus, optimal loading and uniform dispersion of nanoparticles in matrix are the important parameters for a better interaction between the nanoparticles and the matrix.

Further, the effect of adding steel wire mesh (SWM) to jute–polyester composites on mechanical properties of the resulting hybrid composites was studied. The resulting jute–steel wire mesh–polyester hybrid composite (6J4SP) showed improved mechanical properties when compared with plane jute–polyester composite (10JP66) for same volume fraction of polyester. Though this was obvious because of the higher strength and modulus of steel wire, however in order to take full advantage of the steel wire potential, the bonding between steel wire, jute fiber, and matrix is of primary importance. The quality of initial bonding was assessed by conducting pull tests and observed no slippage between steel wire and matrix. This shows good interfacial bonding between steel wire, jute fiber, and matrix resulting in improved mechanical properties. An SEM image of the fractured surface of a 6J4SP specimen is shown in Figure 1.5(a) in which breakage of jute strands and steel wires is seen. An image of higher resolution for the same surface is presented in Figure 1.5(b). A further magnified image of the failure of 6J4SP specimen in Figure 1.5(c) shows the shearing of a steel wire due to tensile loading and damage in the neighboring matrix.

In this study, another hybrid composite laminate (6J4SPN6) was fabricated by using jute and steel wire mesh as reinforcement in nanoclay-loaded (6 wt.%) polyester resin on the basis of initial confidence gained through improved mechanical properties of 6J4SP and 10JPN6. This hybridization of jute–polyester composite resulted in significant improvement in tensile strength and modulus as reported in Table 1.3. The tensile stress–strain curves are compared in Figure 1.6.

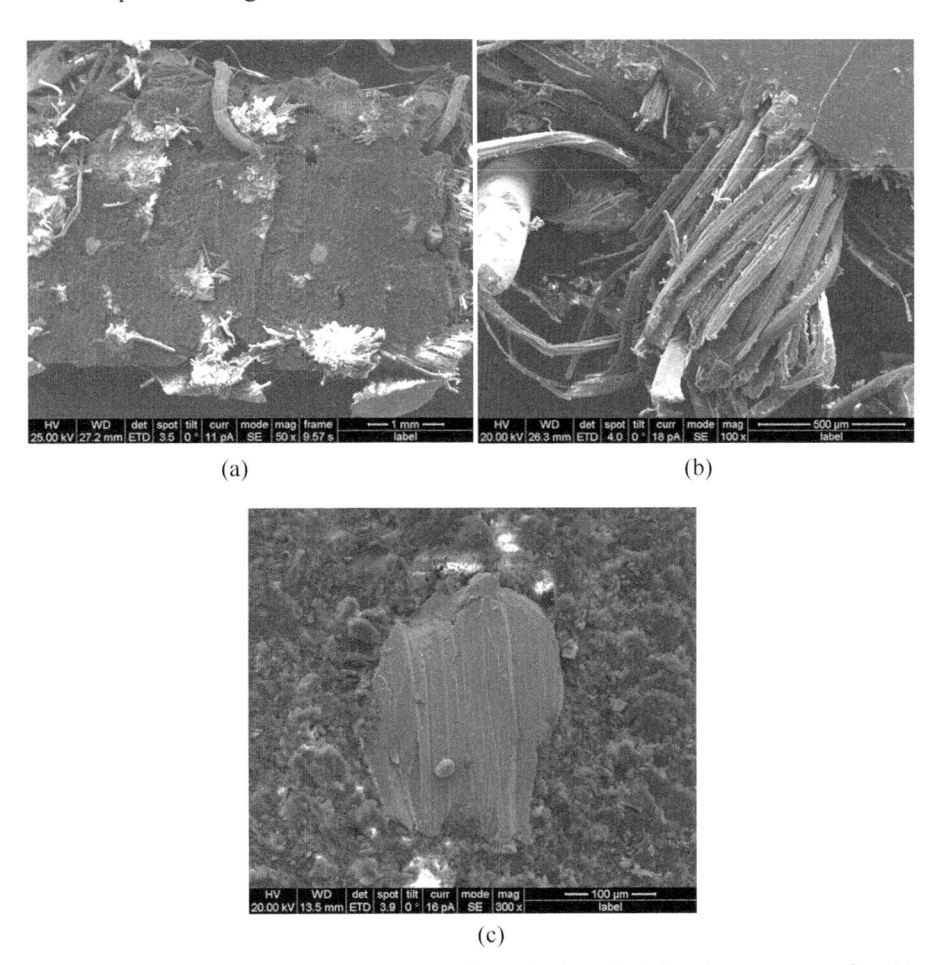

FIGURE 1.5 SEM images of fractured surface of a jute–SWM–polyester composite: (a) broken jute fibers, steel wire mesh, and polyester resin; (b) close-up view of fractured jute fiber strand and steel wire; (c) magnified view of broken single steel wire with cracks in surrounding matrix.

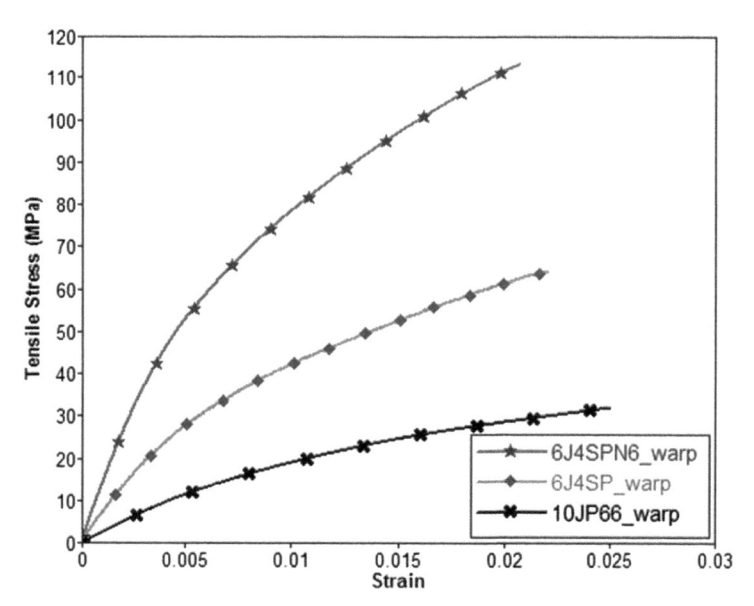

FIGURE 1.6(a) Comparison of tensile stress–strain curves of jute–polyester–steel wiremesh (JSP) and jute–polyester–steel wire mesh-nanoclay (JSPN) hybrid composites in warp direction.

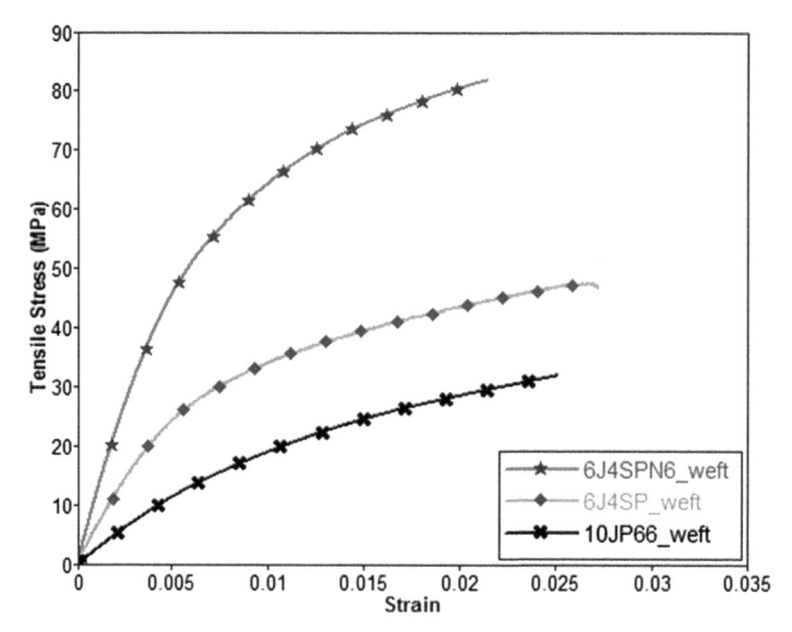

FIGURE 1.6(b) Comparison of tensile stress–strain curves of jute–polyester–steel wire mesh (JSP) and jute–polyester–steel wire mesh–nanoclay (JSPN) hybrid composites in weft direction.

1.4.3 FLEXURAL TEST RESULTS

Flexural stress–strain curves are shown in Figures 1.7 and 1.8. Variations in flexural strength as calculated from eq 1.3 (apparent flexural strength) and 1.7 (tensile and compressive flexural strength) with varying wt.% of nanoclay in polyester resin are presented in Table 1.4. The percentage improvement in flexural strength with respect to the base line composite is reported in Table 1.5. The higher flexural strength was found in the 10JPN6 which showed about 44% (in warp direction) and 49% (in weft direction) improvement in the flexural strength as compared to 10JP60. Also 6J4SPN6 showed up to 30% improvement in flexural strength as compared to 6J4SP. When compared with pure baseline jute–polyester composite (10JP66) for same volume fraction of polyester (or fiber and steel wire mesh), 6J4SPN6 showed about 89% improvement in flexural strength in warp direction and 60% in weft direction. Thus, hybridization of jute–polyester composite laminate with nanoclay and steel wire mesh enhances flexural strength to a great extent.

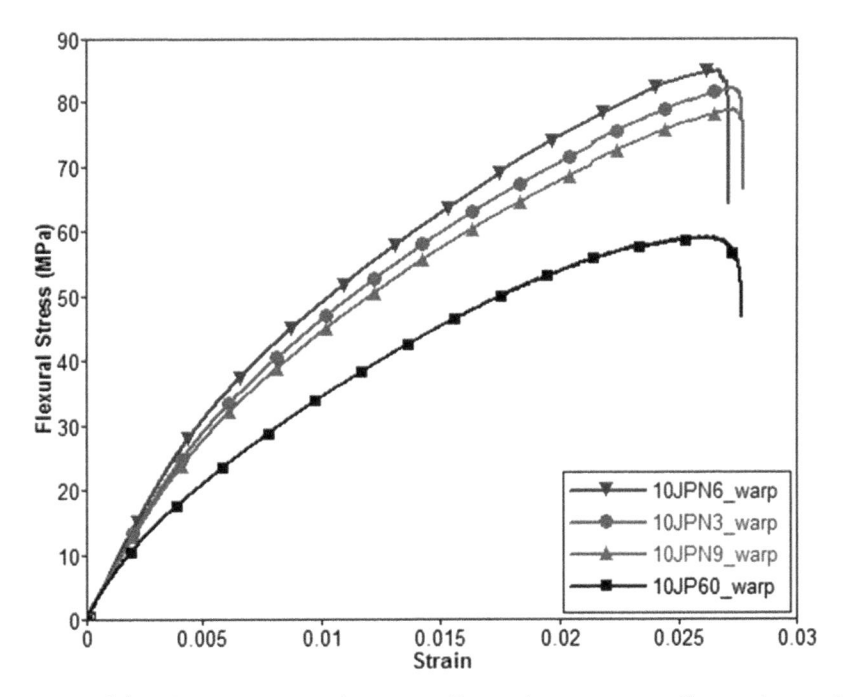

FIGURE 1.7(a) Flexure stress–strain curves of jute polyester curves of jute–polyester (JP) and jute–polyester–nanoclay (JPN) hybrid composites in warp direction.

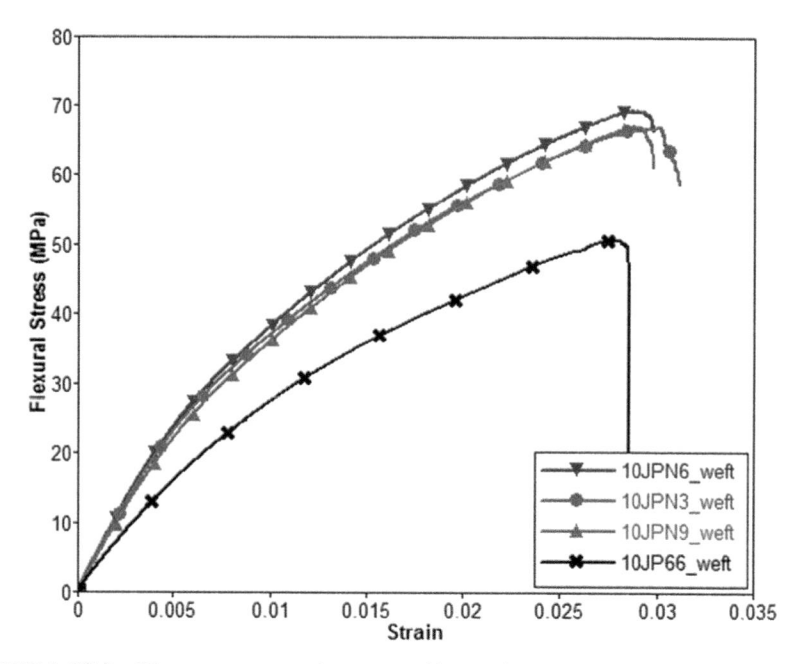

FIGURE 1.7(b) Flexure stress–strain curves of jute polyester curves of jute–polyester (JP) and jute–polyester–nanoclay (JPN) hybrid composites in weft direction.

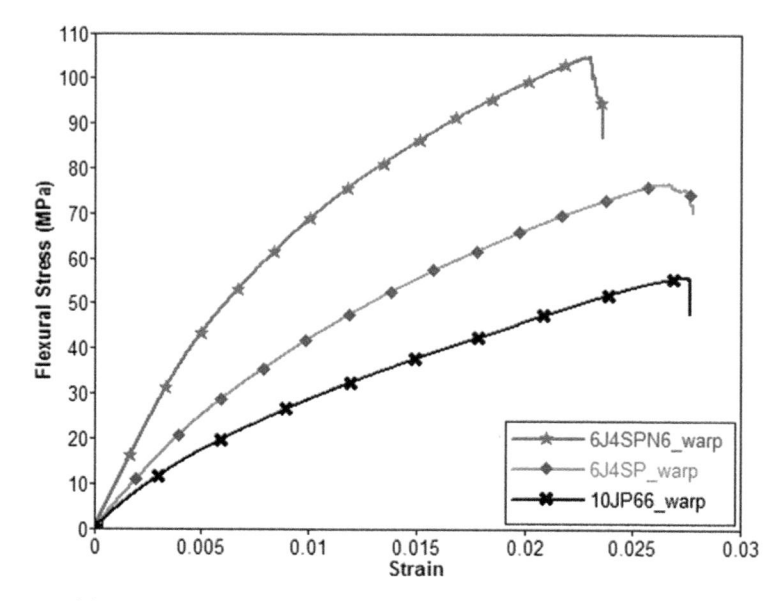

FIGURE 1.8(a) Comparison of flexural stress–strain curves of jute–polyester–steel wire mesh (JSP) and jute–polyester–steel wire mesh–nanoclay (JSPN) hybrid composites in warp direction.

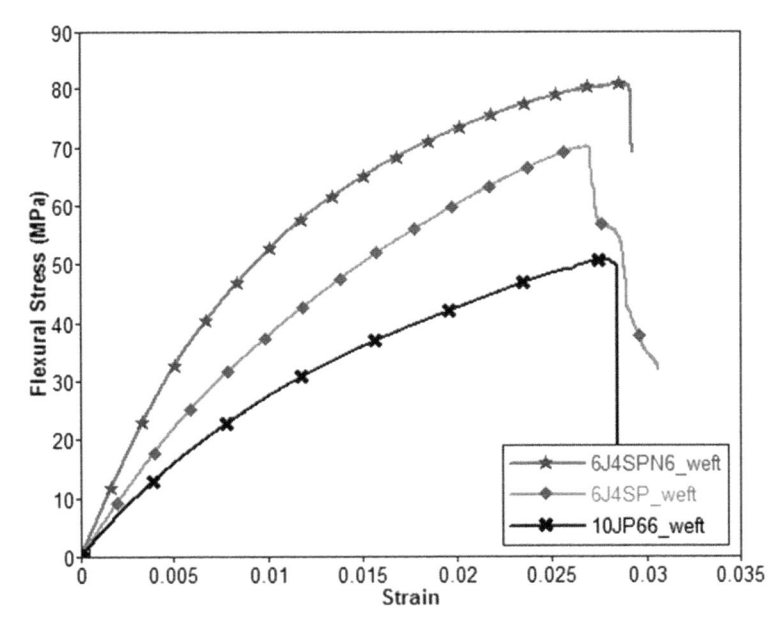

FIGURE 1.8(b) Comparison of flexural stress–strain curves of jute–polyester–steel wire mesh (JSP) and jute–polyester–steel wire mesh-nanoclay (JSPN) hybrid composites in weft direction.

TABLE 1.4 Flexure Test Results of Jute–Polyester, Jute–Polyester–Nanoclay, and Jute–Polyester–Steel Wire Mesh and Jute–Polyester–Steel Wire Mesh–Nanoclay Hybrid Composites.

Composite laminate configuration	Apparent flexural strength (MPa)		Tensile flexural strength (MPa)		Compressive flexural strength (MPa)	
	warp	weft	warp	weft	warp	weft
10JP60	59.97 (2.8)	52.52 (1.8)	77.60	50.31	47.62	40.04
10JPN3	82.13 (0.38)	66.86 (1.65)	125.85	89.11	61.43	55.27
10JPN6	84.86 (1.10)	68.24 (1.84)	155.01	100.71	57.94	51.60
10JPN9	78.81 (1.34)	66.65 (0.15)	132.40	103.63	57.65	49.12
10JP66	55.20 (2.16)	50.65 (2.66)	45.85	38.05	69.35	72.07
6J4SP	76.38 (3.14)	70.07 (3.41)	86.72	67.81	68.24	72.49
6J4SPN6	104.42 (1.82)	80.77 (0.65)	199.20	123.13	69.93	59.79

1.5 CONCLUSIONS

The present study has dealt with the characterization of jute–polyester composite and the effect of hybridization on its mechanical properties. Two different novel approaches of hybridizing jute–polyester composite have been discussed. The first strategy adopted was to additionally reinforce the polymer matrix, viz. polyester, with economically priced alkali-modified Indian nanoclay (montmorillonite), whereas in the second approach hybridization was achieved by incorporating steel wire mesh as reinforcement along with jute fiber in polyester resin to get jute–steel wire mesh–polyester hybrid composite. Both the approaches resulted in considerable improvement in mechanical properties.

TABLE 1.5 Percentage Increase in Tensile and Flexural Properties.

Composite laminate configuration	Percentage increase in tensile strength		Percentage increase in tensile modulus		Percentage increase in flexural strength	
	warp	weft	warp	weft	warp	weft
10JPN3	77	68	106	110	39	46
10JPN6	87	91	106	127	44	49
10JPN9	52	82	95	114	34	46
6J4SPN6	81	78	111	105	37	15

The advantage of replacing some of the jute plies in a 10-ply laminate with steel wire mesh plies has been established by obtaining substantively higher tensile and flexural properties in the hybrid laminates, although compressive strength does not increase as the same appears to be governed primarily by the attributes of the resin, i.e., polyester in the present case. The tensile and flexural strengths of a laminate were found to be higher in the warp direction of jute mats in the laminate which to a great extent is due to the higher count of jute strands in the warp direction per unit width in the weft direction.

KEYWORDS

- **nanoclay**
- **jute**
- **steel wire mesh**
- **polyester**
- **nanocomposite**
- **laminate**
- **tensile modulus strength**

REFERENCES

1. De Rosa, I. M.; Santulli, C.; Sarasini, F. Mechanical and Thermal Characterization of Epoxy Composites Reinforced with Random and Quasi-Unidirectional Untreated Phormium Tenax Leaf Fibres. *Mater. Des.*, **2010**, *31*, 2397–2405.
2. Prasad, A. V. R.; Rao, K. M. Mechanical Properties of Natural Fibre Reinforced Polyester Composites: Jowar, Sisal and Bamboo. *Mater. Des.* **2011**, *32*, 4658–4663.
3. Gowda, T. M.; Naidu, A. C. B.; Chhaya, R. Some Mechanical Properties of Untreated Jute Fabric-Reinforced Polyester Composites. *Composites: Part A* **1999**, *30*, 277–284.
4. Deb, A.; Das, S.; Mache, A.; Laishram, R. A Study on the Mechanical Behaviors of Jute-polyester Composites. *Procedia Eng.* **2017**, *173*, 631–638, https://doi.org/10.1016/j.proeng.2016.12.120
5. Sature, P.; Mache, A. Experimental and Numerical Study on Moisture Diffusion Phenomenon of Natural Fiber based Composites. *Mate. Today: Proc.* **2017**, *4*, 10293–10297, https://doi.org/10.1016/j.matpr.2017.06.367
6. Pingulkar, H.; Mache, A.; Munde, Y.; Siva, I. Synergy of Interlaminar Glass Fiber Hybridization on Mechanical and Dynamic Characteristics of Jute and Flax Fabric Reinforced Epoxy Composites. *J. Nat. Fibers*, 2020, https://doi.org/10.1080/15440478.2020.1856280
7. Mache, A.; Deb, A.; Chou, C. Effect of Strain Rate on Mechanical Responses of Jute-Polyester Composites. *SAE Technical Paper*, **2017**, https://doi.org/10.4271/2017-01-1467.
8. Mache, A.; Deb, A.; Gupta, N. An Experimental Study on Performance of Jute-Polyester Composite Tubes Under Axial and Transverse Impact Loading, *Polym. Compos.* **2020**, *2019*, 1–17,
9. Mache, A.; Deb, A. A Comparative Study on the Axial Impact Performance of Jute and Glass Fiber-Based Composite Tubes. *SAE Int.* **2013**, https://doi.org/10.4271/2013-01-1178
10. Pingulkar, H.; Mache, A.; Munde, Y.; Siva, I. A Comprehensive Review on Drop Weight Impact Characteristics of Bast Natural Fiber Reinforced Polymer Composites. *Mate. Today: Proc.* **2021**. *44*, 3872–3880, https://doi.org/10.1016/j.matpr.2020.12.925
11. Paturkar, A.; Mache, A.; Deshpande, A.; Kulkarni, A. Experimental Investigation of Dry Sliding Wear Behavior of Jute/Epoxy and Jute/Glass/Epoxy Hybrids using Taguchi Approach. *Mate. Today: Proc.* **2018**, *5*, 23974–23983.
12. Ray, D.; Sarkar. B. K.; Das, S.; Rana, A. K. Dynamic Mechanical and Thermal Analysis of Vinylester–Resin–Matrix Composites Reinforced with Untreated and Alkali-Treated Jute Fibres. *Compos. Sci. Techno.* **2002**, *62*, 911–917.
13. Ray, D.; Sarkar, B. K.; Bose, N. R. Impact Fatigue Behaviour of Vinylester Resin Matrix Composites Reinforced with Alkali Treated Jute Fibres. *Compos. A: Appl. Sci. Manuf.* **2002**, *33*, 233–241.
14. Mohanty, A. K.; Khan, M. A.; Hinrichsen, G. Surface Modification of Jute and its Influence on Performance of Biodegradable Jutefabric/Biopol Composites. *Compos. Sci. Techno.* **2000**, *60*, 1115–1124.
15. Fraga, A. N.; Frulloni, E.; de la Osa, O.; Kenny, J. M.; Vazquez, A. Relationship between Water Absorption and Dielectric Behavior of Natural Fibre Composite Materials. *Polym. Test.* **2006**, *25*, 181–187.
16. Sabeel Ahmed, K.; Vijayarangan, S.; Naidu, A. C. B. Elastic Properties, Notched Strength and Fracture Criterion in Untreated Woven Jute–Glass Fabric Reinforced Polyester Hybrid Composites. *Mater. Des.* **2007**, *28*, 2287–2294.

17. Santulli, C. Post-Impact Damage Characterisation on Natural Fibre Reinforced Composites using Acoustic Emission. *NDT and E International* **2001**, *34*, 531–536.
18. Dash, B. N.; Rana, A.; Mishra, H. K.; Nayak, S. K.; Tripathy, S. S. Novel Low-Cost Jute–Polyester Composites. III. Weathering and Thermal Behavior. *J. Appl. Polym. Sci.* **2000**, *78*, 1671–1679.
19. Sever, K.; Sarikanat, M.; Seki. Y.; Erkan, G.; Erdogan, Ü. H. The Mechanical Properties of C-methacryloxypropyltrimethoxy Silanetreated Jute/Polyester Composites. *J. Compos. Mater.* **2010**, *44*, 1913–24.
20. Venkatesh, G. S.; Deb, A.; Karmarkar, A. Characterization and Finite element Modeling of Montmorillonite/Polypropylene Nanocomposites. *Mater. Des.* **2012**, *35*, 425–433.
21. Venkatesh, G. S.; Deb, A.; Karmarkar, A. Synthesis and Effects of Nanoclay and Compatibilizer on Viscoelastic Properties of Montmorillonite/Polypropylene Nanocomposites. *Mater. Des.* **2012**, *37*, 285–291.
22. Venkatesh, G. S.; Deb, A.; Karmarkar, A.; Srinivas, G. R.; Shivakumar, N. D. Viscoelastic, Mechanical and DOE-Based Study on PP-Nanocomposites. *Polym.-Plast. Technol. Eng.* **2012**, *51*, 832–839.
23. Subramaniyan, A. K.; Sun, C. T. Enhancing Compressive Strength of Unidirectional Polymeric Composites using Nanoclay. *Composites Part A* **2006**, *37*, 2257–2268.
24. Bensadoun, F.; Kchit, N.; Billotte, C.; Bickerton, S.; Trochu, F.; Ruiz, E. A Study of Nanoclay Reinforcement of Biocomposites Made by Liquid Composite Molding. *Int. J. Polym. Sci.* **2011**, *2011*, 1–10.
25. Behera, K. A.; Avancha, S.; Sen. R.; Adhikari, B. Development and Characterization of Nanoclay-Modified Soy Resin-Based Jute Composite as an Ecofriendly/Green Product. *Polym.-Plast. Technol. Eng.*, **2013**, *52*, 833–840.
26. Dewan, M. W.; Hossain, M. K.; Hosur, M.; Jeelani, S. Thermomechanical Properties of Alkali Treated Jute-Polyester/Nanoclay Biocomposites Fabricated By VARTM Process. *J. Appl. Polym. Sci.* **2013**, *128* (6), 4110–4123
27. Chan, M. L.; Lau, K. T.; Wong, T. T.; Ho, M. P.; Hui, D. Mechanism of Reinforcement in a Nanoclay/Polymer Composite. *Compos. Part B: Eng.* **2011**, *42* (6), 1708–1712.
28. Chan, M.-L.; Lau, K.-T.; Wong, T. T.; Cardona, F. Interfacial Bonding Characteristic of Nanoclay/Polymer Composites. *Appl. Surf. Sci.* **2011**, *258* (2), 860–864.
29. Subramaniyan, A. K.; Sun, C. T. Enhancing Compressive Strength of Unidirectional Polymeric Composites Using Nanoclay. *Compos. A: Appl. Sci. Manuf.* **2006**, *37* (12), 2257–2268.
30. Hakamy, A.; Shaikh, F. U. A.; Low, I. M. Characteristics of Hemp Fabric Reinforced Nanoclay–Cement Nanocomposites. *Cem. Concr. Compos.* **2014**, *50*, 27–35.
31. Suprakas, S. R.; Bousmina, M. Biodegradable Polymers and Their Layered Silicate Nanocomposites: In Greening The 21st Century Materials World. *Prog. Mater. Sci.* **2005**, *50* (8), 962–1079.
32. Pavlidou, S.; Papaspyrides, C. D. A Review on Polymer–Layered Silicate Nanocomposites. *Progress in Polymer Science* **2008**, *33* (12), 1119–1198.
33. Fengge, G. Clay/Polymer Composites: The Story. *Materials Today* **2004**, *7* (11), 50–55.
34. Akelah, A.; Moet, A. Polymer-Clay Nanocomposites: Free-Radical Grafting of Polystyrene on to Organophilic Montmorillonite Interlayers. *J. Mater. Sci.* **1996**, *31* (13), 3589–3596.
35. Greenland, D. J. Adsorption of Polyvinyl Alcohols by Montmorillonite. *J. Colloid Sci.* **1963**, *18* (7), 647–664.

CHAPTER 2

Effect of Epoxidized Soyabean Oil on Mechanical and Structural Properties of Sepiolite-Filled Polypropylene/Polyolefin Elastomer Composites

M. B. KULKARNI[1], R. KUMBHAKARNA[1], D. S. BHUTADA[1],
BHUSHAN HAZARE[1], S. THORAT[1], S. RADHAKRISHNAN[1] and
Y. S. MUNDE[2]

[1]*School of Petroleum, Polymer and Chemical Engineering, Dr. Vishwanath Karad MIT World Peace University, Pune, India*

[2]*MKSSS's Cummins College of Engineering for Women, Karvenagar, Pune, Maharashtra, India*

ABSTRACT

In the present work, polypropylene/sepiolite/polyolefin elastomer (POE)/ epoxidized soyabean oil (ESBO)-based composites were extruded and injection molded in order to obtain the standard testing specimens. The sepiolite percentage was kept in between 0–30 wt% and the changes after adding 20 wt% POE as an impact modifier and 7.5wt% and 15wt% loading of ESBO on structure and properties of composites were investigated. The microstructure evaluation was done using scanning electron microscopy (SEM) and various properties of the composite such as thermal, mechanical were investigated using different methods. The melting and crystallization characteristics were evaluated using differential scanning calorimetry (DSC). From the results, the observation was made that the addition of sepiolite increases the crystallinity. The evaluation of the structure development was done with the help of wide-angle X-ray diffraction (XRD) and FTIR was used for the examination of molecular interaction. XRD measurements revealed that PP crystals were in the α phase, but in the presence of sepiolite and ESBO, the relative intensities of 110, 040, and 131 reflections altered, implying that preferred growth. Improvement in crystallinity with the addition of sepiolite was confirmed by XRD results. Thermogravimetric analysis was performed and observed that addition of ESBO in filled composites improved the thermal stability. FTIR analysis showed good interaction of sepiolite in the presence of ESBO and can be used as a dual additive for filled composites. As a result of the inclusion of optimum loading of POE, ESBO in sepiolite-filled PP-based composites helps to achieve considerable improvements in impact strength, thermal stability, and stiffness. According to the finding from SEM, it can be seen that the adequate sepiolite content can be safely added to the PP matrix, but on the other hand higher sepiolite content affects the interfacial compatibility of sepiolite and PP, and with the addition of POE and ESBO helped to encapsulate the filler particles of sepiolite particles with improvement in dispersion. The filler properties such as wettability affect the final properties of sepiolite/PP composites.

2.1 INTRODUCTION

Sepiolite is also known by a name Meerschaum which dictates as "Seafoam" in German. It is a hydrated magnesium silicate and holds a formula $Mg_4Si_6O_{15}(OH)_2 \cdot 6H_2O$. It varies in colors such as white, opaque, gray, or cream. Another identifying characteristic of sepiolite is that it can be compared to the bones of the cuttlefish sepia, which also gives it the same Sepiolite. The silicon-oxygen sheets are structurally embedded in sepiolite which makes it fall under layer silicate family. The fibrous characteristics of the sepiolite are a result of sheets containing tetrahedral SiO4 group. Single fiber of the mineral is said to have dimensions which are of length 6.2–4 µm, width 10–30 µm, and thickness 5–10 µm. It is also highly porous in structure. High surface area (300 m^2/g) and elevated sorption capacity are other distinctive features of sepiolite.[1–4]

In order to make SEP useful for high end applications, work is currently being developed to establish SEP as a valid alternative to other mineral fillers which are used in composites such as high-density polyethylene,[3] polypropylene[5] etc.

Polypropylene is one of the thermoplastic polymers, perhaps the lightest polymeric material with low cost, outstanding moldability, and easy processing. Polypropylene is having many commodity applications such as producing packaging films, pipes, storage tanks, seat covers, chairs, tiffin boxes etc. Polypropylene also having high crystallinity exhibits high stiffness, hardness, tensile strength, and high strength-to-weight ratio, due to the fact that it found applications in automobile such as car bumper, instrumental panels, seat covers, and door trims.[6–9] Polypropylene has been developed by compounding with several other materials, to improve its performance characteristic and gets several grades of polypropylene. Impact strength of polypropylene compound is needed to enhance for automotive application.[6,9] It can be achieved by physically blending polypropylene. For this purpose various elastomers are used such as Ethylene-Propylene Diene Monomer (EPDM),[9] Ethylene Propylene Rubber,[7] Styrene-butadiene-styrene rubber[8,10] and Poly(ethylene-octene) (POE)[1,9,11] and Standard Malaysian Rubber.[12] POE presented good toughing efficiency characterized by low molecular weight distributions with advantage of good compatibility with polypropylene and good processability[6,13] Poly(ethylene-octene) a polyolefin elastomer, commercially available with trade name of Engage. Incorporation of POE led to the elevation of impact properties of polypropylene. Finer dispersion of minor component is resulted after adding 10% POE to the blend and such fine dispersing trend is observed even after increasing POE content.[13] Incorporation of elastomer in polypropylene improves impact resistance, but it also shows decrease in stiffness, tensile, and flexural strength. To achieve balancing between impact and stiffness properties of impact modified polypropylene, third component filler is added in it which often acts as crystallization nuclei[1,11] Influence of fillers such as talc,[14] calcium carbonate,[15] fly ash,[16–19] and organic phosphates[11] is investigated on impact modified polypropylene by researchers.

This study focuses on improving the balance between two major properties of polypropylene and POE blend which are impact strength and flexural stiffness using sepiolite filler. Reinforcement of sepiolite in blend enhanced mechanical properties of blend especially flexural strength and modulus, which may decrease due to incorporation POE in polypropylene.[1] The study has been made on the effects of various SEP contents on different characteristics of PP/POE/SEP/ESBO composites. Mechanical and thermal properties were studied as well along with the morphological characteristics, melting and crystallization properties. The interaction between PP and SEP, as well as the distribution of SEP, was investigated.

2.2 EXPERIMENTAL METHODS AND MATERIALS

2.2.1 MATERIALS

Homopolymer polypropylene (PP Repol commercial grade H110MA) was supplied by Reliance Polymers, having a density of 0.90 g/cc, MFI (210C/2.16 kg) of 11 g/10 min. Sepiolite was supplied by Astrra Chemicals, Chennai. Its density was 2.26. Polyolefin elastomer (Engage) Grade-8450 was purchased from Dow Chemical Company. Its density was 0.902 g/cc and MFI (190°C/2.16 kg) of 3 g/10 min. Epoxidized Soyabean oil was supplied by S. V. Plastochem Pvt. Ltd., Nashik.

2.2.2 SAMPLE PREPARATION

Prior to preparation, to minimize the effects of moisture, the drying of PP pellets is done at 80°C for 8 h using an oven. All the raw material were mixed physically as per the formulations given in Table 2.1; twin screw extruder was used for melt blending of mixed batches using temperature range of 170°C to 220°C. Screw rotation was 60 RPM. Extruded strands were pelletized using pelletizer; the test specimens were made from the resulting pellets, which were then injection molded at temperature from 200 to 220°C.

TABLE 2.1 Blends Composition

Batch Code	Polypropylene (wt.%)	POE (wt.%)	Epoxidized Oil (wt.%)	Sepiolite Filler (wt.%)
PP	100	-	-	-
PP/POE/SEP0/ESBO7.5	72.50	20	7.50	0
PP/POE/SEP10/ESBO7.5	62.50	20	7.50	10
PP/POE/SEP20/ESBO7.5	52.50	20	7.50	20
PP/POE/SEP30/ESBO7.5	42.50	20	7.50	30
PP/POE/SEP20/ESBO15	45	20	15	20
PP/POE/SEP30/ESBO15	35	20	15	30

2.2.3 CHARACTERIZATION

FTIR analysis is done using a Bruker Alpha spectrometer, USA (wave no 4000–500 cm^{-1}); XRD of samples were carried out 2Θ from 5° to 70°.

Intensity of the samples was plotted against 2Θ. This analysis was actually performed for evaluating % crystallinity increment because of introducing filler to the blend. Crystallinity was found out by calculating area under crystalline and amorphous peaks by using eq 2.1.

$$\varphi_C = \frac{S_c}{S_c + S_a} \quad (2.1)$$

where φ_C is crystallinity, Sc is area under the crystalline peaks, and Sa is area under the amorphous peak.[20]

In order to evaluate the thermal stability of the given samples, TGA analysis was used. The analysis included heating of a weighed quantity of the sample (9 mg) in the range of 25–800°C with continuous rate of heating (10°C/minute). For origination of various reactions along with oxidation heating was done electrically. Nitrogen gas atmosphere was used for accurate results. As for the results the weight reductions were noted at given time.

The DSC curves were recorded using an HITACHI DSC 7000 differential scanning calorimeter. The rate of heating and cooling was maintained to be at 10°C/min in the temperature range of 30–200°C. As used in TGA analysis, nitrogen atmosphere was used here as well. The resulted melting and cooling curves were studied to evaluate the heat of fusion and crystallization.

Using different magnifications of ZEISS SIGMA FE-SEM, phase morphology and filler dispersion in the blend were studied.

2.3 RESULTS AND DISCUSSION

2.3.1 *THE INTERFACIAL AND DISTRIBUTION CHARACTERISTICS OF SEP IN PP/POE/ESBO COMPOSITES*

Figure 2.1 is a result of scanning electron microscope used to study impact fracture surface. Figure 2.1(a) indicates that PP/SEP20 composition has rough surface and irregularity. Figure 2.1(b) shows proper dispersion of sepiolite in PP matrix when kept at 10% concentration. Furthermore, as discussed, ESBO is well compatible in the composite which results in proper interfacial adhesion of PP and sepiolite. As the concentration of sepiolite increases it is observed agglomeration at PP-POE interface shown in Figure 2.1(c) and (d). It is showing plenty of Void formation causing poor interfacial adhesion.

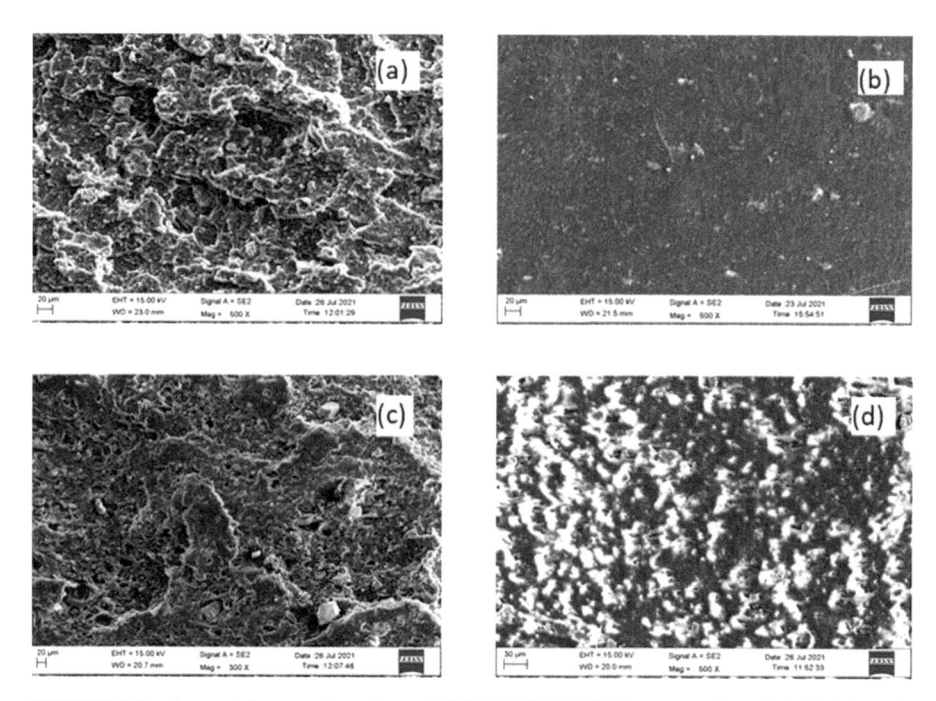

FIGURE 2.1 Impact fractured surface of PP/POE/SEP/ESBO composite: (a) PPUS-20, (b) SEP-10, (c) SEP-20, and (d) SEP-30 (SEM images).

2.3.2 CRYSTALLIZATION AND MELTING PROPERTIES OF PP/POE/SEP/ESBO COMPOSITES

Figure 2.2 indicates melting and crystallization curves. The onset temp of crystallization (T_{onset}), crystallization temp (T_c), and the melting temperature (T_m) can be found in Table 2.2. As per values it can be observed that increasing SEP content increases onset temp of crystallization and crystallization temp. The ability of SEP to act as a nucleating agent in crystallization process is the reason behind this increase in temperature. As for the T_m of composites, there is slightly reduction in melting temperature. The reason is that PP crystals were in α phase, but the relative intensities of 110, 040, and 131 reflections changed in the presence of sepiolite and ESBO suggesting preferential nucleation/growth direction of the crystallites. This is confirmed by XRD results.

The XRD scans were thoroughly analyzed shown in Figure 2.3 and the various peak positions have been tabulated in Table 2.3. Pure PP has four major peaks in the XRD while the samples containing sepiolite exhibit two additional peaks. Since the peak positions for the main peaks remain more or less same, the crystalline form is same, i.e. α composition of PP. However, according to the graph, relative intensities vary acutely which are due to the nucleation effects and preferential growth direction of PP in the presence of the sepiolite. Such varying intensities of XRD peaks of PP have been reported before for example in the presence of wollastonite,[21] calcium carbonate,[22] calcium phosphate,[23] etc. In all these cases PP crystals were nucleated by the additives.

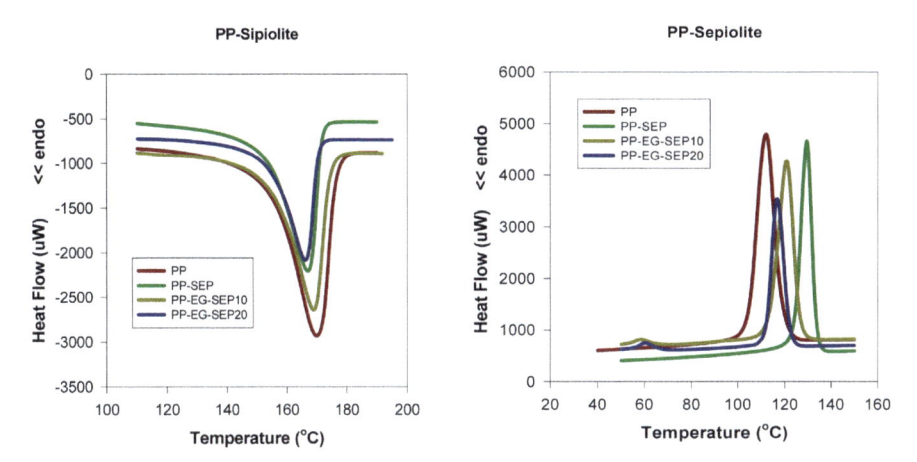

FIGURE 2.2 DSC scans for heating and cooling cycles of PP, PP/SEP, PP+POE+ESBO with 10,20 % sepiolite

TABLE 2.2 DSC Data of PP/POE/SEP/ESBO Composites

Composites	T_{onset} (°C)	T_c (°C)	T_m (°C)
PP	125.4	112.2	170
PP-SEP	137.9	129.3	167
PP-POE-SEP10	131.4	120.3	168.9
PP-POE-SEP20	126.6	116.9	166

TGA analysis was introduced to evaluate the thermal properties of composites, and the TGA curves are displayed in Figure 2.4. Table 2.4 shows temperatures at different mass losses namely 10% is (T10), 50% is (T50), and 80% is (T80) obtained from TGA curves. Despite the presence of POE and SEP in composites, the decomposition process only has one step. This can be explained by two factors. On the one hand, POE's breakdown mechanism is similar to that of PP. After all, the molecular linkages between POE and

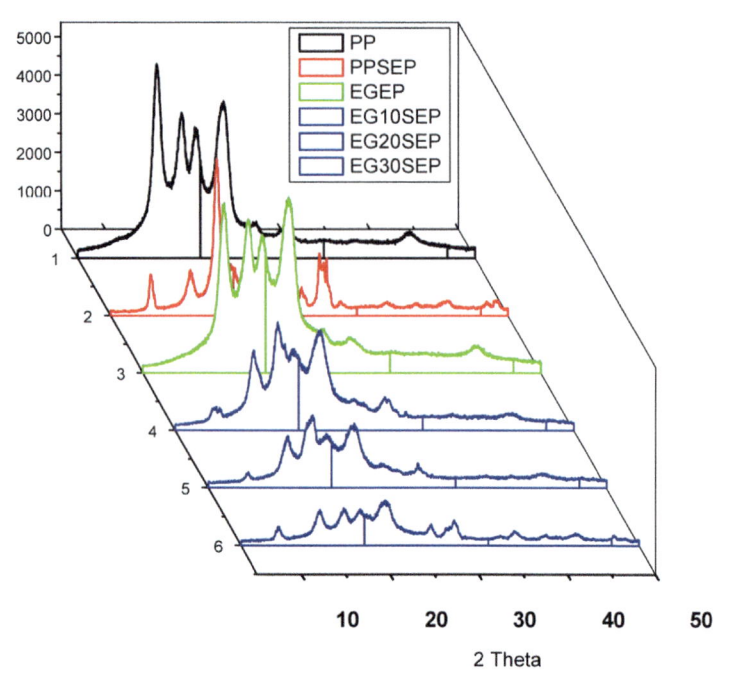

FIGURE 2.3 XRD scans of PP, PP+20%sepiolite, PP+POE+ESBO, PP+POE+ESBO with 10,20 and 30% sepiolite 3.3 The thermal stability of PP/SEP/ESBO composites

TABLE 2.3 Blend Composition and % Crystallinity.

Composition	% Crystallinity	2Θ Position
H110 MA	60.05%	14.17°, 16.84°,18.56°,21.50°
PP/SEP20	64.77%	7.98°,9.62°,14.14°,17.12°,18.09°, 21.83°, 25.70°,29.53°
PP/POE20/SEP0/ESBO7.5	54.24%	14.27°,16.90°,18.77°,21.59°
PP/POE20/SEP10/ESBO7.5	63.46%	9.32°,13.80°,16.57°,18.49°,21.29°, 25.60°,28.80°
PP/POE20/SEP20/ESBO7.5	64.70%	9.51°,14.04°,16.52°,18.68°,21.49°, 24.65°,28.69°
PP/POE20/SEPF30/ESBO7.5	69.57%	9.29°,13.94°,16.63°,18.63°,21.36°, 26.50°,35.91°,39.41°

PP, C–C, and C–H are comparable. As presented in Table 2.4, T_{10}, T_{50}, and T_{80} of composites increase as compared with neat PP value. The T_{10} is seen to increase progressively with the addition of ESBO. On the other hand, T_{50}, T_{80}

increase with the addition of ESBO up to 7.5% but decrease slightly at 15% ESBO. The reason for this phenomenon is thermally stable epoxy group linkage to PP matrix increases its thermal stability. ESBO can attach to POE elastomer also acting as compatibilizer. At a higher concentration of ESBO, the T50 and T80 of composites containing 15% ESBO oil show lower value compared to 7.5% containing ESBO. Since the POE content was fixed at 20% with filler content also at 20%, the 15% ESBO interacts with both these dispersed domains through epoxy groups. T_{10} is the initial onset temperature while T50 and T80 are on the higher temperature region where the ESBO can detach from the domains and the excess oil will give more flexibility for main polymer chains. This leads to faster thermal degradation. Even so, the thermal stability is much higher than neat PP. Thus, it appears that at 7.5% ESBO the tight bound network is formed while at 15% ESBO there is loose network suggesting that the optimum composition for this system is 20% POE and 7.5% ESBO.

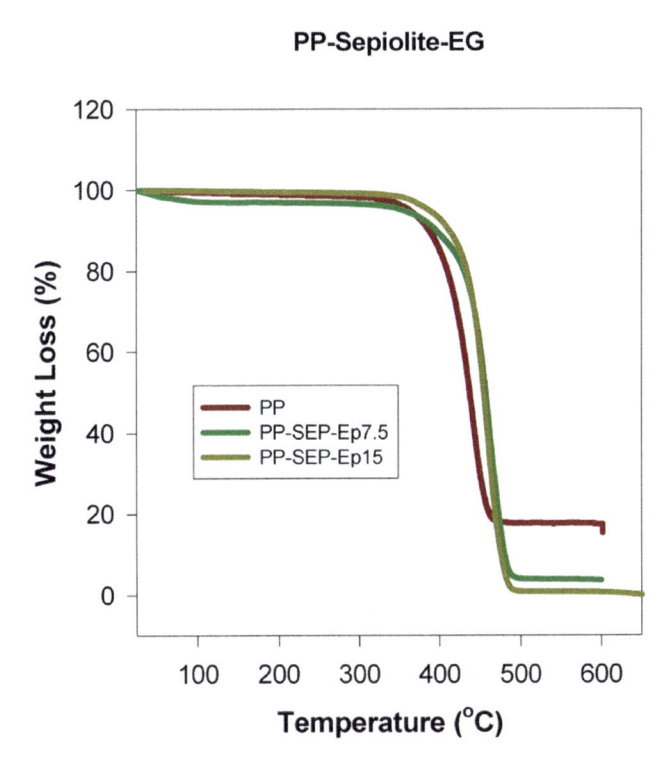

FIGURE 2.4 TGA curves of PP/SEP/POE/ESBO composites.

TABLE 2.4 TGA data of PP/SEP/ESBO composites

Composites	T_{10} (°C)	T_{50} (°C)	T_{80} (°C)
NEAT PP	395	425.2	457.4
PP/SEP/ESBO7.5	401.4	462.0	483.1
PP/SEP/ESBO15	413.2	456.9	473.8

2.3.3 MECHANICAL PROPERTIES OF PP/POE/SEP/ESBO COMPOSITES

Table 2.5 contains the mechanical properties of composites, whereas Figure 2.5 represents the interchanging variations of mechanical properties in an elaborated manner. According to Table 2.5 neat PP has a tensile strength if 35 MPa. The addition of 20% POE and 7.5% ESBO to the PP reduces the tensile strength of composites to 25 MPa. Further adding SEP to the blend at 10% increases tensile strength 26 MPa. This can be referred to composites' greater crystallinity, which results in increased stiffness. As previously stated, as the percentage of SEP is increased, the harmonious properties between SEP and PP tend to decrease which in turn results into reduced tensile strength of the composite. Therefore, in the beginning the increasing of tensile strength can be observed followed by a slight reduction. Increase in flexural strength is obtained with the incorporation of 10% sepiolite in PP/POE/ESBO blend, but as percentage of sepiolite increases reduction in flexural strength is recorded. As for the toughness of composites, Izod impact strength significantly improves at 10% Sepiolite loaded composites. Compared with neat PP from the above results it can be seen that optimum loading of Sepiolite is about 10%.

2.4 CONCLUSION

Composites of PP/POE/ESBO were made using molding processes and various SEP percentages were added. According to the said experimental outcomes, the SEP proportion has a significant impact on the structure as well as on the characteristics of PP/POE/SEP/ESBO composites. A modest amount of SEP can be well dispersed in composites when compared to composites without SEP, but after adding a high amount of SEP, the interfacial harmony of composites deteriorates. Furthermore, a noticeable dissociation of interface in the impact fracture surface is seen. As the SEP content is increased it was seen that there is an increasing trend in the T_{onset}, T_c, and X_c of composites. Another change in thermal properties can be seen as the melting properties are reduced after boosting the SEP percentage. Thermal stability greatly improves containing ESBO in composite. From the mechanical results, it is observed that 10% Sepiolite loading to the PP/POE/ESBO composite balances stiffness as well as toughness.

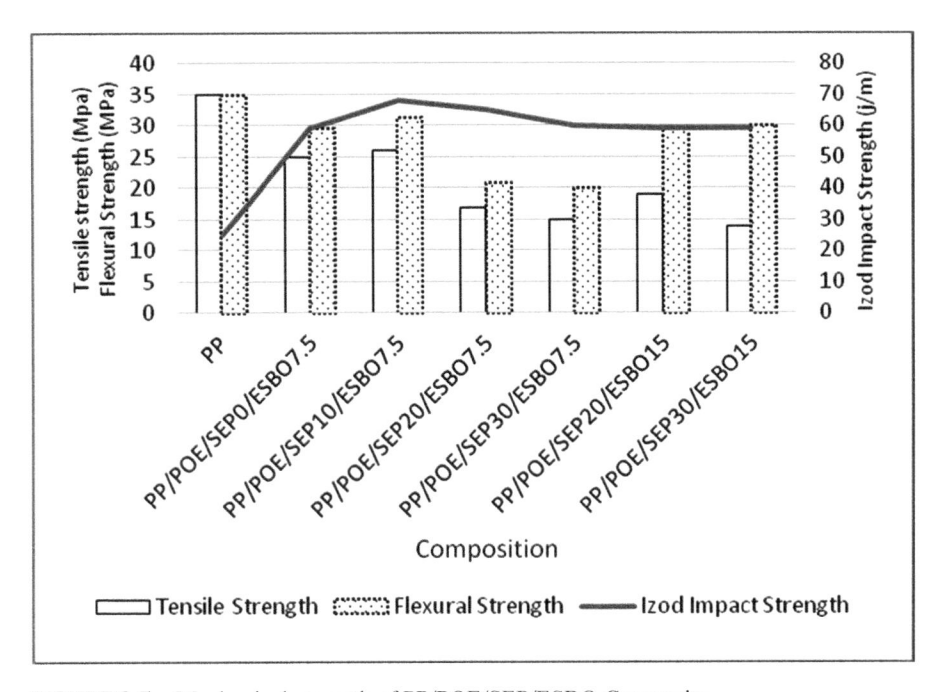

FIGURE 2.5 Mechanical strength of PP/POE/SEP/ESBO Composite.

TABLE 2.5 Mechanical Strength of PP/POE/SEP/ESBO Composite.

Composites	Tensile strength (MPa)	Flexural strength (MPa)	Impact strength (J/m)
PP	35 ± 0.82	35.00 ± 0.55	25 ± 0.45
PP/POE/SEP0/ESBO7.5	25 ± 0.67	29.51 ± 0.19	59 ± 1.01
PP/POE/SEP10/ESBO7.5	26 ± 0.72	31.38 ± 0.29	68 ± 1.31
PP/POE/SEP20/ESBO7.5	17 ± 1.03	20.79 ± 0.39	65 ± 1.57
PP/POE/SEP30/ESBO7.5	15 ± 1.80	20.44 ± 0.24	60 ± 1.22
PP/POE/SEP20/ESBO15	19 ± 1.24	29.32 ± 0.05	59 ± 0.95
PP/POE/SEP30/ESBO15	14 ± 1.49	30.00 ± 0.39	59 ± 1.34

KEYWORDS

- polypropylene (PP)
- sepiolite
- polyolefin elastomer (POE)
- epoxidized soyabean oil (ESBO)

REFERENCES

1. Morales, E.; White, J. Injection-Moulded Sepiolite-Filled Polypropylene: Mechanical Properties and Dimensional Stability. *J. Mater. Sci.* **1988**, *23*, 4525–4533.
2. Ruiz-Hitzky E.; Aranda, P.; Álvarez, A.; Santarén, J.; Esteban-Cubillo, A. Advanced Materials and New Applications of Sepiolite and Palygorskite. In *Developments in Clay Science;* Galàn, E., Singer, A., Eds.; 2011; pp 393–452.
3. Bilott, E.: Fischer, H. R.; Peijs, T. Polymer Nanocomposites based on Needle-Like Sepiolite Clays: Effect of Functionalized Polymers on the Dispersion of Nanofiller, Crystallinity, and Mechanical Properties. *J. Appl. Polym. Sci.* **2008**, 1116–1123.
4. Alvarez, A.; Santaren, J.; Esteban-Cubillo, A.; Aparicio, P. Current Industrial Applications of Palygorskite and Sepiolit. *Dev. Clay Sci. Elsevier* **2011**, 281–298.
5. Moritomi, S.; Watanabe, T.; Kanzaki, S. Polypropylene Compounds for Automotive Applications, **2010**, 10.
6. Liu, G.; Qiu, G. Study on the Mechanical and Morphological Properties of Toughened Polypropylene Blends for Automobile Bumpers. *Polym. Bull.* **2013**, *70*, 849–857.
7. Naiki, M.; Matsumura, T.; Matsuda, M. Tensile Elongation of High-Fluid Polypropylene/Ethylene–Propylene Rubber Blends: Dependence on Molecular Weight of the Components and Propylene Content of the Rubber. *J. Appl. Polym. Sci.* **2002**, *83*, 46–56.
8. Abreu, F.; Forte, M.; Liberman, S.; SBS and SEBS Block Copolymers as Impact Modifiers for Polypropylene Compounds. *J. Appl. Polym. Sci.,* **2005**, *95*, 254–263.
9. Bouchart, V.; Bhatnagar, N.; Brieu, M.; Ghosh, A.; Kondo, D. Study of EPDM/PP Polymeric Blends: Mechanical Behavior and Effects of Compatibilization. *C. R. Mec.* **2008**, *336*, 714–721.
10. Gupta, A.; Purwar, S. Crystallization of PP in PP/SEBS Blends and its Correlation with Tensile Properties. *J. Appl. Polym. Sci.* **1984**, *29*, 1595–1609.
11. Diani, J.; Liu, Y.; Gall, K. Finite strain 3D Thermoviscoelastic Constitutive Model for Shape Memory Polymers. *Polym. Eng. Sci.* **2006**, *46*, 486–492.
12. Norzalia, S.; Surani, B.; Ahmad Fuad, M. Properties of Rubber-Modified Polypropylene Impact Blends. *J. Elastomers Plast.* **1994**, *26*, 183–204.
13. Paul, S.; Kale, D. Impact Modification of Polypropylene Copolymer with a Polyolefinic Elastomer. *J. Appl. Polym. Sci.* **2000**, *76*, 1480–1484.
14. Shri Kant, U.; Kumar. J.; Pundir, G. Study of Talc Filled Polypropylene—A Concept for Improving Mechanical Properties of Polypropylene. *IJRET* **2013**, *15* (20).

15. Premphet, K.; Horanont, P. Improving Performance of Polypropylene Through Combined use of Calcium Carbonate and Metallocene-Produced Impact Modifier. *Polym.-Plast. Technol. Eng.* **2001,** *40,* 235–247.
16. Kulkarni, M.; Radhakrishnan, S.; Samarth, N.; Mahanwar, P. Structure, Mechanical and Thermal Properties of Polypropylene Based Hybrid Composites with Banana Fiber and Fly Ash. *Mater. Res. Express* **2019,** *6,* 075318.
17. Radhakrishnan, S.; Kulkarni, M.; Samarth, N.; Mahanwar, P. Melt Rheological Studies of Polypropylene Filled with Coconut Water Treated and Untreated Fly Ash. *J. Appl. Polym. Sci.* **2016,** 133.
18. Kulkarni, M. B.; Mahanwar, P. A. Studies on Effect of Titanate-Coupling Agent (0.5, 1.5, and 2.5%) on the Mechanical, Thermal, and Morphological Properties of Fly Ash–Filled Polypropylene Composites. *J. Thermoplast. Compos. Mater.* **2016,** *29,* 344–365.
19. Kulkarni, M.; Mahanwar, P. Studies on the Effect of Maleic Anhydride–Grafted Polypropylene with Different MFI on Mechanical, Thermal and Morphological Properties of Fly Ash-Filled PP Composites. *J. Thermoplast. Compos. Mater.* **2014,** *27,* 1679–1700.
20. Weidinger, A.; Hermans, P. H. On the Determination of the Crystalline Fraction of Isotactic Polypropylene from X-ray Diffraction. *Die Makromolekulare Chemie* **1961,** *50,* 98–115.
21. Ding, Q.; Zhang, Z.; Dai, X.; Li, M.; Mai, K. Crystalline Morphology and Mechanical Properties of Isotactic Polypropylene Composites Filled by Wollastonite with β-Nucleating Surface. *Polym. Compos.* **2014,** *35,* 1445–1452.
22. Jing, Y.; Nai, X.; Dang, L.; Zhu, D.; Wang, Y.; Dong, Y.; Li, W. Reinforcing Polypropylene with Calcium Carbonate of Different Morphologies and Polymorphs. *Sci. Eng. Compos. Mater.* **2018,** *25,* 745–751.
23. Saujanya, C.; Radhakrishnan, S. Structure Development and Crystallization Behaviour of PP/Nanoparticulate Composite. *Polymer* **2001,** *42,* 6723–6731.

CHAPTER 3

Process Parameter Optimization by an Ant Lion Algorithm of Austenitic Stainless Steel (SS 304) for Cutting Force in Turning Using PVD-Coated Tools Deposited with TiAlN/TiSiN Coating Materials

CHRISTOPH SCHIFFERS[1], MANISH ADWANI[2], OMKAR KULKARNI[3], ATUL KULKARNI[4], and GANESH KAKANDIKAR[3]

[1]*Product Manager, CemeCon AG, Germany*

[2]*Sales Manager-India & South East Asia, CemeCon AG, India*

[3]*Department of Mechanical Engineering, MIT World Peace University, Pune, India*

[4]*Department of Mechanical Engineering, Vishwakarma Institute of Information Technology, Pune, India*

ABSTRACT

Austenitic treated steel has amazing properties like high erosion opposition, high temperature obstruction, and high sturdiness, settling on it an alluring decision for a wide scope of utilizations. However, it is among the "difficult-to-cut" material due to work hardening, high strength, and low thermal conductivity. This paper focuses on experiments that were carried out using multilayer TiAlN/TiSiN, coated tools. The coatings were stored utilizing DC magnetron sputtering (DCMS) PVD procedures on (K-grade)

Smart Innovations and Technological Advancements in Civil and Mechanical Engineering.
Satish Chinchanikar, Ashok Mache, Shardul Joshi, & Preeti Kulkarni (Eds.)
© 2025 Apple Academic Press, Inc. Co-publis hed with CRC Press (Taylor & Francis)

uncoated established carbide embeds. The influence of coating architecture, coating properties, and cutting parameters on the machinability of SS 304 was investigated during dry turning. The cutting parameters selected for the study include cutting speed, feed, and depth of cut. The response variable was selected to be cutting force. A regression model was developed to predict the values of different response variables and also validated experimentally. Bio-inspired and Nature-inspired algorithm plays a very crucial role in optimizing real-life problems. In this research work Ant Lion Optimizer was used which predicted a much better result as compared to the experimental one. It has been seen that feed has a significant influence on cutting force compared to the depth of cut and cutting speed.

3.1 INTRODUCTION

Manufacturing with machine is one of the methods in which the workpiece is formed into the desired shape by material removal. Usually, a suitable cutting tool is used to remove the excess material on the workpiece in the form of chips. For a given material, the achievement of machining depends on the correct choice of material used in tool and geometry. The expression "processability" is frequently used to depict the simplicity or intricacy of handling materials. There are an assortment of cutting instrument materials and handling materials to browse, with various attributes, execution, and cost blends. Stainless steel[1] is the one of the potential work materials and commonly used in a variety of applications. Contrasted with different grades of steel, it is devoured in huge volumes (72%). However, stainless steel is also considered to be one of the difficult-to-cut materials.[2]

3.1.1 STAINLESS-STEEL (SS)

Stainless steel has been discovered accidentally by British metallurgist Harry Brearley in 1913. By including chromium to low-carbon steel, a consumption safe material was acquired, which later became stainless steel. "Today's stainless steel also contains other elements, such as titanium, nickel, niobium and molybdenum, as well as iron, carbon and chromium." By adding nickel, molybdenum, niobium, and chromium, the corrosion resistance of stainless steel is improved. Low carbon steel's antifouling or rust prevention begins with the addition of at least 12% of chromium. The chromium present in steel associates with atmospheric oxygen to create a thin and nonvisible layer of chromium oxide, called a passivation film. Stainless steel stands as a widely recognized alloy, distinguished for its consistent performance enhancements and expanding range of applications. This growth is possible because of the integral characteristics of hardened steel, for example, resistance to corrosion, high strength-to-weight proportion, and exquisite appearance. Because of its recyclability it is the first choice because of its reliability, reusability, long service life, low maintenance and product safety, and minimal production emissions.

Based on the extensive literature, it is noticed that the cutting force depends on cutting speed in large percentage and feed and depth of cut marginal percentage. In this research work cutting force is the response variable which is selected and the parameters selected are speed, feed, and depth of cut. As this material is hard to machine so multilayer TiAlN/TiSiN,[2] coated tools are used and experiments are performed with this. The data is recorded and further analyzed. The response surface methodology (RSM) is used to generate an equation for the cutting force. To optimize the parameters of process to minimize cutting force required Ant Lion Optimizer is used which ends up in better results.

Bio-inspired, nature-inspired, and socio-inspired[3] algorithms recently have gained an important role in optimizing engineering real-life problems. These algorithms now have proven their importance in mechanical engineering problems as well. Many algorithms such as Genetic Algorithm (GA),[4] Particle Swarm Optimization (PSO),[5] Grey wolf optimizer (GWO),[6] Firefly algorithm,[7] cuckoo search algorithm,[8] Grasshopper optimization algorithm,[9–11] Salp swarm algorithm,[12,13] Bat inspired algorithm,[14] Cohort intelligence (CI)[15] have already proven their applications to mechanical engineering problems. One such algorithm Ant lion optimizer is used in this research to obtain minimum value of cutting force and optimum values of the parameters, i.e., spindle speed, feed and depth of cut.[16]

3.2 METHODOLOGY

It is understood from the literature survey that most of the researchers have used multiple-layer coating of TiAlN/TiSiN[17] for turning of SS 304 stainless steel. PVD coating and its application technology have been continuously improved to obtain coatings with enhanced thermal stability, wear resistance, and hardness. Generally, it is observed that materials with high wear resistance have low toughness.

Characterization of the work material (SS 304)[18] before machining was carried out to confirm the actual composition of the selected material.

Spectroscopic analysis has been done to determine chemical composition of SS 304. It is shown in Table 3.1. Table 3.2 shows the mechanical and physical properties of SS 304.

TABLE 3.1 Chemical Composition of the SS 304 Steel.

Elements	PNi	Al	V	C	Cr	Si	Mn	S	Mo	Fe
% by Wt	0.034	8.75	0.001	0.03	0.065	18.53	0.59	1.197	0.024	0.21 Balance

TABLE 3.2 Mechanical and Physical Properties of SS 304 Steel.[2-4]

Physical properties		Mechanical properties	
Melting point (°C)	1455	Elongation (%)	45
Young's modulus (KN/mm^2)	208	Hardness (Rockwell B)	92
Thermal conductivity (W/m K) at 100°C	16.7	Tensile strength (MPa)	520
		Compressive strength (MPa)	210

However, cemented carbide is renowned for its balanced combination of properties like higher wear resistance and adequate toughness.[19] There are a variety of tool shapes and geometric shapes to choose from, as well as corresponding attribute combinations. The coatings of materials selected are shown in Table 3.3.

TABLE 3.3 Details of PVD Coatings Designated for Performance Evaluation.

Coating type	Coating technique	Coating materials	Tool name
Multilayer	CAE	TiAlN/TiSiN	TM1

Cathodic arc evaporation technique is used to apply coating on the tip of tool and then used for machining. The used machine for turning process SS 304 is ACE CNC LATHE JOBBER XL. Evaluate tool performance with SS 304 dry turning and a variety of cutting conditions. Cutting parameters (especially) are determined based on literature reviews, machine characteristics, and cutting tool manufacturers' recommendations.

3.2.1 CUTTING FORCES

In a turning operation, there are three component forces: cutting force (CF), feed force (F), and radial force (RF). The key cutting force and feed force are mainly observed in the direction of cutting speed and tool feed respectively. The main cutting force plays a major role in determining the machinability and hence these are discussed.[20] The force of cutting is restrained with a three-piece dynamometer (Kistler 9257A) and related (5019 B130) charge amplifier. The working of Kistler dynamometer is based on the principle of piezoelectricity.

3.3 DESIGN OF EXPERIMENTS

Experimentation work was conducted for different cutting speed,[21] feed, and depth of cut. The arrays of cutting speed, feed, and depth of cut are selected based on the performance of machine, literature and cutting tool reviews, and references from coating manufacturers. The nominated parameters and its levels are shown in Table 3.4.

TABLE 3.4 Levels Selected of Cutting Parameters.

Run no.	Parameters	Low level	High level
1	Cutting speed (m/min)	140	320
2	Feed (mm/rev)	0.08	0.26
3	depth of cut (mm)	0.5	1

In statistics, the response surface method (RSM) examines the relationship between multiple independent factors and at least one reaction factor. Response surface methodology is an effective tool to optimize the performance.[22] RSM is a set of statistical and mathematical methods used to create empirical models. The models developed using this method have a high degree of reliability. Response surface methodology is used for this research work and experimental design matrix is as given in Table 3.5.

3.4 MODELING MATHAMATICALLY

During turning of SS304 using TM1 multifaceted cutting tool with different cutting conditions, cutting power, surface roughness and interface temperature was determined. Investigation of the trial results was performed utilizing standard response surface methodology (RSM) procedure.[2] The relapse condition is created depending on exploratory clarifications utilizing Design Expert 7.0 programming. The condition which is quadratic in nature, is

TABLE 3.5 Matrix Used for the Experiments.

Run no	Dept of cut (mm)	Cutting speed (m/min)	Feed (mm/rev)
1	0.75	140	0.17
2	0.75	320	0.17
3	0.75	230	0.08
4	0.75	230	0.26
5	0.5	230	0.17
6	1	230	0.17
7	0.9	283	0.22
8	0.6	283	0.22
9	0.9	283	0.12
10	0.6	283	0.12
11	0.9	176	0.22
12	0.6	176	0.22
13	0.9	176	0.12
14	0.6	176	0.12
15	0.75	230	0.17
16	0.75	230	0.17
17	0.75	230	0.17
18	0.75	230	0.17
19	0.75	230	0.17
20	0.75	230	0.17

utilized to mimic three factors, viz. cutting pace (V), feed (f), and profundity of cut (ap), is communicated as follows eq. 3.1.

$$y = K + a1 \times V + a2 \times f + a3 \times ap + a4 \times V f \quad (3.1)$$

where y is the quadratic capacity, and a1 to a4 are the relapse coefficients of the cutoff variable. The consistent qualities (K) and relapse coefficients (a1 to a4) acquired for various mishaps during turning of SS 304 with various covered devices are shown in Table 3.6.

TABLE 3.6 Values of Constants and Coefficients Obtained.

Tool coating	Tool name	Output response	K	V a1	f a2	ap a3	Vf a4
TiAlN/TiSiN	TM1	Cutting force (Fc)	138.36	−0.49	2187.22	168.08	−4.45

The final regression equation for cutting forces turns down to be as follows in eq 3.2.

$$CF = 138.36 - 0.49V + 2187.22f + 168.08a_p - 4.45f. \qquad (3.2)$$

3.5 RESULTS AND DISCUSSION

TiAlN/TiSiN, multi-layered[1] coated tools have been used for performance evaluation. Experiments with cutting speed, feed, and depth of cut were performed at five different levels using the central rotating composite structure (CCD) method as planned. In the present study, twenty experiments were carried out at each coating level of the tool in the order shown in Table 3.5 to evaluate and mathematically simulate the cutting force. Hereinafter, the temperature of the chip-tool edge is referred to as the cutting temperature. Cutting conditions are anticipated dependent on beginning investigations, past experience, cutting instrument/covering producer's suggestions. Perform analysis of variance on the obtained experiment, see Table 3.7.

TABLE 3.7 ANOVA Results for Various Parameters for TiAlN/TiSiN Coated Tool.

Elements	Cutting force	% Contribution
V	100.10	48.95
f	87.18	42.63
a_p	14.04	6.87
V f	2.14	1.05
V^2	–	–
f^2	–	–
Residual	0.79	0.39
Total	204.49	100.00

3.5.1 EFFECT OF CUTTING PARAMETERS ON CUTTING FORCE

Cutting force is one of the most critical parameters in machining. He helps to determine power requirement in machining. Figures 3.1–3.3 show the variation of cutting force with respect to cutting speed, feed, and depth of cut, respectively. The main cutting force is derived from formula. 1', substitute the constant value in Table 3.6. When drawing a curve, one of the input parameters changes, while the other parameters remain unchanged.

It is worth noting that as the cutting speed increases, the key cutting force inclines to decrease and remains almost unchanged. This is because as the cutting speed increases, the energy of the shear surface and friction energy increase. As a result, the temperature in the shear plane rises and the material becomes softer. This will lead to a decrease in resistance to deformation and ultimately a decrease in cutting force.

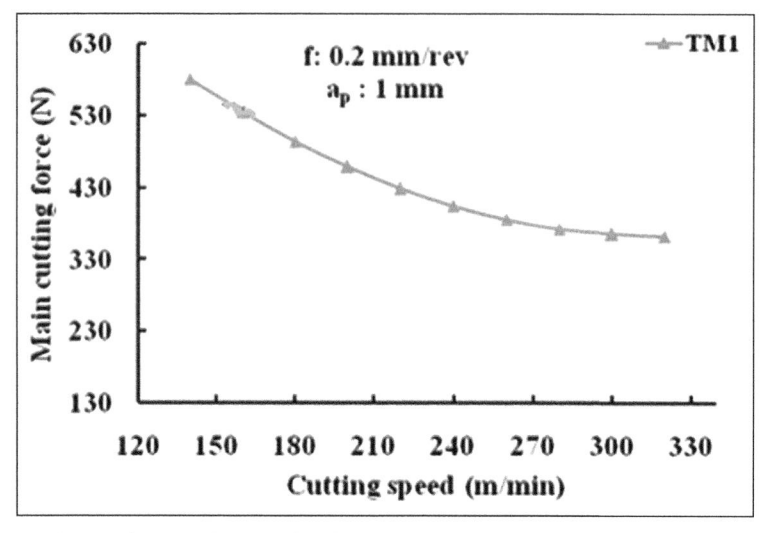

FIGURE 3.1 Cutting speed v/s cutting force.

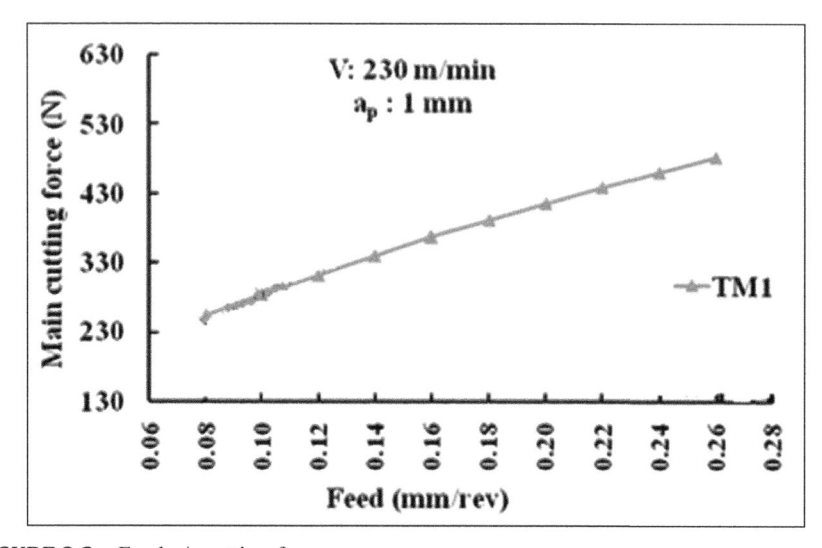

FIGURE 3.2 Feed v/s cutting force.

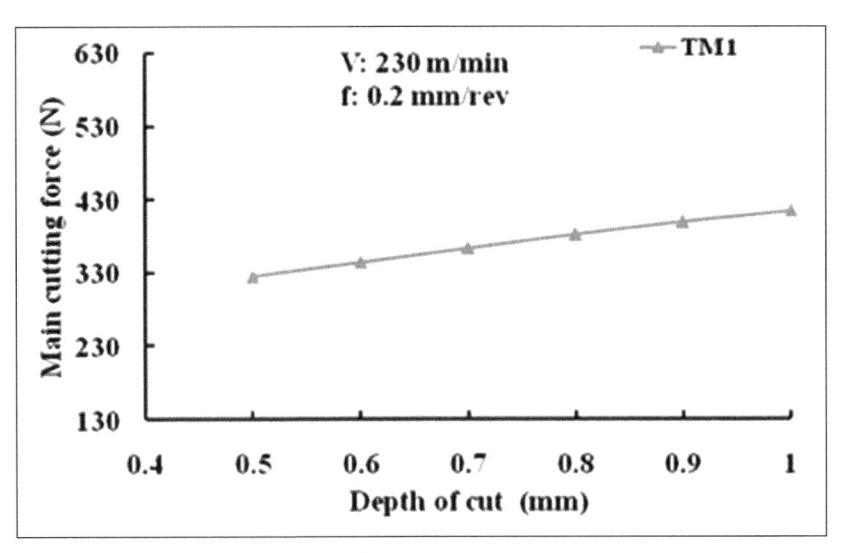

FIGURE 3.3 Depth of cut v/s cutting force.

3.6 OPTIMIZATION—ANT LION OPTIMIZATION ALGORITHM

The technique of hunting of ant lions in nature is mimicked in ant lion optimization (ALO)[23] algorithm for finding the best solutions. Five basic steps of hunting prey are realized, such as randomly walking ants, building traps, letting ants enter the trap, catching the prey, and recovering the trap.

3.6.1 INSPIRATION

Ant lions derive their moniker from their distinctive hunting behavior and tetraborate production. The larva of the ant lion drills a cone-shaped hole in the sand, moves in a circular route, and throws the sand out with its massive jaw. The larva hides at the bottom of the cone (like a sedentary anteater) after digging the trap, waiting for the insect (ideally an ant) to be captured in the hole. The edges of the cone are sharp enough that insects can easily fall to the bottom of the trap. Once the lion ant realizes that the victim is trapped, he will try to catch it. When the prey hits the jaw, it is dragged into the ground and eaten. After the ant lion has eaten its prey, it throws the wreck out of the pit and adjusts the pit for the next hunt. Another interesting behavior observed in the lifestyle of ant lions is that they tend to dig big traps when they feel hungry and/or full moon. To increase their chances of survival, they have evolved and adapted. The activity of ant lion larvae hunting for food is the main source of inspiration for the ALO algorithm."[23] The ALO algorithm simulates lion ants interacting with imprisoned ants. To further imitate this relationship, the ants must wander about the search area, allowing the ants and lions to chase them down and utilize traps. Because ants travel at random when searching for food, random walks are used to imitate their movement:

$$X(iter) = [0, cumsum(2r(1)-1), cumsum(2r(2)-2), \ldots, (2r(n)-1)]$$

The cumulative total is calculated with cumsum, where n is the maximum number of repetitions. $iter$ shows the iteration of random walk, and $r(t)$ is a stochastic function defined as follows:

$$r(t)i = 1 \text{ if } rand > 0.5$$
$$r(t)i = 0 \text{ if } rand \leq 0.5$$

The positions of the ants, as well as their associated objective functions, are stored in matrices, MAnt and MOA, respectively.[23] Suppose that in addition to the ants, the ant lion hides somewhere in the search area. The matrices MAntlion and MOAl are used to store their location and applicability values. In the ALO algorithm, the pseudo code is as follows:

Step 1: Randomly populate the first ant and ant lion population, initially. Calculate ants and lion ant's fitness.

Step 2: Discover the best lion anthill and mistake it for the elite. This research retains the best lion ant found in each iteration so far and is considered an elite.

Step 3: Using roulette wheel, choose an ant lion for each ant.
Step 3.1: Make a walk that is completely random.
Step 3.2: Normalize them to keep the random walks contained inside the search space.[23]

$$X_i^{iter+1} = \left\{ \left[\left(X_i^{iter} - a_i \right) \times \left(d_i - a_i \right) \right] \div \left(b_i - a_i \right) \right\} + c_i$$

where
a_i is the min random walk of the ith variable
b_i is the max random walk in the ith variable
c_i and d_i are the min and max of the ith variable at the current iteration

Step 3.3: Update the position of ant

$$Ant_i^{iter+1} = \left(R_A^{iter} + R_E^{iter} \right) \div 2$$

where

R_A^{iter} is the roulette wheel's choice of a random tour around the ant lion
R_E^{iter} is the random walk around the elite
Ant_i^{iter+1} indicates the position of the *i*th ant at the iteration iter + 1

Step 3.4: Update c and d using the following:

$$C^{iter+1} = C^{iter} \div I$$
$$d^{iter+1} = d^{iter} \div I$$

where

$$I = 10^w \div \frac{iter}{n}$$

and *w* is a constant defined based on the current iteration (*w* = 2 when *iter* > 0.1*n*, *w* = 3 when *iter* > 0.5*n*, *w* = 4 when *iter* > 0.75*n*, *w* = 5 when *iter* > 0.9*n*, and *w* = 6 when *iter* > 0.75*n*)[23]

Step 4: Determine the fitness of all ants.
Step 5: Replace an ant lion with its corresponding ant if it becomes fitter.
Step 6: Update elite if an ant lion becomes fitter than the elite.
Step 7: Repeat from Step 3 until a stopping criterion is satisfied.

3.6.2 RANDOM WALKS OF ANT

At each level of optimization, ants update their positions by traveling randomly. The equation cannot be used directly to update the ant's position because each search space has a boundary (varying range). They are normalized using the following equation (minimum-maximum normalization) in order to preserve a random walk in the search space. (minimum-maximum normalization):[23]

$$X_i^t = \frac{(X_i^t - a_i) \times (d_i^t - c_i^t)}{(b_i - c_i)} + c_i^t$$

where

a_i is the min. random walk of the ith variable
b_i is the max. random walk in the ith variable
c_i^t is the min. of the ith variable at the tth iteration
d_i^t denotes the max. of the ith variable at the tth iteration

3.6.3 TRAPPING IN ANT LIONS' PITS

During random walks, ants will fall into the ant-lion trap. In order to mathematically model this hypothesis, the following equation is proposed:

$$c_i^t = Antlion_j^t + c^t$$
$$d_i^t = Antlion_j^t + d^t$$

where
c^t min. of all variables at the tth iteration
d^t denotes the vector including the max. of all variables at the tth iteration
c_i^t is the min. of all variables for the ith ant
d_i^t is the max. of all variables for the ith ant

Ant lion$_j^t$ shows the position of the selected j^{th} ant lion at the tth iteration. According to the equations, ants move about a selected ant lion in a hypersphere defined by the vectors c and d at random.

3.6.4 BUILDING THE TRAP

The ant lion's hunting abilities are simulated using a roulette wheel approach. Let us pretend the ants are only locked in one lion ant mound. During optimization, the ALO algorithm must employ the roulette operator to select the ant lion based on the ant lion's fitness. This method makes it very easy to gather lions in order to catch ants.

3.6.5 SLIDING ANTS TOWARD ANT LION

Ant lions can set traps based on their health, and ants must move sporadically, according to the currently proposed mechanism. When the ant lion realizes an ant is caught, however, it sprays sand from the pit's core. This activity will get in the way of trapped ants who are attempting to flee. The hyper spherical radius of the ant's random walk is adaptively lowered for mathematical modeling of this behavior. The following equation is presented in this regard:[23]

$$c^t = \frac{c^t}{I}$$

$$d^t = \frac{d^t}{I}$$

3.6.6 CREATING PREY AND REBUILDING THE PIT

When the ant reaches the bottom of the hole and collides with the lion ant's jaw, the search is over. The lion ant then drags the ant into the sand and consumes its body. To model this process, it is considered that the ant becomes better than the lion ant when it captures the victim (into the sand). The lion ant must then adjust its position to that of the hunted ant in order to increase its chances of catching new prey. In this regard, the following equation is proposed:[23]

$$Antlion_j^t = Ant_i^t \text{ if } f(Ant_i^t) > f(Antlion_j^t)$$

where

t shows the current iteration

$Antlion_j^t$ shows the position of the selected j^{th} antlion at the tth iteration

Ant_i^t indicates the position of the ith ant at the tth iteration

3.6.7 ELITISM

Elitism is a crucial feature of evolutionary algorithms, enabling them to keep the best solution found at any point during the optimization process. In the present study, the best lion ant obtained in each iteration so far was retained and considered the elite.[23] Since elites are the most suitable lion ants, they should be able to influence the movement of all ants in the iterative process. Therefore, it is assumed that each ant randomly walks around the roulette and the lion ant selected by the elite at the same time, as shown in the following figure:

$$Ant_i^t = \frac{R_A^t + R_E^t}{2}$$

where

"R_A^t is the random walk about the ant lion nominated by the roulette wheel at the tth iteration

R_E^t is the random walk around the elite at the tth iteration

Ant_i^t indicates the position of the ith ant at the tth iteration"

Program Logic:

Begin

  Initialize the first population of ants and ant lions randomly
  Calculate the fitness of ants and ant lions
  Find the Best Antlion and assume it as the elite (Determined optimum)

While,

  End criterion is not satisfied
  For, every ant, select as antlion using Roulette Wheel
  Update 'c' and 'd' using equation: $c^t = \dfrac{c^t}{I}$ and $d^t = \dfrac{d^t}{I}$
  Create a random walk and normalize it using equation:

  $$X(iter) = [0, cumsum(2r(1)1), cumsum(2r(2)2),...., cumsum(2r(n)1)$$

and

$$X_i^t = \dfrac{(X_i^t - a_i) \times (d_i^t - c_i^t)}{(b_i - c_i)} + c_i^t$$

  Update the position of and using equation: $Ant_i^t = \dfrac{R_A^t + R_E^t}{2}$

end for

  Calculate the fitness of all ants
  Replace an ant lion with its corresponding ant if it becomes fitter by equation:

  $$Ant\,lion_j^t = Ant_i^t \text{ if } f(Ant_i^t) > f(Ant\,lion_j^t)$$

  Update elite if ant lion becomes fitter than the elite

End while

Return elite

3.6.8 OPTIMIZATION RESULTS

After performing Ant lion optimizer algorithm over 500 iterations following optimal results are obtained for cutting force at the dry as shown in Table 3.9. The algorithm was run for over 20 times for each time 500 iterations were performed and then the best results were selected.

TABLE 3.8 Optimization Results.

Method	Cutting speed (m/min)	Feed rate (m/min)	Depth of cut (mm)	Optiman cutting force values (N)
Experimentally	320	0.08	1	334
Ant lion algorithm	320	0.08	0.5	240.22

3.7 CONCLUSION

The optimal values obtained for cutting force with the help of experimental results were observed in dry run; the main influencing parameter for cutting force was found to be cutting speed. The mathematical modeling of the cutting force is performed by the RSM method, i.e., response surface methodology. The range of coating thickness for multilayer coatings TiAlN/TiSiN (TM1) was 2.65 μm to 4.2 μm. At the low cutting speeds, high level of cutting forces were detected for all tools. This is because as the chips move slowly along the rake face, the contact length between the chips and the chips increases, thereby increasing the cutting force. As the tool chip contact length increases, the cutting force increases. The equation defined was single objective function with lower and upper bounds for the parameters. The objective was to minimize the cutting force with the help of ant lion optimizer which was achieved successfully; the cutting force was reduced by 28%.

KEYWORDS

- **austenitic treated steel**
- **multilayer coating**
- **machinability**
- **cutting parameters**
- **regression model**

REFERENCES

1. Fernández-Abia, A. I.; Barreiro, J.; De Lacalle, L. N. L.; Martínez, S. Effect of Very High Cutting Speeds on Shearing, Cutting Forces and Roughness in Dry Turning of Austenitic

Stainless Steels. *Int. J. Adv. Manuf. Technol.* **2011**, *57* (1–4), 61–71. doi: 10.1007/s00170-011-3267-9.
2. Liew, W. Y. H. Low-Speed Milling of Stainless Steel with TiAlN Single-Layer and TiAlN/AlCrN Nano-multilayer Coated Carbide Tools Under Different Lubrication Conditions. *Wear* **2010**, *269* (7–8), 617–631. DOI: 10.1016/j.wear.2010.06.012.
3. Patel, S.; Kakandikar, G. M.; Kulkarni, O. Applicability and Efficiency of Socio-Cultural Inspired Algorithms in Optimizing Mechanical Systems—A Critical Review. *Rev. Comput. Eng. Stud.* **2020**, *7* (2), 31–41. DOI: 10.18280/rces.070203.
4. Bhoskar, T.; Kulkarni, O. K.; Kulkarni, N. K.; Patekar, S. L.; Kakandikar, G. M. Genetic Algorithm and Its Applications to Mechanical Engineering: A Review. *Mater. Today Proc.* **2015**.
5. Kulkarni, N. K.; Patekar, S.; Bhoskar, T.; Kulkarni, O.; Kakandikar, G. M.; Nandedkar, V. M. Particle Swarm Optimization Applications to Mechanical Engineering—A Review. *Mater. Today Proc.* **2015**.
6. Kulkarni, O.; Kulkarni, S. Process Parameter Optimization in WEDM by Grey Wolf Optimizer. *Mater. Today Proc.* 2018, *5* (2), 4402–4412. DOI: 10.1016/j.matpr.2017.12.008.
7. Kakandikar, G. M.; Kulkarni, O.; Patekar, S.; Bhoskar, T. Optimising Fracture in Automotive Tail Cap by Firefly Algorithm. *Int. J. Swarm Intell.* **2020**, *5* (1), 136. DOI: 10.1504/ijsi.2020.106396.
8. Joshi, A. S.; Kulkarni, O.; Kakandikar, G. M.; Nandedkar, V. M. Cuckoo Search Optimization—A Review. *Mater. Today Proc.* **2017**, *4* (8), 7262–7269. DOI: 10.1016/j.matpr.2017.07.055.
9. Neve, A. G.; Kakandiar, G. M.; Kulkarni, O. Application of Grasshopper Optimization Algorithm for Constrained and Unconstrained Test Functions. *Int J Swarm Intel Evol Comput.* **2017**, *6* (3), 165. DOI: 10.4172/2090-4908.1000165.
10. Jawade, S.; Kulkarni, O. K.; Kakandikar, G. M. Parameter Optimization of AISI 316 Austenitic Stainless Steel for Surface Roughness by Grasshopper Optimization Algorithm. *J. Mech. Eng. Autom. Control Syst.* **2021**, 1–11. DOI: 10.21595/jmeacs.2021.22149.
11. Neve, A. G.; Kakandikar, G. M.; Kulkarni, O.; Nandedkar, V. M. Optimization of Railway Bogie Snubber Spring with Grasshopper Algorithm. *Data Eng. Commun. Technol. Adv. Intell. Syst. Comput.* **2020**, *1079*. DOI: 10.1007/978-981-15-1097-7_80.
12. Mhatugade, S. P.; Kakandikar, G. M.; Kulkarni, O. K.; Nandedkar, V. M. Development of a Multi-objective Salp Swarm Algorithm for Benchmark Functions and Real-world Problems. *Optim. Eng. Probl.* **2019**, 101–130. DOI: 10.1002/9781119644552.ch5.
13. Kulkarni, O.; Kakandikar, G. M. Nandedkar, V. M. Application of Salp Swarm Algorithm to Solve Constrained Optimization Problems with Dynamic Penalty Approach in Real-Life Problems. In *Metaheuristic Algorithms in Industry 4.0*; CRC Press, 2021; p. 10.
14. Burande, C. G.; Kulkarni, O. K.; Jawade, S.; Kakandikar, G. M. Process Parameters Optimization by bat Inspired Algorithm of CNC Turning on EN8 Steel for Prediction of Surface Roughness. *J. Mechatronics Artif. Intell. Eng.* **2021**, 1–13. DOI: 10.21595/jmai.2021.22148.
15. Kulkarni, O.; Kulkarni, N.; Kulkarni, A. J.; Kakandikar, G. M. Constrained Cohort Intelligence Using Static and Dynamic Penalty Function Approach for Mechanical Components Design. *Int. J. Parallel. Emergent Distrib. Syst.* **2016**, 570–588. DOI: 10.1080/17445760.2016.1242728.
16. Hovsepian, P. E.; Ehiasarian, A. P.; Petrov, I. TiAlCN/VCN Nanolayer Coatings Suitable for Machining of Al and Ti Alloys Deposited by Combined High Power

Impulse Magnetron Sputtering/Unbalanced Magnetron Sputtering. *Surf. Eng.* **2010**, *26* (8), 610–614. DOI: 10.1179/026708408X336337.
17. Chauhan, K. V.; Rawal, S. K. A Review Paper on Tribological and Mechanical Properties of Ternary Nitride based Coatings. *Procedia Technol.* **2014**, *14*, 430–437. DOI: 10.1016/j.protcy.2014.08.055.
18. Rizzo, A.; et al. Improved Properties of TiAlN Coatings Through the Multilayer Structure. *Surf. Coatings Technol.* **2013**, *235*, 475–483. DOI: 10.1016/j.surfcoat.2013.08.006.
19. Lattemann, M.; Ehiasarian, A. P.; Bohlmark, J.; Persson, P. Å. O.; Helmersson, U. Investigation of High Power Impulse Magnetron Sputtering Pretreated Interfaces for Adhesion Enhancement of Hard Coatings on Steel. *Surf. Coatings Technol.* **2006**, *200* (22–23) *Spec. Iss.* 6495–6499. DOI: 10.1016/j.surfcoat.2005.11.082.
20. Lembke, M. I.; Lewis, D. B.; Münz, W. D.; Titchmarsh, J. M. Significance of Y and Cr in TiAlN Hard Coatings for Dry High Speed Cutting. *Surf. Eng.* **2001**, *17* (2), 153–158. DOI: 10.1179/026708401101517656.
21. Joshi, G.; Kulkarni, A. P. Machinability Aspects in Dry Turning of Ti6Al4V Alloy with HiPIMS Coated Carbide Inserts. *Ind. J. Engg. Mat. Sci.* **2021**, *28*, 533–41; *Tribol. Int.* **2009**, *42* (11–12), 1758–64. DOI: 10.1016/j.triboint.2009.04.026.
22. Erkens, G.; Alami, J.; Vetter, J.; Müller, J. About Unique Micro-Alloyed Coating Solutions and the Related Innovative Pvd Process Technology. no. October, 2008; pp. 1–3.
23. Mirjalili, S. The Ant Lion Optimizer. *Adv. Eng. Softw.* **2015**, *83*, 80–98. DOI: 10.1016/j.advengsoft.2015.01.010.

CHAPTER 4

Electrical Discharge Machining Process Optimization During Machining of EN19 Alloy Steel Using a Desirability Concept

VIJAY KUMAR S JATTI[1], VINAY KUMAR S. JATTI[2], SAVITA V. JATTI[1], PAWANDEEP DHALL[1], and SATISH CHINCHANIKAR[3]

[1]DY Patil College of Engineering, Akurdi, Pune, India

[2]Symbiosis Institute of Technology, Pune, India

[3]Vishwakarma Institute of Information Technology, Pune, India

ABSTRACT

Owing to the increasing demand, EN19 alloy steel in the automotive industry calls for machining difficult-to-cut material using an electrical discharge machine. Based on the complexity of the electric discharge machining process, this study focuses on determining the optimal setting of input process parameters to maximize material removal rate and minimize tool wear rate and surface roughness. A central composite design has been employed for designing the experimental layout using Minitab 16 statistical software. Desirability concept and optimizer function were used to get the optimum value of responses, namely material removal rate (MRR), tool wear rate (TWR), and surface roughness (Ra). The best MRR obtained was 44.332 mm^3/min, TWR 0.678 mm^3/min, and Ra 4.34 µm. This study provides a guideline to shop-floor operators, to set the input parameters according to the required values of responses.

4.1 INTRODUCTION

Electric discharge machining (EDM) is an unconventional thermoelectric machining process that uses rapid and repetitive spark discharges between an electrode and the workpiece to wear conductive materials. The polarities of a dielectric liquid are contained at a short distance from each other. Sparks produce enough heat to vaporize and melt the material after being generated by measured pulses. In small batches or even in workshops, electrodischarge machining is used to perform work on difficult machine tool materials, high strength-to-weight alloys, and high-temperature alloys.

The composition of steel alloys varies widely. By weight, they range from 1% to 2% Cr or Ni in some low alloy steels, which is up to 15–18% by weight Cr content. Alloy steels are available in a variety of conventional and exotic alloys, including 4140, 316, and Hadfield's steel. These alloys take advantage of the properties of their microstructure to give them many different properties. As a result of their wide range, many industries use them. Depending on the situation, this alloy might be the only one capable of providing all the necessary characteristics. Using it in the automotive industry is vital to developing safer vehicles and improving fuel efficiency.

The ability to machine some of the more complex and harder materials requires the use of discharge machining in a manufacturing environment. Because of the complexity of this process, researchers have difficulty determining the optimal process parameters. Research was conducted with the aim of developing an efficient system that would remove material efficiently, provide superior surface finishes, and minimize tool wear. Due to the materials' high strength, hardness, and impact strength, conventional processing methods cannot be used to process these materials. Processes for processing solid materials, such as discharge, are among the most widely used. In order to maximize productivity, different process parameters must be optimized for processing different materials.

Durairaj et al.[1] performed an experimental study of the discharge treatment of stainless steel wire (SS304) using 0.25 mm copper wire, and the following conclusions were made: The minimum surface roughness (Ra) is a standby voltage of 40 V, a wire feed of 2 mm/min, a pulse duration of 6 µs, a pulse pause time of 10 µs, and similar optimized conditions for obtaining the desired kerf width. Gray's relational analysis was used to optimize the input parameter combinations to obtain both the minimum Ra and the nominal kerf width based on a standby voltage of 50 V and a wire feed rate of 2 mm/min. In both Taguchi optimization and Gray's relational analysis, the variance leading to pulses has a significant impact on Ra (*m) and kerf width (mm). The numerical results of the Taguchi method and gray-scale analysis have been compared with experimental results for determining Ra and kerf width. Studies on the electrical parameters of EDM performance utilizing a variety of modeling and optimization techniques have revealed that significant work has been done to simulate and enhance those effects. In order to optimize the performance of the processor, extensive research has been conducted; however, only a small amount of attention has been paid to various nonelectrical factors.

Manivannan et al.[2] used TOPSIS and COPRAS algorithms. COPRAS was applied to control many drilling parameters simultaneously to minimize all reactions when drilling Mg AZ91 carbide tools. This method improved drilling accuracy and reduced production costs. By combining TOPSIS and COPRAS drilling methods, all feedbacks were simultaneously minimalized at a spindle speed of 4540 rpm and a feed rate of 0.076 mm/revolution. DFA determines the optimum input by using the same process variable. With minimal error rates in the confirmation tests, the recommended independent parameter has been confirmed. Spindle speed has been found to result in higher slitting speeds when added to drilling speed, increasing machine capacity. When chip movement was taken into account in the study, it was found that the rate of percolation is related to material removal. Increased feed rates result in faster penetration of the tool, resulting in less feedback. By combining both responses, experiments can be run faster. A minimum inlet bezel thickness is obtained by combining both feedbacks at the bottom. In order to achieve the lowest entry burr, the spindle speed is lowered and the feed rate is increased.

High spindle speed and low feed rates are combined to achieve the highest surface quality. Using above-average feed rates alongside intermittent spindle speeds minimizes exit burr height and exit clearance thickness. An assessment of the influence of tool size on material removal rate (MRR), tool wear rate (TWR), and Ra is presented in this study while maintaining all other electrical parameters, tool material, and workpiece parameters constant.

Singh et al.[3] treated 316 L stainless steel by discharge treatment with nanopowder TiO_2 to make it surface-modifiable. It was found that TiO_2 nanoparticles in the dielectric medium and current strength had the most influence on microhardness, which increased by 233% over the nontreated metal. On the surface of stainless steel 316 L treated with TiO_2, layers without cracks, carbides, and other intermetallic layers were discovered, which increased wear resistance. In addition to improved wear resistance and microhardness, new phases formed such as titanium carbide, silicon iron carbide, chromium, cobalt silicide, and titanium can be considered for the promotion of biological activity.

Engineering process engineers pose machining optimization tasks that improve the efficiency of the overall machining process. Multicriteria decision-making (MCDM) is a form of discrete processing optimization developed to solve discrete processing problems using the PSI method. Several case studies, related to material machinability and the performance of various cutting fluids, are examined to validate the PSI method. The PSI MCDM method does not require criterion weight selection, as is common with other MCDM methods. The method calculates the criteria weights objectively by using only components of the decision matrix. Using this method, a decision maker can easily follow a relatively simple calculation process. Because the PSI method can be implemented easily in MS Excel, solving many MCDM problems does not require the use of specialized software packages. A complete classification of the alternatives is used when calculating preferred selection index values. An alternative's assessment is not influenced by its implementation or the introduction of additional parameters, as in the case of other MCDM methods, such as λ in the WSPAS method. It may also be possible to update the PSI method in future research and compare it with other MCDM techniques to resolve MCDM problems in a production environment.[4]

Prabhu et al.[5] combined Taguchi's experimental engineering technique with TOPSIS to optimize the EDM machining of AISI D2 tool steel. The best experimental combination based on TOPSIS analysis was a pulse current of 8 A, a pulse duration of 5 s, and 80 V. The impulse voltage was having a greater influence on the combined response. Combining pulsed current and pulse width with level 3 and level 3 pulse voltage will minimize Ra, maximize material removal, and minimize crack size. The regression model was validated using variance analysis (ANOVA) and the F-test, and both Ra and crack size were significantly influenced by a single parameter. In a close comparison to TOPSIS multipurpose optimization, ANOVA showed pulse current, pulse width, and pulse voltage to be 42.42%, 11.13%, and 44.17%, respectively. F-test values above 1 indicate greater changes in EDM processing in response to a change in impulse voltage. The RMS roughness of surface morphology was used for the quantitative analysis of AFM images.

When nanoelectrodes are mechanically processed, their surface finish, slope, and fractal dimensions increase. Therefore, nanoelectrode liquid processing can produce a high surface quality. For a machined piece of AISI D2 tool steel, GAP was used to optimize the process parameters. GA is used in conjunction with the developed mathematical model to determine the ideal conditions leading to minimal Ra.

Gadakh et al.[6] utilized the TOPSIS method to select the optimal WEDM process parameters. Any selection problem can be solved by using the proposed method, irrespective of the number of selection criteria. The application is demonstrated with three illustrative examples validated by TOPSIS. There is no difference between the classes of alternative alternatives that are best classified and those previously suggested. The TOPSIS method takes into account the attributes as well as the rank and gives a more precise assessment of the alternatives. It combines both quantitative and qualitative attributes in an efficient and computationally simple manner and provides a reasonable, objective method of selection.

Dastagiri et al.[7] used TOPSIS, GRA, and the Taguchi method for optimizing discharge machining with EN31 wire cutting with multiple performance characteristics. Hence, performance characteristics such as MRR and SR can be improved with this approach. Based on the experimental results and the confirmatory test, the test results for optimal configurations show significant improvements in performance, that is, metal removal rates and Ra. The most important factors affecting the reliability of the WEDM process have been identified as on-time pulse, off-pulse, servo power, and servo voltage. These methods are more convenient and economical for predicting optimal processing parameters.

Gopalakannan et al.[8] performed an experiment on Al7075-B4C MMC by modeling an equation with the help of the central composite design of response surfaces. For MRR, TWR, and Ra, a mathematical model was developed. ANOVA was used to test MRR, TWR, and SR effects on process parameters and their interactions. In this study, we found that increasing SiC percentage decreased the MRR, increased the TWR and SR.

Rajmohan et al.[9] observed the effect of pulse-on time, pulse-off time, voltage, and current on MRR during 304 L stainless steel machining. As per the design, the experiments were carried out. The MRR of 304 L stainless steel can be increased by using different combinations of EDM process parameters. 304 L stainless steel MRR is primarily affected by current and pulse-off time, according to a study. Abbas et al.[10] studied the dielectric, which turns into waste after certain cycles and is found to be dangerous to the environment. An average EDM manufacturer in Malaysia will dispose of 4393 L of dielectric. Multiplying the data against the number of manufacturers, we see an alarming state of pollution. Therefore, environmentally friendly dielectrics are needed.

Vishwakarma et al.[11] investigated the effect of input machining parameters on the MRR. In order to conduct experiments on EN19 alloy steel, a tungsten–copper electrode was used. An optimal analysis of copper electrode input parameters was used to determine the effect of these parameters on EN19 MRRs. MRR of EN19 alloy steel was affected the most by input current, voltage gap, and flushing pressure, according to the results. It has been shown that the developed model shows high accuracy in the experimental region. Among all the parameters affecting EN19's machinability, pulse current was identified as the most significant. Rajesh et al.[12] used multiple regressions and modified genetic algorithm models to determine the optimum machining parameters of electrical discharge machining. An empirical model for MRR and Ra based on gray relational analysis was developed by a designed experiment based on working current, working voltage, oil pressure, spark gap, pulse-on time, and pulse-off time. In this process, the Ra of the test piece has been minimized and the MRR increased.

From the literature review, it is evident that considerable work has been done on modeling and optimizing the various parameters that affect EDM's performance characteristics. To maximize machining performance, various materials of workpieces and tools have been investigated. Despite this, little but worthwhile work has been done on nonelectric parameters. It is noted that there is a lot of potential in discovering how geometry affects performance, in particular the elements of the geometry of the tools. This study focuses on evaluating how the size of the tool affects MRR, TWR, and Ra, having electrical parameters like pulse on-off times and voltage, and keeping other electrical parameters and tool materials constant while the discharge current varies simultaneously.

4.2 EXPERIMENTAL DETAILS

For the studies, a commercial-grade oil EDM electrode was used along with a Die Sink Model C with an NC Z-axis control and negative polarization (specific gravity and freezing point, at 0.763 and 94°C, respectively). This dielectric fluid had a dielectric strength of 12 kV, and the side flush pressure was 0.5 kg/cm² in the experiment. Figure 4.1 shows the electric discharge machine used for the present experimental work.

The alloy steel EN19 specifications are: A material density of 7.85 g/cm³ (EN19) was selected for the workpiece. Despite its good ductility and impact resistance, it is an alloy steel with high strength and wear resistance. 30 mm × 30 mm × 14 mm were the dimensions used for the test. A billet is loaded into the induction furnace, then calcined for 30 min and cooled in an oil bath to 45 HRc, after which the temperature stabilizes at 820°C. In the following process, the parts were ground on a surface grinder to produce a flat surface. During the experiment, copper electrodes (99.9% purity) were used in five different diameters (φ12 mm, φ15 mm, φ18 mm, φ21 mm, and φ24 mm), each 50 mm in length. As a matter of weight percent, 99.9% copper is contained in it as well as 0.04% oxygen, and another 8.97 g per cubic centimeter is its density.

FIGURE 4.1 Electric discharge machine.

Table 4.1 displays the process constraints that were used in the experiment to determine how they affected the performance.

TABLE 4.1 Design Parameters Used in the Study.

Process constraints	Performance measures	Constant constraints
1. Discharge current (I_g)	1. Material removal rate.	1. Flushing pressure
2. Gap voltage (V_g)	2. Tool wear rate	2. Polarity
3. Pulse-on time (T_{on})	3. Surface roughness	
4. Pulse-off time (T_{off})	4. Over cut (OC)	
5. Diameter of electrode		

A suitable experimental design for data collection is used to estimate model parameters more efficiently. The Response Surface Methodology (RSM) design fits curved surfaces to the dataset. It is used to define an area of element space that defines specific performance characteristics for a system or to identify the optimal operating conditions for a system. Even though RSM can help gain these insights, it cannot explain the physical mechanism of the system. Moreover, RSM uses optimization to produce a mathematical model that represents a quadratic polynomial response surface. Using five input variables, three levels of coded values, and 32 experimental runs, a central composite design was constructed.

An electronic balance is used to weigh workpieces and electrodes with a minimum quantity of 0.001 g. To measure the average Ra of a surface, a contact probe Ra tester is used. It has an estimated length of 4 mm and a cut length of 0.8 mm.

4.3 RESULTS AND DISCUSSION

For developing mathematical models that are based on experimental data, experiments need to be carefully planned. When combined with well-designed experiments, mathematical modeling can provide a robust equation that reduces the total number of experiments when compared to classical experiments. As part of the current research, productivity is affected by parameters such as discharge current (I), gap voltage (V), pulse-on time (T_{on}), and pulse-off time (T_{off}). The Ra of the material removed during submerged arc EDM is measured by calculating the MRR, TWR, and Ra. Each variable is divided into five levels.

TABLE 4.2 Experimental Matrix.

Factors	Levels				
	−2	−1	0	+1	+2
Gap current, I, Amp	4	6	8	10	12
Gap voltage, V, Volts	35	45	55	65	75
Pulse-on time, T_{on}, μs	4	27	50	73	96
Pulse-off time, T_{off}, μs	4	5	6	7	8
Tool diameter, D, mm	12	15	18	21	24

Using the encoders and vivo in the table, the central composite design (CCD) matrix was created with 32 encoders. With 16 block points, 6 block centers, 10 pivot points, and 2 alpha points for each of the parameters, there are 24 factorials on each of the 5 levels. At intermediate level (0), all machining parameters form center points. In addition to the input data, Table 4.3 includes the response tables for MRR, TWR, and Ra.

4.3.1 MATERIAL REMOVAL RATE

A total of 32 experiments were carried out, as shown in Table 4.3, according to the experimental plan, each with a combination of several process variable values.

A regression analysis was performed on the input data to determine the relationship between factors and responses, assuming linearly related factors and responses. The error surface was determined using Minitab 16 to ensure the least squares of the error surface. As variables are added to the model, the R^2adj statistics do not always increase. It is often the case that adding unnecessary terms will reduce R^2adj. The presence of a significant difference between R^2 and R^2adj indicates that important terms are not included in the model. An R^2 value of 86.79% was obtained for this experiment, indicating that the predictors explained the variation in the response. In the model with a coefficient of 84.25%, the data are well-adjusted.

Figure 4.2 shows the residual plot of MRR, and in a normal probability plot of residuals, the theoretical percentiles of the normal distribution are plotted on the x-axis and the sample percentiles of the residuals are plotted on the y-axis.

It is possible to see whether the actual and theoretical percentiles are linear by tracing the diagonal line. It is apparent from the scatterplot that the sample percentiles approximate the theoretical percentiles. Based on this

TABLE 4.3 Experimental Matrix and Performance Measurements.

Run no.	I	V	T_{on}	T_{off}	D	MRR	TWR	Ra
8	8	55	50	6	18	17.656	0.223	8.150
32	10	65	27	7	15	34.140	1.672	10.410
9	6	45	27	7	15	14.688	0.401	6.280
26	10	45	27	5	15	29.095	0.803	8.09
25	8	55	50	8	18	18.522	0.089	6.840
11	8	55	50	6	24	18.624	0.089	10.010
15	8	55	50	6	18	17.733	0.111	8.360
29	10	65	27	5	21	25.456	1.468	7.960
22	6	65	73	5	21	9.087	0.037	9.020
16	6	65	27	7	21	12.059	0.669	7.170
1	6	45	73	7	21	5.596	0.178	6.430
14	6	65	27	5	15	13.350	0.123	6.090
18	8	55	50	6	18	19.388	0.089	9.560
24	8	55	50	4	18	18.471	0.468	10.02
12	6	45	73	5	15	1.771	0.045	1.760
13	8	35	50	6	18	14.369	0.178	6.980
19	10	45	73	7	15	19.350	0.134	7.550
27	8	55	50	6	18	20.127	0.156	10.580
7	10	45	73	5	21	18.195	0.102	7.480
5	8	55	96	6	18	13.146	0.223	4.880
28	8	55	50	6	18	17.325	0.167	7.960
10	4	55	50	6	18	3.006	0.156	3.230
6	8	75	50	6	18	22.420	0.089	10.130
20	8	55	50	6	12	15.083	0.089	6.620
4	6	45	27	5	21	14.140	0.087	6.540
2	10	65	73	5	15	26.102	0.279	10.850
17	8	55	4	6	18	14.975	0.203	5.560
30	8	55	50	6	18	20.255	0.123	9.210
3	10	65	73	7	21	28.280	0.100	10.810
23	6	65	73	7	15	8.611	0.167	5.400
31	12	55	50	6	18	35.592	0.357	10.880
21	10	45	27	7	21	32.378	1.115	9.360

data, the residuals follow a normal probability distribution, corresponding to a usually distributed error term. A normal distribution of residuals can be seen in the histogram of residuals. There is, however, one extreme outlier. There is a random bounce between residuals. There is no serial correlation when residuals in the residual = 0 region exhibit normal random noise.

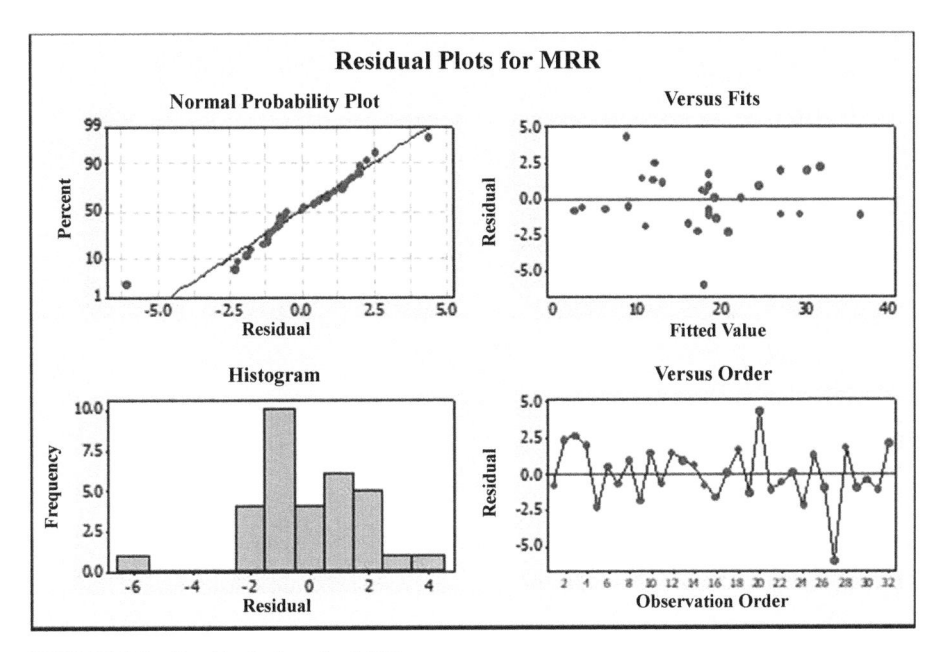

FIGURE 4.2 Residual plots for MRR.

As a result of regression or analysis of variance, the residual plot is used to determine the goodness-of-fit. Assumptions of ordinary least squares are checked by examining residual plots. These assumptions will help to provide an unbiased estimate of the coefficients with the least variance with ordinary least squares regression.

Figure 4.3(a) depicts low MRRs at low input current. MRRs do increase with increasing input current, creating a linear relationship, but then increase with increasing current. Figure 4.3(b) shows a similar relationship between MRR versus T_{on} and V, showing that a high MRR is achieved at the V peaks. Figure 4.3(c) shows the association between I and V. Plotting the peaks shows that the MRR increases with increasing I and V.

The aim now is to maximize the MRR response while maintaining other responses in the EDM process. Table 4.4 shows the general input parameter solution obtained by the response optimizer using the maximize descriptor function.

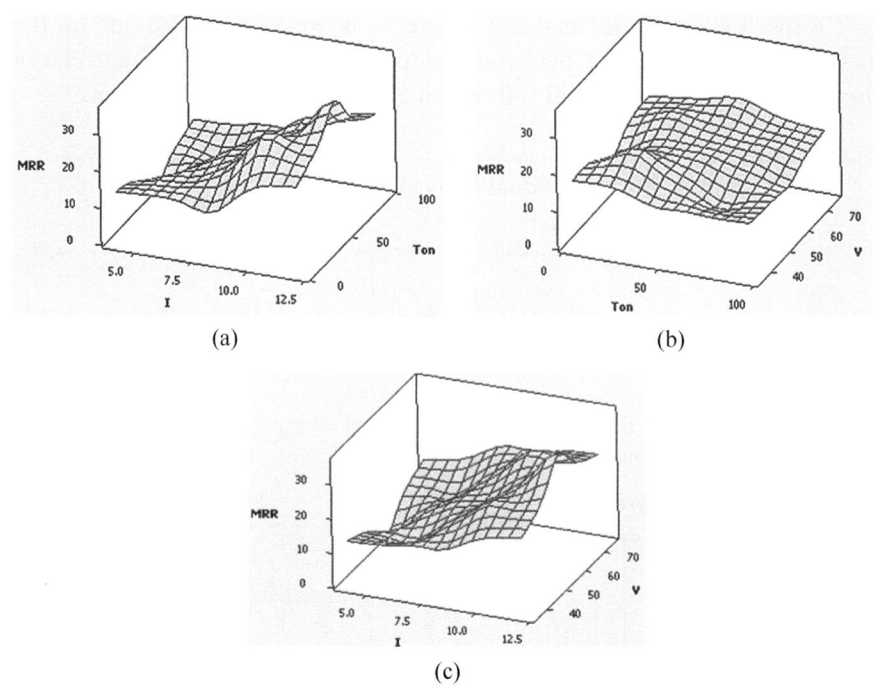

FIGURE 4.3 Surface plots for (a) MRR Vs T_{on} and I, (b) MRR Vs T_{on} and Vg, and (c) MRR Vs V and I.

TABLE 4.4 Response Optimizer by Maximization Desirability Function for MRR.

	Parameters					
	Goal	Lower	Target	Upper	Weight	Import
MRR	Maximum	10	40	40	0.1	1
Starting Point						
	$I_g = 8$	$V_g = 55$	$T_{on} = 50$	$T_{off} = 6$	D = 18	
Global Solution						
	$I_g = 12$	$V_g = 75$	$T_{on} = 4$	$T_{off} = 8$	D = 24	
Predicted Responses						
MRR = 44.3319		Desirability = 1.0000		Composite Desirability = 1.000000		

Determining the overall solution to the input variables to satisfy the previous MRR maximization criterion is solved using the desired likelihood maximization function of the response optimizer in the Minitab 16 environment. Individual MRR material removal requirements are a global solution. Figure 4.4 depicts the response optimization plot for MRR.

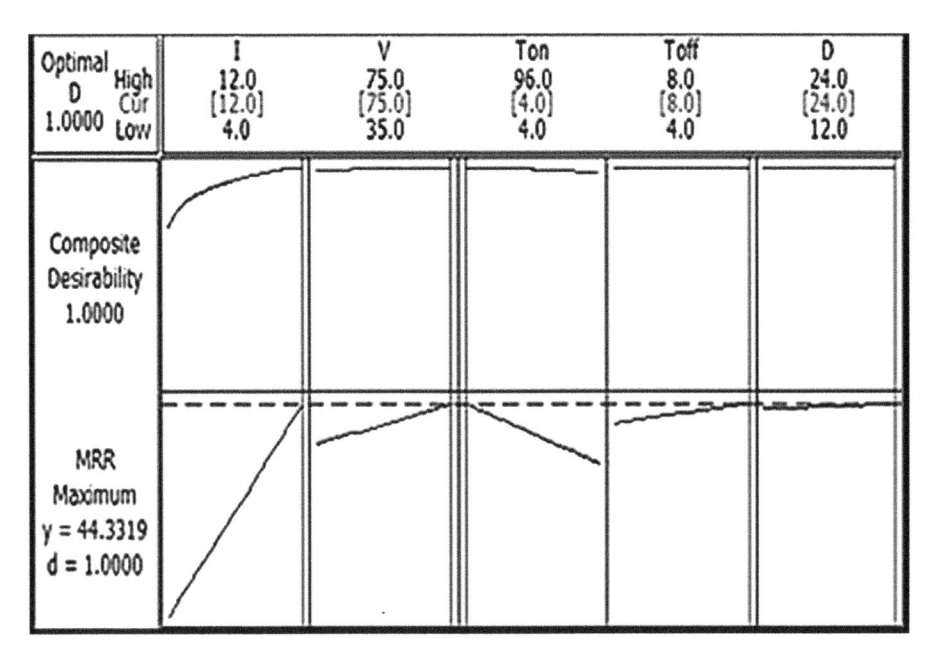

FIGURE 4.4 Response optimization plot for material removal rate.

4.3.2 TOOL WEAR RATE

A total of 32 experiments were conducted according to the experimental design, each with combinations of the process variables. A linear equation can be used to represent the solutions to the experiments.

Minitab 16 was used to compute the least squares of the error surface using regression analysis. In order to determine the relationship between input data and TWR, regression analysis was performed. The factors and responses are assumed to have a linear relationship in the process of regression analysis. An increase in R^2 adj stat does not always result from the addition of additional variables. Aside from that, adding unnecessary terms usually decreases the R^2 adj value. The R^2 adj value is likely to differ significantly from R^2 and is more indicative of important terms not being included in the model. An R^2 score of 72.80% indicates that the predictors explained a large percentage of the variation in response. A different set of data was compared. Figure 4.5 depicts the residual plot for TWR, and in a normal probability plot of residuals, the theoretical percentiles of the normal distribution are plotted on the x-axis and the sample percentiles of the residuals are plotted on the y-axis.

Residual Plots for MRR

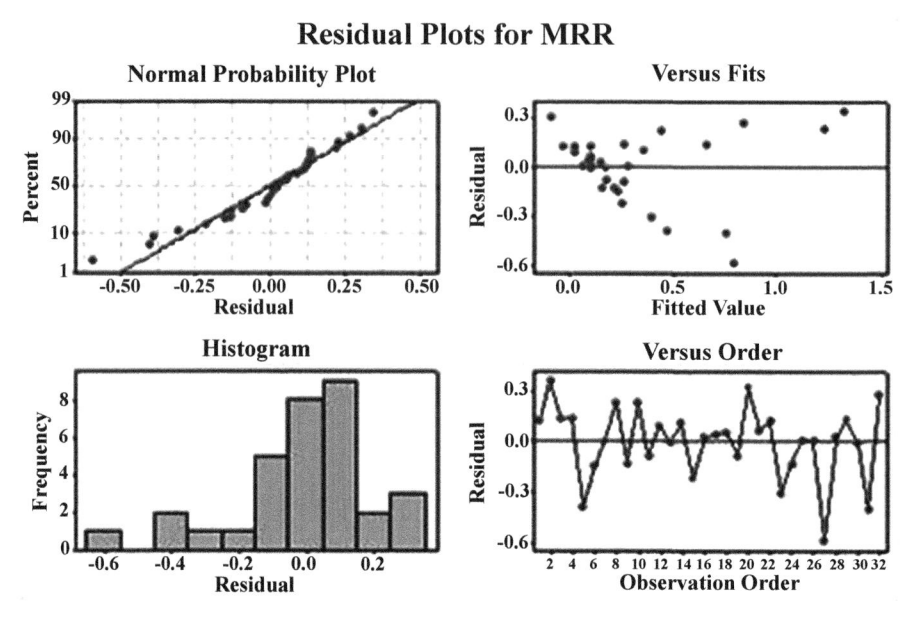

FIGURE 4.5 Residual plot for TWR.

It has been observed that the electrode wear rate increases with increasing input current, although a low TWR at initial low I_g values and a low T_{on} create a linear relationship, but then it increases with increasing current. Figure 4.6(a) shows the relationship between I_g and V. In this graph, the peaks show an increase in TWR with increasing V. Figure 4.6(b) shows the relationship between TWR versus T_{on} and V, where it is clear that there is a linear relationship between V and TWR. The influence of V on the electrode wear rate is minimal.

Each reaction inside the study is expressed one at a time as linear and nonlinear features of the input variables including I, T_{on}, T_{off}, V, and D. It is preferred to limit the reaction TWR and concurrently preserve different responses in the EDM process. Table 4.5 indicates the global solution of input parameters that's acquired by means of a reaction optimizer. To decide the global solution of input variables so that it will fulfill the above standards of TWR minimization, it has been solved via the reaction optimizer desirability minimization feature in the Minitab 16 environment. The user desirability

Electrical Discharge Machining Process Optimization

for TWR is 1. To acquire this desirability, the most efficient value component ranges may be set as proven beneath the global solution. Figure 4.7 depicts the response optimization plot for TWR.

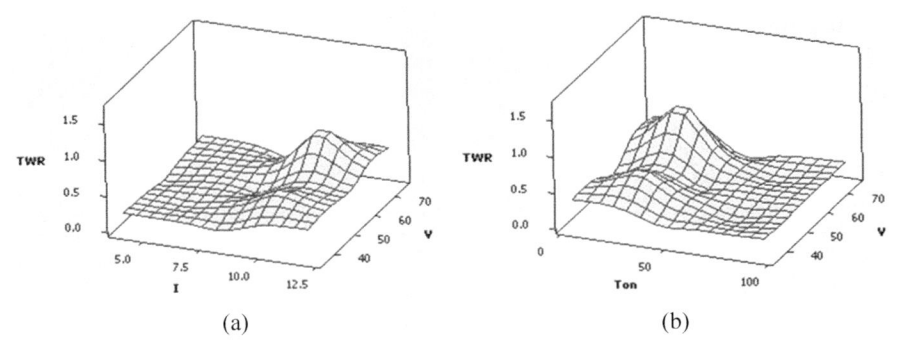

FIGURE 4.6 Surface plot for (a) TWR vs Vg and Ig and (b) TWR vs T_{on} and V_g.

FIGURE 4.7 Response optimization plot for tool wear rate.

4.3.3 SURFACE ROUGHNESS (RA)

Several variations of each process variable were used in the 32 experiments, according to the experiment plan. The results are provided as a linear equation as follows.

TABLE 4.5 Response Optimizer by Minimization Desirability Function for TWR.

	Parameters					
	Goal	Lower	Target	Upper	Weight	Import
TWR	Minimum	0.1	0.1	1	10	1
Starting point						
	$I_g = 8$	$V_g = 55$	$T_{on} = 50$	$T_{off} = 6$	$D = 18$	
Global solution						
	$I_g = 4$	$V_g = 35$	$T_{on} = 96$	$T_{off} = 4$	$D = 12$	
Predicted responses						
TWR = −0.678396		Desirability = 1.0000		Composite Desirability = 1.000000		

Minitab 16 was used to provide the least squares error surface for the regression analysis of the input data and Ra. An analysis of regression assumes factors and responses have linear relationships. As variables are added to the model, the R^2 adj stat will not always increase. R^2 adj values are usually decreased when terms are added that are unneeded. R^2 indicates that 83.47% of the response variable was explained by the predictors in this experiment. A different dataset was then used to validate the predictive model. The residual plot for Ra is shown in Figure 4.8.

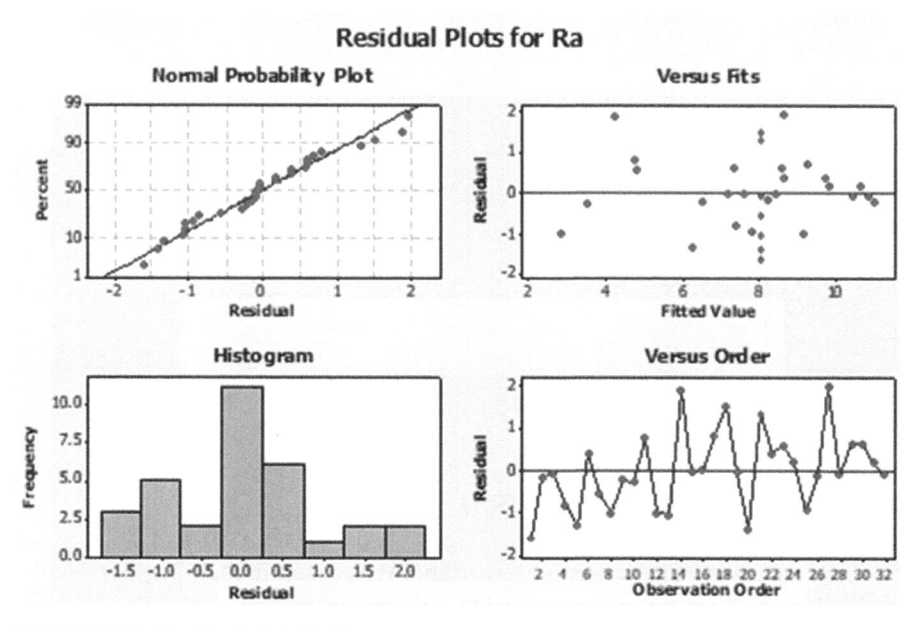

FIGURE 4.8 Residual plots for Ra.

Ra increases as input current increases, according to Figure 4.9(a). It increases with increasing voltage and decreases with increasing current, even though the initial low Ig values have a low Ra and a low Vg, giving a linear relationship. Ra versus T_{on} and Vg is shown in Figure 4.9(b), and there is a linear relationship between Ra and Vg in this context. Ra is minimally affected by Vg. Figure 4.9(c) shows Ig's relationship with T_{on}. Ra reaches its optimum value with an increasing gap current. Ra increases as T_{on} increases on this graph. Ra is determined by the interaction of Ig and T_{on}.

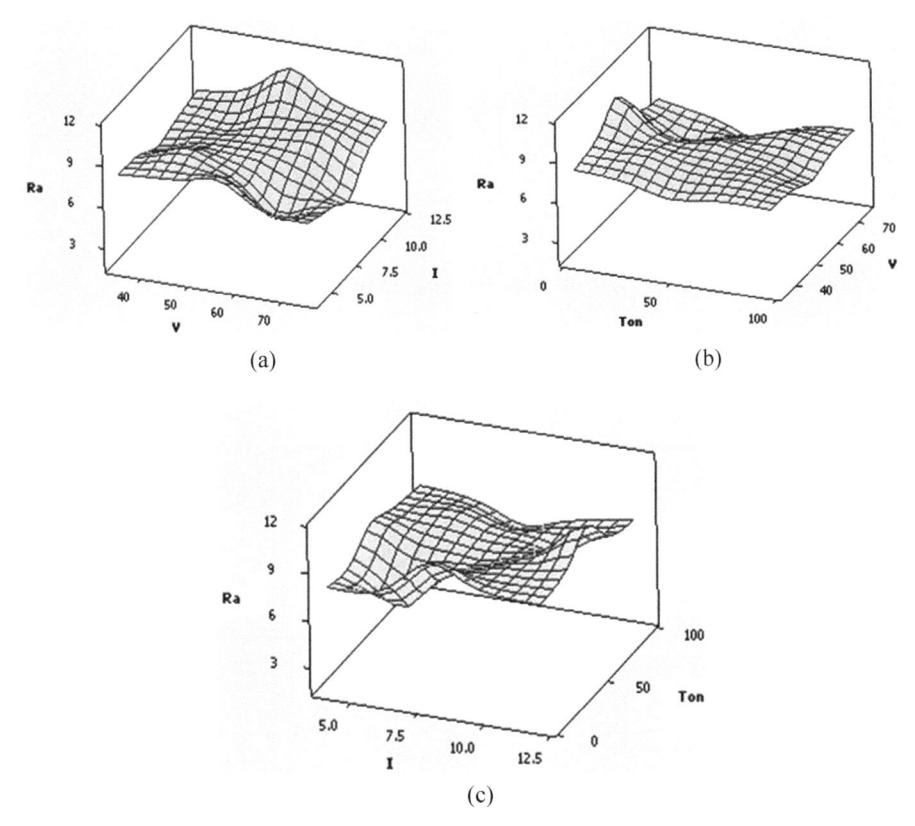

FIGURE 4.9 Surface plots for (a) Ra Vs V_g and I_g, (b) Ra Vs T_{on} and V_g, and (c) Ra Vs I_g and T_{on}

There are separate linear and nonlinear functions for each response to each input variable in this study, such as I, T_{on}, T_{off}, V, and D. Table 4.6 shows the overall input parameter solution obtained by the response optimizer. In order to determine the general solution of the input variables that satisfies the above criteria for minimizing Ra, it was solved using the response optimization function in Minitab 16, which is required separately for Ra. For this convenience, you can set the optimal levels of the coefficients, as in the general solution.

TABLE 4.6 Response Optimizer by Minimization Desirability Function for Ra.

	Parameters					
	Goal	Lower	Target	Upper	Weight	Import
TWR	Minimum	1	4.34	10	0.1	1
Starting Point						
	$I_g = 4$	$V_g = 35$	$T_{on} = 19$	$T_{off} = 4$	$D = 12$	
Global Solution						
	$I_g = 4$	$V_g = 35$	$T_{on} = 15$	$T_{off} = 4$	$D = 12$	
Predicted Responses						
Ra = 4.34		Desirability = 1.0000		Composite Desirability = 1.000000		

The response optimization plot for Ra is shown in Figure 4.10.

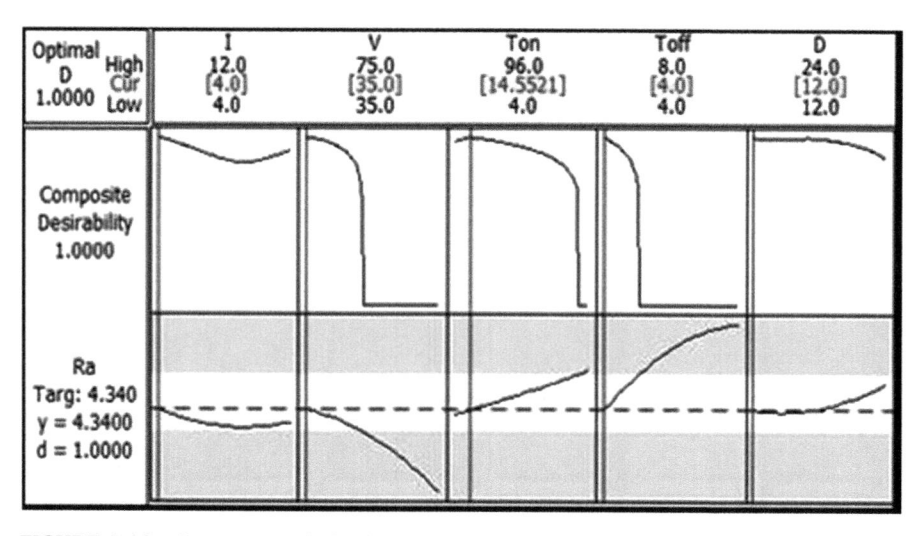

FIGURE 4.10 Response optimization plot for surface roughness.

4.4 CONCLUSIONS

In this work, on the basis of experimental results, a parametric analysis of the process of electrical discharge machining was carried out. Copper electrodes were used for machining EN19 alloy steel, and commercial EDM oil was used as the dielectric. Response surface methodology was employed to design the experimental layout, and the observed responses in this study are MRR, TWR, and Ra. From experimental response surface plots, it is possible to satisfactorily predict the values of Ra, TWR, and MRR from experimental data. Using the composite design, experiments can be efficiently and economically carried out by generating graphs and statistical models. Based on the desirability concept and Minitab optimizer function, the best MRR is 44.332 mm^3/min at 12 A gap current, 75 V gap voltage, 4 μs pulse duration, 8 μs pulse-off time, and 24 mm diameter. A gap current of 12 A, a voltage of 75 V, on and off times of 8s, and a 24 mm diameter produced the best MRR. The best TWR obtained from the desirability concept and Minitab optimizer function is 0.678 mm^3/min at 4 A gap current, 35 V gap voltage, 96 μs pulse-on time, and 4 μs pulse-off time for 12 mm diameter. The best Ra obtained from the desirability concept and Minitab optimizer function is 4.34 μm at 4 A gap current, 35 V gap voltage, 15 μs pulse-on time, and 4 μs pulse-off time for 12 mm diameter. EN19 full-hardening steel is a material of importance in the automotive sector. The manufacturers need a system for setting parameters that is fast and consistent. Manufacturers will be able to respond to new demands with better surface finish quality using the developed model. Mass production can benefit greatly from this conclusion. A process parameter can be set so that production time can be shortened without sacrificing surface quality.

KEYWORDS

- **electrical discharge machining**
- **optimization**
- **En19 alloy steel**
- **desirability**
- **copper electrodes**

REFERENCES

1. Durairaj, M.; Sudharsun, D.; Swamynathan, N. Analysis of Process Parameters in Wire EDM with Stainless Steel using Single Objective Taguchi Method and Multi Objective Grey Relational Grade, *Procedia Eng.* **2013,** *64,* 868–877.

2. Manivannan, R.; Pradeep, K. M. Multi-attribute Decision-making of Cryogenically Cooled Micro-EDM Drilling Process Parameters Using TOPSIS Method, *J. Mater. Manuf. Proc.* **2017,** *32* (2), 209–215.
3. Singh, G.; Sidhu, S, S.; Bains, P. S.; Bhui, A. S. Improving Microhardness and Wear Resistance of 316L by TiO2 Powder Mixed Electro Discharge Treatment. *Mater. Res. Express.* **2019,** 6086501.
4. Dušan, P.; Miloš, M.; Miroslav, R.; Valentina, G. Application of the Performance Selection Index Method for Solving Machining Mcdm Problems. *Mech. Eng.* **2017,** *15,* 97–106.
5. Prabhu, S.; Vinayagam, B. K. Multiresponse Optimization of EDM Process with Nanofluids Using TOPSIS Method and Genetic Algorithm. *Arch. Mech. Eng.* **2016,** 45–71.
6. Gadakh, V. S.. Parametric Optimization of Wire Electrical Discharge Machining Using Topsis Method. *Adv. Prod. Eng. Manag.* **2012,** *7* (3), 157–164.
7. Dastagiri, M.; Srinivasa, R. P.; Valli, P. M. In *TOPSIS, GRA Methods for Parametric Optimization on Wire Electrical Discharge Machining (WEDM) Process*, Design Research Conference (AIMTDR–2016) College of Engineering; 2016.
8. Gopalakannan, S.; Senthilvean, T.; Ranganathan, S. Modeling and Optimization of EDM Process Parameter on Machining of Al 7075-B$_4$C MMC Using RSM. *Procedia Eng.* **2012,** *38,* 685–690.
9. Rajmohan, T.; Prabhu, R.; Rao, S. G.; Palanikumar, K.; Optimization of Machining Parameter in EDM of 304 Stainless Steel. *Procedia Eng.* **2012,** *38,* 1030–1036.
10. Norliana, M. A.; Yusoff, N.; Mahmod. R.; Electrical Discharge Machining: Practices in Malaysian Industries and Possible Change towards Green Manufacturing. *Procedia Eng.* **2012,** *41,* 1684–1688.
11. Vishwakarma, M.; Parashar, V.; Khare. V. K. Regression Analysis and Optimization of Material Removal Rate on Electric Discharge Machine for EN-19 Alloy Steel. *Int. J. Sci. Res. Public.* **2012,** *2* (11).
12. Rajesh, R.; Anand, M. D. The Optimization of the Electro-Discharge Machining Process Using Response Surface Methodology and Genetic Algorithms. *Procedia Eng.* **2012,** *38,* 3941–3950.

CHAPTER 5

Some Studies on Hole Feature Recommendation for Additive Manufacturing Processes

DAMA Y. B., BHAGWAN F. JOGI, and R. S. PAWADE

Department of Mechanical Engineering, Dr. Babasaheb Ambedkar Technological University, Lonere-Raigad, Maharashtra, India

ABSTRACT

Additive manufacturing (AM) is developing as a major manufacturing technology in today's world. Colossal development has been observed in recent decades and that too at a much faster pace. Additionally, it has matured from simple prototyping to actual end-use products and production tooling. Various manufacturing processes are developed by using AM techniques such as SLS, FDM, DMLS, SLA, PolyJet, LTP, FDM, LENS, and binder jet printing. Layer-by-layer addition of materials is a crucial method in all these processes, and hence this technology is referred to as AM. There are many equivalent terms used like rapid prototyping, 3D printing, digital manufacturing, and so on. AM is used in various industries like aerospace components, automotive, consumer goods, medical equipment parts and devices, fashion industry, jewelry, etc. All parts can be manufactured by using either subtractive manufacturing or AM techniques. However, several design features have manufacturing challenges in metal substation manufacturing and metal AM techniques. Since their has been an increase in the number of parts manufactured directly using AM techniques, it is essential to list the best design for manufacturing principles suitable for different AM processes. This will help the design community ensure parts are designed from AM instead of general strategy and printed. The fundamental point of this work is to consider and comprehend the distinctive additive processes and discover the best design for manufacturing practices to be followed. Consequently, one should be ready to pick a proper AM process for printing based on the design features of a component to be printed.

5.1 INTRODUCTION

The additive manufacturing (AM) technique[1] is a promising technology that will be widely accepted across significant manufacturing industries worldwide in the coming future. Colossal development has been seen in recent decades and is at a much faster pace.[2] It has matured from simple prototyping to actual end-use products and production tooling. Various manufacturing processes are developed by using AM techniques like stereolithography (SLA), selective laser sintering (SLS), fused-deposition modeling (FDM), PolyJet (3DP), liquid thermal polymerization (LTP), direct metal-laser sintering (DMLS), ballistic particle manufacturing (BPM), binder jet printing (BJP), and laser-engineered net shaping (LENS),[2–9] etc. Layer-by-layer deposition of materials is the critical method in all these processes, and hence this technology is referred to as AM. There are several equivalent terms used, such as rapid prototyping, 3D printing, etc. AM is used in various aerospace, automotive, consumer goods, medical equipment parts and devices, fashion industry, jewellery,[10–14] etc. Wohlers Report, 2017, shows percentagewise use of AM in different domains, as is shown in Figure 5.1.[15]

As per the Wohlers Report, April 2019, the AM process is extensively used by the USA, followed by China and Germany.[16] Looking into the initial research and development investment and process adoption in the USA and China, other countries adopted this technology. Future tread can be seen in Figure 5.2. Also, Wohlers Report, August 2019, shows that the use of AM has increased for end-use parts compared to functional prototype parts. This indicates that the AM process is slowly moving toward stabilization and industry adoption for functional use rather than prototyping.[17]

Following typical challenges are the main reasons for the slower pace of adopting AM technique:[18–20]

- High initial investment and maintenance cost
- Increased processing and raw material cost
- Layer-by-layer build creation slows the part printing
- Lack of design standards
- Design for manufacturing—first-time right component

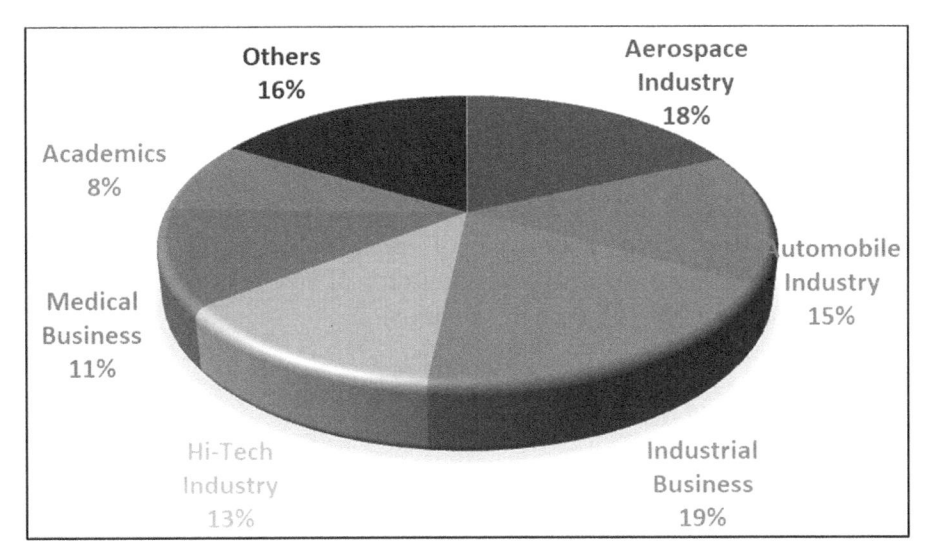

Some Studies on Hole Feature Recommendation 79

FIGURE 5.1 Additive manufacturing application in different domains.
Source: Adapted from Ref. [15].

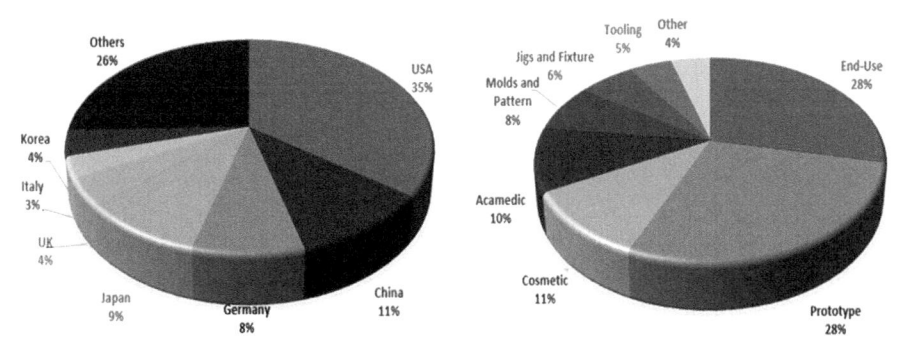

FIGURE 5.2 AM research and developments in different countries and process adoption in different applications.
Source: Adapted from Refs. [16,17].

Many of the 3D-printing manufacturers are working on the first three challenges to capture most of the market share. These first three challenges will be overcome in the coming few years, as predicted by the current trend and growth of AM. However, to cope with the last two challenges, researchers and industry R&D need to come forward to work together to share their 3D printing experience and knowledge. This would help establish the design standards for an AM process and promote the further adoption and use of this technique.

3D printing of a final component in the first build is another crucial challenge. One needs to get command on the 3D printing of all parts irrespective of particular industry domain components. To overcome this challenge, it is necessary to conduct 3D-printing exercises and understand different factors to achieve part quality and accuracy. Based on the recent 3D-printing experiences, it is observed that additional design features created on the part to be produced gave rise to different challenges to print.

This paper focuses on the 3D-printing study of hole features for different AM processes and presents recommendations.

5.2 MATERIALS AND METHODS

Different CAD softwares were used for the 3D-printing study of hole features for the various AM processes. These included PTC—Creo Parametric 6.0,[21] DS—SolidWorks 2019,[22] and Siemens—NX 1899.[23] The part with hole features was modeled using their CAD software and standard triangle language (STL) file preparation and subsequent printing of a few of the sample parts.

The following AM machines were used for the 3D-printing study of hole features for a different AM process; these are—fused-deposition method,[1] selective laser apparatus,[4] and selective laser sintering.[10]

5.2.1 DESIGN REQUIREMENTS

To achieve some functional requirements, design engineers first carry out component design and then assembly design. Once a design is ready from the operating point of view, a prototyping model is created to check how part/assembly functions. A further possibility of any modifications concerning other aspects is also considered; it might be the aesthetic look, strength, etc. Figure 5.3 exhibits some of the features used while designing the component to achieve the functional needs. Different types of features can also be created using traditional manufacturing processes. Table 5.1 depicts some of the features of the conventional manufacturing process.

The features mentioned in Table 5.1 for the different manufacturing processes are well practiced in the conventional manufacturing processes. There are some standard guidelines available for designing the design with such features based on the manufacturing process. For 3D printing of parts, it is observed that designers are directly using similar traditional techniques for the AM process due to a lack of experience in manufacturing/printing components using AM. Besides, there are no well-defined AM design guidelines for building the right ingredients and assemblies for the first time. Hence, it is imperative that designers also understand the different AM processes and working principles. In addition, one needs to understand the advantages and limitations of each approach in terms of printing material, application of the printed component, and life cycle in actual use.

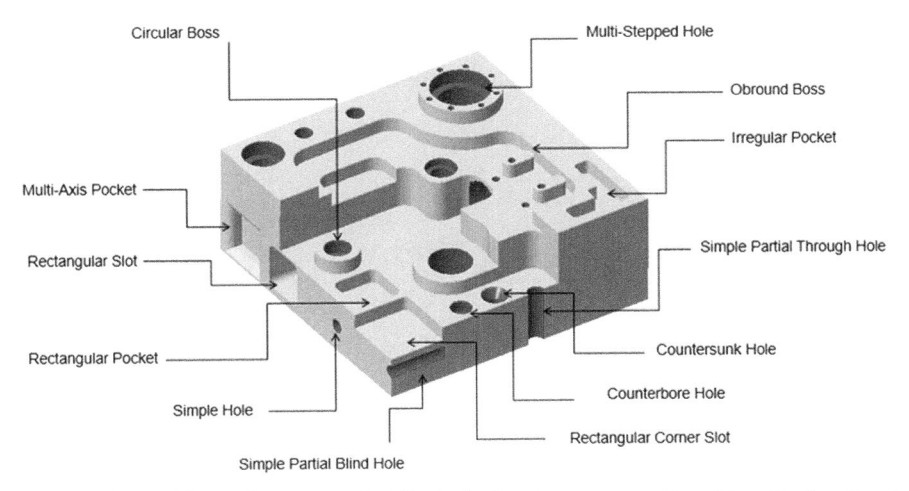

FIGURE 5.3 Different features used while designing the component to achieve the functional needs.

TABLE 5.1 Typical Features Created for Traditional Manufacturing Processes.

Machining	Sheet metal	Injection molding	Casting	Forging
Pocket	Holes	Ribs	Corner radius	Bosses
Cutouts	Cutouts	Bosses	Holes	Ribs
Slots	Gusset	Corners	Ribs	Fillets
Holes	Emboss	Holes	Bosses	Tapers
Grooves	Hem	Pin	Draft	Corners
Side radius	Curl	Wall thickness	Projections	Webs
Bottom radius	Notch	Snap	Wall thickness	Planes

5.2.2 AM PROCESSES

There are seven AM processes according to ASTM committee F42 AM.[18,25]

1. Material extrusion

2. Material jetting
3. Powder bed fusion
4. Sheet lamination
5. Binder jetting
6. Directed energy deposition or laser metal deposition
7. Hybrid technologies

A popular approach is to classify according to baseline technology, like using lasers, printer technology, extrusion technology, etc. Another method is to collect the processes together according to the type of raw material input.

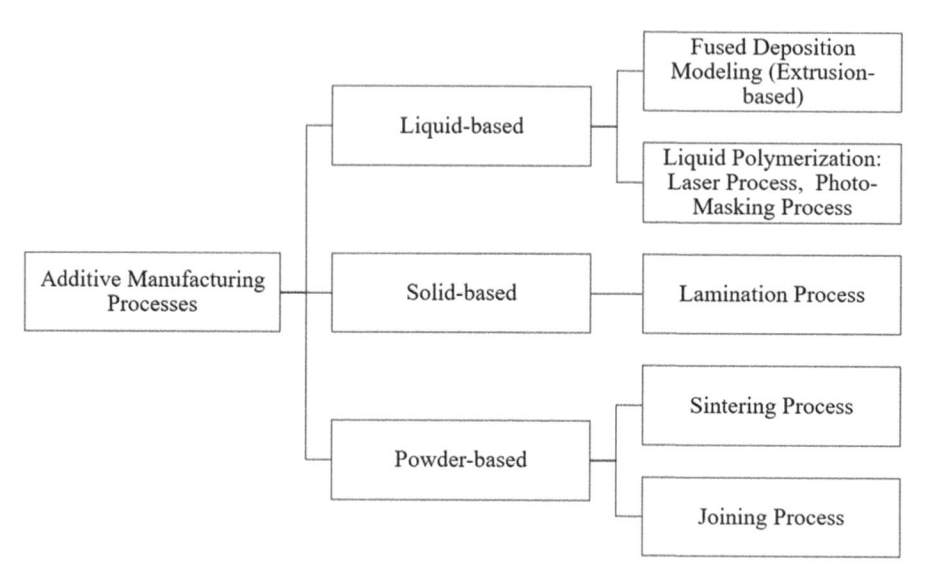

FIGURE 5.4 Classification of additive manufacturing processes.

5.3 RESULTS AND DISCUSSION

5.3.1 CHALLENGES DURING THE 3D PRINTING OF HOLE FEATURES

The results are shown in Figures 5.5–5.12. A detailed study was conducted to understand different AM process behavior while building components with varying design features. The initial focus was on the hole features on the part, and AM process might be more suitable for printing holes. Hole features are generally used in different component designs, for example, machining, sheet metal, plastic, casting, and die-casting. In conventional methods, mostly hole features are produced using a material removal process depending on the accuracy requirement. Also, hole features are generated by using cores in the casting and die-casting process. Therefore, developing hole features on parts is challenging for the AM process when accuracy is essential.

5.3.2 HOLE FEATURES AND ITS TYPES

There are several types of hole features created on part design depending on the functional requirements. Following are some of the typical holes features available in CAD software which are most commonly used (Figure 5.5).

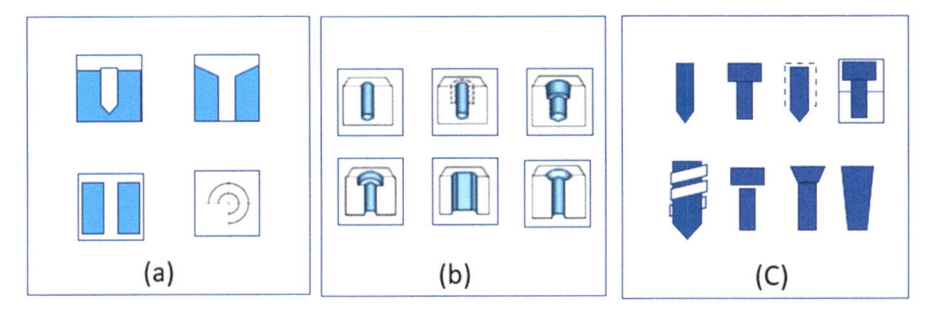

FIGURE 5.5 (a) Creo parametric—hole features, (b) solidworks—hole features, and (c) NX-hole features.

Among different hole features, a simple hole feature is chosen to study manufacturability using an AM process and 3D-printing challenges.

5.3.3 SIMPLE HOLE FEATURE

A hole feature is defined as an opening in something, usually a solid body. The hole feature was found very useful for various engineering/non-engineering purposes. The hole can be a through-hole or blind spot, depending on the functional requirements. A blind hole can be generated using reaming, drilling, or milling operation to a specific depth and not to the other side of the workpiece. Through-hole can be caused by using drilling or milling holes throughout the material of a solid object.[26] Figure 5.6 depicts a few types of spots.

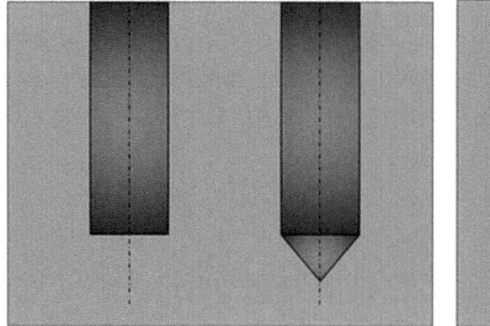

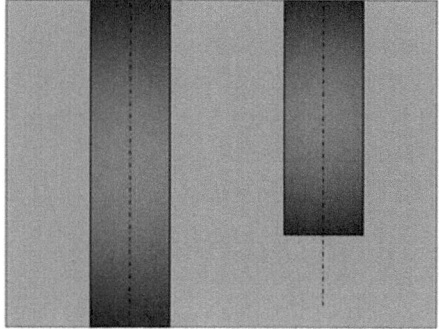

FIGURE 5.6 Types of simple holes—(a) flat bottom hole and conical bottom hole, and (b) through-hole and blind hole.

Following are the few factors which has a reasonable impact on the accuracy of hole features (Figure 5.6).
Design for manufacturing approach

  i. Hole size and parameters
 ii. Hole axis concerning the build direction
iii. Impact of tensile load on printed part hole feature
 iv. Tessellation parameters used to generate the STL file

Design for Manufacturing Approach

i. Hole size and parameters

Smaller hole diameter exhibits less accuracy as compared to larger diameters. However, hole size below a particular value is challenging to achieve as hole features are filled with printing material, as shown in Figures 5.8 and 5.9. These capabilities varices with different AM processes.

ii. Hole axis concerning the build direction

In an experimental study, further observations were made for hole feature design—holes when parallel to build direction and spots when perpendicular to create movement (Figure 5.9).

Case A: Hole Axis Parallel to Build Direction: In this case, 3D-printer builds the part as a series of layers on top of one another. Since the previous layer supports the new layer, part of this method generates the relatively good quality holes.

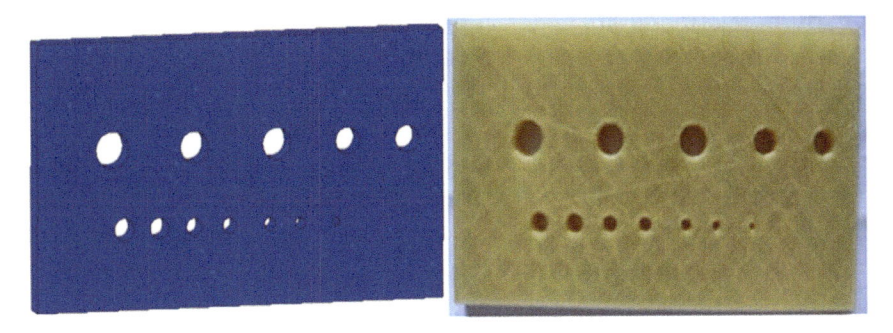

Some Studies on Hole Feature Recommendation 85

FIGURE 5.7 Photographs hole size in CAD vs. actual 3D-printed.

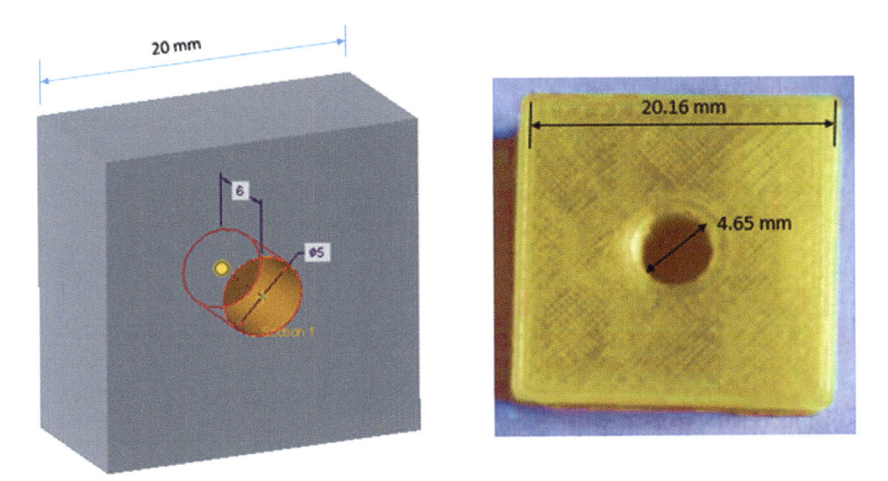

FIGURE 5.8 Photographs of the test artifact hole size in CAD vs. actual 3D-printed.

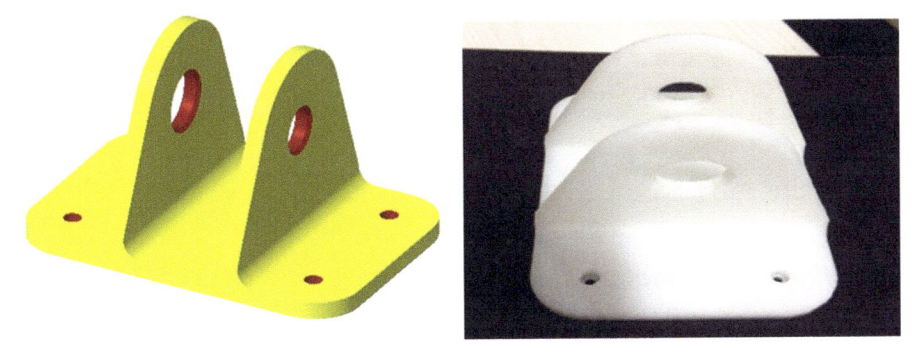

FIGURE 5.9 Photographs of the test artifact—(a) CAD file with different hole axis concerning the build direction, and (b) 3D-printed part to study the hole features.

Case B: Here, the hole axis is perpendicular to the build direction. If the part with the hole axis is oriented perpendicular to the build direction, the amount will be built as a series of layers' top of each other. In the hole feature, the region support structure is required in the FDM and SLA process, which is a wastage of material and cycle time increases for building support structures. Therefore, the overall cost will increase for holes perpendicular to the build direction. Further, the accuracy of the spot is impacted. The comparison of accuracy achieved between 3D-printed dilemmas in the case of A and B shows that the holes with axis parallel to build direction are more accurate than holes with axis perpendicular to creating demand.

iii. Impact of Tensile Load on 3D-Printed Part with Hole Feature

3D-printing layers and tensile loading directions are shown in Figure 5.11. There are two possibilities: tension load should be normal to the layers as per case A, and the tensile load will be parallel to the layers as per case B. The effect of the loading on the layers is as follows:

A: Tension load average to layers—Part is weak
B: Tension load parallel to layers—Part is strong

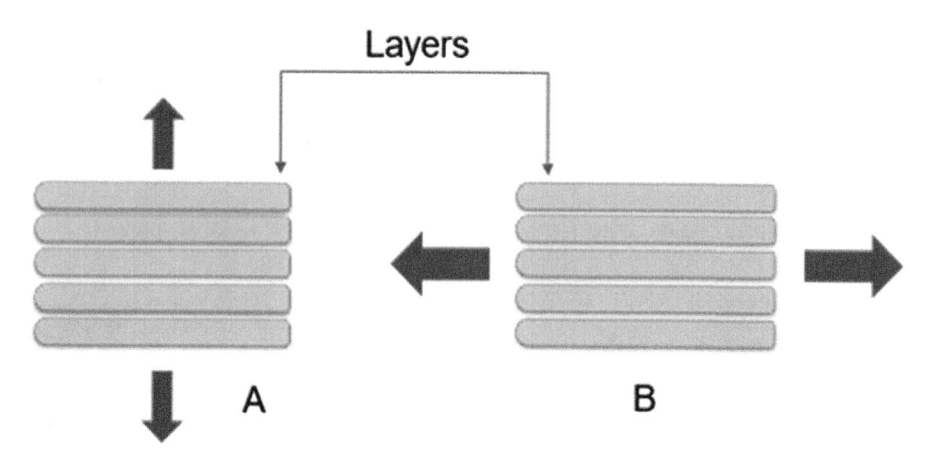

FIGURE 5.10 3D-printing layers and tensile loading directions.

Tessellation Parameters of STL File

STL file is input format to AM process machine. Once the CAD design is ready to build the physical model, the CAD file needs to be translated into STL file format. All the major CAD software has an in-built facility to convert the CAD file into STL file format. During the CAD file conversion, it tessellates the complete model into triangulation format as shown in Figure 5.9(a), and surface geometry shape information of the object is stored in this file. STL file is a universal file format accepted as an input file by all 3D printers which read the file and slice the 3D object into 2D layers information. Further, this slicing information is stored as G-code language for building the physical parts.

Tessellation parameters like chord height, angle of contact, and step size play an essential role while 3D-printing the designed parts. Figure 5.11(b)–(d) exhibit how tessellation size varies after changing the chord height. It is observed that the accuracy of hole size increases with acceptable tessellation parameters like chord height which should be as minimum as possible.

The hole feature should be reasonable in size to properly get 3D-printed. The hole size should be greater than the minimum size specified in the machine manufacturer configuration details.

One should ideally avoid creating a hole with an axis perpendicular to build direction. If it is not possible to avoid creating a hole with an axis perpendicular to build direction, modify the hole cross-section to a pear or cone shape as shown below in Figure 5.12.

As shown above, the alternate ways to create holes on vertical faces or holes with axis perpendicular to build direction will help to achieve the functional need. Even if hole gets created in smaller size or not much accurate in size, such methods will be useful to accommodate the functional need.

A—Nonpreferred for hole preparation when accuracy is very important.

B, C, and D—Preferred for hole preparations with 3D-printed parts.

C and D—More preferred when some other components needs to be assembled.

B—Type hole preferred where much flexibility is not available for hole method and accuracy of 3D-printed hole feature is important.

5.4 CONCLUSION

To improve the hole features 3D-printing, the following are recommended in this study.

The hole feature should be reasonable in size to properly get 3D print. The hole size should be greater than the minimum size specified in the machine manufacturer configuration details. One should ideally avoid creating a hole with an axis perpendicular to build direction. If it is not possible to avoid

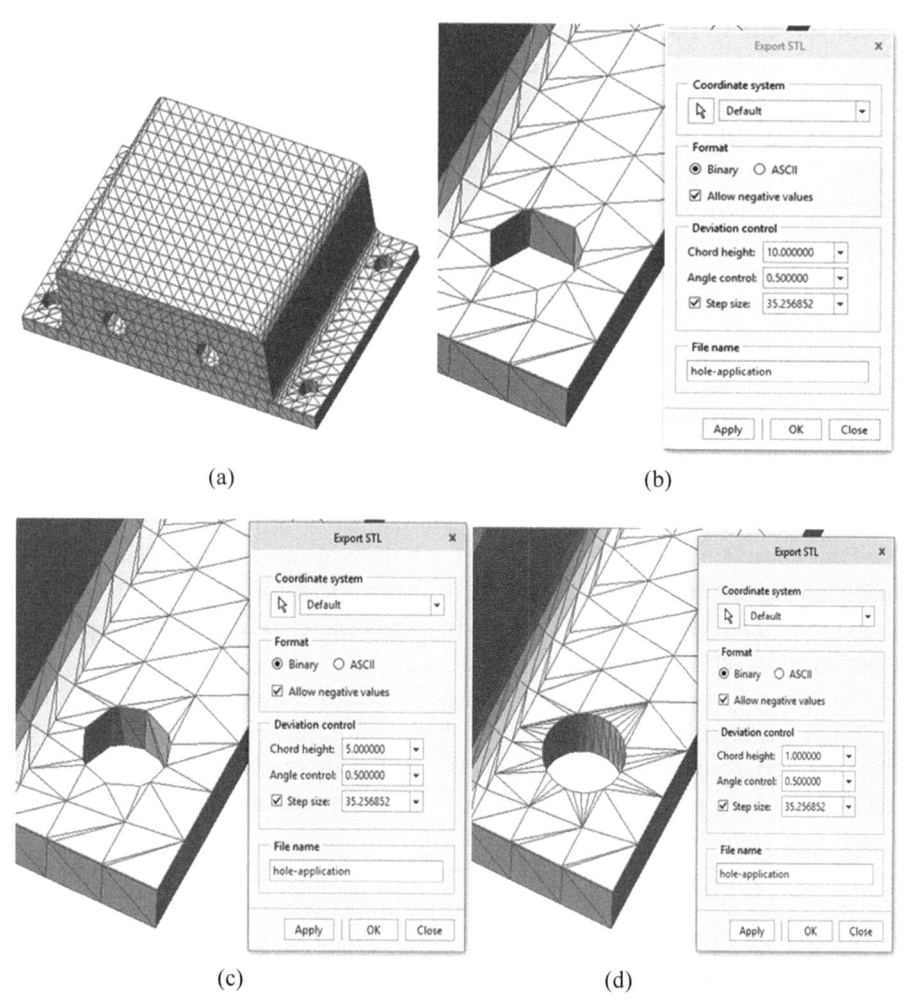

FIGURE 5.11 Screenshot of the test artifact—(a) tessellation CAD file, and (b), (c), and (d) impact of chord height for generating STL file.

creating a hole with an axis perpendicular to build direction, modify the hole cross-section to a pear or cone shape, as shown in Figure 5.12.

A—Non-preferred for hole preparation when accuracy is very important.

B, C, and D—Preferred for hole preparations with 3D printed parts.

C and D—More preferred when some other components needs to be assembled.

B—Type hole preferred where much flexibility is not available for hole method and accuracy of 3D Printed hole feature is important.

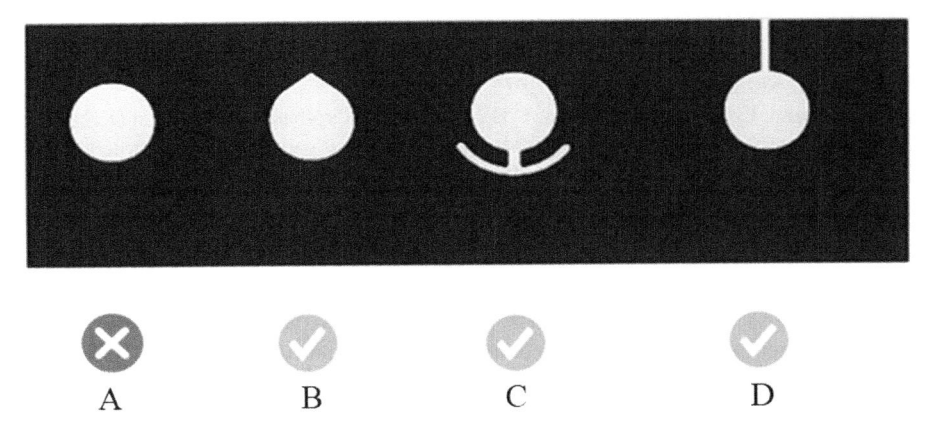

FIGURE 5.12 Alternate methods to create hole features perpendicular to build direction.

No error was found in the B type of hole as compared to other hole preparation methods.

Hence, B method is very useful for making holes in 3D-printing parts.

This research work is very useful for researchers, academicians, and people working in 3D-printing-based manufacturing domain.

KEYWORDS

- **additive manufacturing**
- **design for manufacturing**
- **FDM**
- **hole feature**

REFERENCES

1. Gibson, L.; Rosen, D. W.; Stucker, I. B. *Additive Manufacturing Technologies Rapid Prototyping to Direct Digital Manufacturing*; Springer; 2010. ISBN: 978-1-4419-1119-3
2. Wohlers, A. Wohlers Report, 3D Printing and Additive Manufacturing State of the Industry. *Annual Worldwide Progress Report*, Associates Wohlers; 2010. ISBN 978-0-9913332-6-4
3. Thompson, M. K.; Vaneker, G. M. T.; Fade, G.; Campbell, R. I.; Gibson, I.; Bernard, A.; Schulz, J.; Graf, P.; Ahuja, B.; Martina, F. Design for Additive Manufacturing: Trends, Opportunities, Considerations, and Constraints. *CIRP Ann.-Manuf. Techn.* **2016,** *65,* 737–760.

4. Olakanmi, E. O.; Cochrane, R. F.; Dalgarno, K. W. A Review on Selective Laser Sintering/Melting (SLS/SLM) of Aluminum Alloy Powders: Processing, Microstructure and Properties. *Prog. Mater. Sci.* **2015**, *74*, 401–477.
5. Uhlmann, E.; Kersting, R.; Klein, T. B.; Cruz, M. F. Wire-based Laser Metal Deposition for Additive Manufacturing of TiAl6V4: Basic Investigations of Microstructure and Mechanical Properties for Buildup Parts. *Procedia CIRP,* **2015**, *35,* 55–60.
6. Herzog, D.; Seyda, V.; Wycisk, E.; Emmelmann, C. Additive Manufacturing of Metals. *Acta Mater.* **2016**, *117,* 371–392.
7. Annoni, M.; Giberti, H.; Strano, M. Feasibility Study of an Extrusion-based Direct Metal Additive Manufacturing Technique. *Procedia Manuf.* **2016**, *5,* 916–927.
8. Thompson, S. M.; Bian, L.; Shamsaei, N.; Yadollahi, A. An Overview of Direct Laser Deposition for Additive Manufacturing; Part I: Transport Phenomena, Modeling and Diagnostics. *Addit. Manuf.* **2015**, *8,* 36–62.
9. Morvan, S.; Fadel, G. M.; Keicher, D.; Love, J. In *Manufacturing of a Heterogeneous Flywheel on a LENS Apparatus*, Solid Freeform Fabrication Conference, Austin TX Published in Conference Proceedings, 2001.
10. R. Liu; et al. *Aerospace Applications of Laser Additive Manufacturing A2—Brandt, Milan, Laser Additive Manufacturing*; Woodhead Publishing, 2016; pp 351–371.
11. Sreehitha, V. Impact of 3D Printing in Automotive Industry. *Int. J. Mech. Product. Eng.* **2017**, *5* (2), 91–94.
12. Parthasarathy, J.; Starly, B.; Raman, S. A Design for the Additive Manu-facture of Functionally Graded Porous Structures with Tailored Mechanical Properties for Biomedical Applications. *J. Manuf. Process.* **2011**, *13* (2), 160–170.
13. Swaelens, B.; Vancraen, W. In *Laser Photopolymerisation Models Based on Medical Imaging: A Development Improving the Accuracy of Surgery*, Proceedings of the Seventh International Conference on Rapid Prototyping; 1997; pp 130–131.
14. Ferreira, T.; et al. In *Additive Manufacturing in Jewellery Design. In: Volume 4: Advanced Manufacturing Processes; Biomedical Engineering; Multiscale Mechanics of Biological Tissues; Sciences, Engineering and Education; Multiphysics; Emerging Technologies for Inspection*, ASME 11th Biennial Conference on Engineering Systems Design and Analysis, Nantes, France, 2012; pp 187–194.
15. Wohlers, A. Wohlers Report, 3D Printing and Additive Manufacturing State of the Industry. *Annual Worldwide Progress Report*; Wohlers Associates: Fort Collins, CO, 2017.
16. Wohlers, A. Wohlers Report, 3D Printing and Additive Manufacturing State of the Industry. *Annual Worldwide Progress Report*, Wohlers Associates: Fort Collins, CO, 2019.
17. Wohlers, A. Wohlers Report 3D Printing and Additive Manufacturing State of the Industry. *Annual Worldwide Progress Report*, Wohlers Associates: Fort Collins, CO., August 2019.
18. Syed, A. M. T.; Elias, P. K.; Amit, B.; Susmita, B.; Lisa, O.; Charitidis, C. Additive Manu-facturing: Scientific and Technological Challenges, Market Uptake and Opportunities. *Mater. Today* **2017**, *1,* 1–16.
19. Ford, S. J. Additive Manufacturing Technology: Potential Implications for U.S. Manu-facturing Competitiveness. *Int. Commerce Econ.* **2014**, 2–35.
20. Allison, A.; Scudamore, R. Additive Manufacturing: Strategic Research Agenda [Online]. http://www.rm-platform.com/linkdoc/AM%20SRA%20-%20February%202014.pdf.2014
21. PTC Creo Parametric CAD Software [Online]; 2021. https://www.ptc.com/

22. SolidWorks CAD Software [Online]; 2021. https://www.solidworks.com/
23. Siemens NX CAD Software [Online]; 2021. https://www.plm.automation.siemens.com
24. Cunningham, J. J.; Dixon, J. R. In *Designing With Features: The Origins of Features*, Proceedings of the ASME Conference on Computers in Engineering, 1988; pp 237.
25. Sever AM Techniques. 2021.
26. https://www.lboro.ac.uk/research/amrg/about/the7categoriesofadditivemanufacturing/
27. Hole feature https://en.wikipedia.org/wiki/Hole

PART II
Design Engineering, Automation, and Electric Vehicle Technology

CHAPTER 6

Comparative Evaluation of Modal and Harmonic Analysis of Stepped Horn for Ultrasonic Vibration Assisted Turning

GOVIND S. GHULE[1], SUDARSHAN SANAP[2], SATISH CHINCHANIKAR[3], and AVEZ SHAIKH[3]

[1]Department of Mechanical Engineering, MIT-SOE, MIT-ADTU, Pune, India

[2]MIT-SOE, MIT-ADTU, Pune, India

[3]Department Mechanical Engineering, VIIT, Pune, India

ABSTRACT

The demand for sustainable machining with innovative technologies is continuously growing due to ecological and environmental aspects. Ultrasonic vibration-assisted turning (UVAT), wherein ultrasonic vibrations are superimposed on cutting tools, is used for sustainable machining of hard-to-cut materials. This paper discusses the theoretical aspects of the design of a stepped horn used to transfer high-frequency vibrations from the source to cutting tools for different materials. Further, this study presents a modal and harmonic analysis of a stepped horn for UVAT. The harmonic analysis is carried out to estimate the working frequency and the vibration amplitude using finite element analysis (FEA) in ANSYS®. This study finds the lowest deformation under defined conditions with the titanium alloy stepped horn followed by aluminum alloy, mild steel, and stainless-steel stepped horns. Further, the outcomes of equivalent stress (von Mises stress) have been discussed in this study.

Smart Innovations and Technological Advancements in Civil and Mechanical Engineering.
Satish Chinchanikar, Ashok Mache, Shardul Joshi, & Preeti Kulkarni (Eds.)
© 2025 Apple Academic Press, Inc. Co-publis hed with CRC Press (Taylor & Francis)

6.1 INTRODUCTION

The key to survival in the emerging and competitive market of the manufacturing world for accomplishing reliability and improved productivity, better efficiency, and highest customer satisfaction with the best quality product designing is only the acknowledgment of sustainable machining with innovative technologies.[1] Machining operations on a workpiece with a hardness worth over 45 HRC is considered hard turning. It involves a substantial rise in cutting forces, accelerated tool wear, high-temperature generation, and heavy power consumption which are the major hindrance in smooth machining.[2,3] The utilization of coolants to decrease the machining temperature obliging the interaction a bit is yet a significant block. Thus, dry machining activities are financially successful and climate-amicable as they oblige the drawdown of carbon impressions.[4,5] Ultrasonic comes under the umbrella of acoustics with a grouping of uses in the arena of non-destructive testing (NDT), ultrasonic vibration-assisted milling,[6] ultrasonic vibration-assisted welding,[7] ultrasonic vibration-assisted drilling,[8] and in clinical applications. Ultrasonic vibration-assisted turning (UVAT) comprehends ultrasonic generators along with an ultrasonic vibratory tool (UVT).[9–11] UVAT is categorized as per the course of direction of vibrations as tangential vibration turning, radial vibration turning, and axial (feed) vibration turning.

In UVAT, the cutting tool and workpiece are occasionally confined and brought in touch with each other. Subsequently, the summation of tool–workpiece contacts and separation is reduced from the customary measure of turning operation. The significant outcomes of this technique result in lesser cutting forces, lower tool wear, lower residual stresses, improved surface finish, and less power consumption. The UVAT setup is shown in Figure 6.1. It consists of a frequency generator, piezoelectric transducer and booster assembly, and sonotrode, which also acts as a tool holder with an insert attached.

The UVT is a component working in a longitudinal mode utilized for the effective transmission of energy in the form of vibration from a component called a piezoelectric transducer to a sonotrode. An ultrasonic horn or an acoustic horn functions as a tool holder. The horn is a solid metallic component used to alter the displacement as vibrational amplitude. It is given by an ultrasonic transducer working at an ultrasonic recurrence range, in the middle, of 16 kHz–100 kHz.

Amini et al.[14] proposed the FEM investigation of the ultrasonic horn and the cutting tool highlighting the alteration in machining forces and stress for conventional turning and UVAT. In this study, various combinations, for example, (a) cylindrical horn with long and eccentric tool holder, (b) cylindrical horn with concentric tool insert, and (c) conical–cylindrical horn with concentric tool insert, have been taken into consideration. Researchers have additionally viewed the impact of cutting speed, the amplitude of ultrasonic vibration, the cutting tool's rake angles, and clearance angles. Furthermore, it is stated that an aluminum horn with a conical–cylindrical shape and a cutting insert concentrically connected to its lower portion outperforms a cylindrical with an eccentrically attached fairly lengthy tool holder or steel horns.

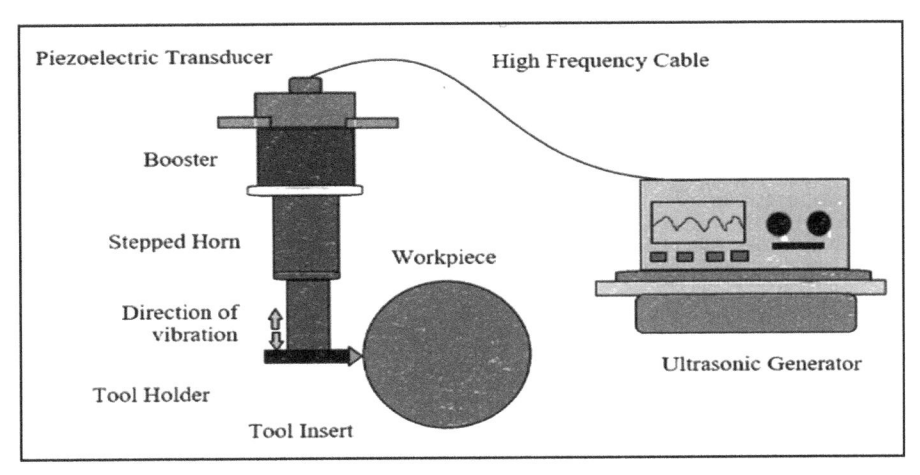

FIGURE 6.1 Assembly of UVAT.[12,13]
Source: Adapted from Ref. [12,13]

Sharma et al.[15] presented a modal, as well as harmonic, analysis using ANSYS®. Researchers have compared conventional turning and UVAT utilizing flat and textured cutting inserts (CNMA 120404) with a stepped horn made up of SS316. It is observed that the highest stress concentration at the stepped part of the horn was in the order of 71.44 MPa at a natural frequency of 20 kHz. The harmonic analysis was used to assess the variation in vibration amplitude with natural frequency. At a transducer end input of 3 mm, a maximum amplitude of vibration of 13.7 mm was obtained. The amplitude at the booster end was altered from 3 to 5 mm during simulation, and the amplitude at the cutting insert was measured.

Kukkala et al.[16] used ANSYS to perform modal and harmonic analysis on a stepped horn made of titanium alloy. According to modal analysis, the generated frequency is found to be 19,653 Hz. This frequency is equivalent to other components of the ultrasound system and is magnified with four times the resonating frequency of the horn under the generator frequency and with the UVT. It is concluded that the UVAT procedure was determined to be a good approach for creating high-quality surfaces and decreasing the demand for cutting force not only for hard materials but also for standard materials such as stainless steel.

From the literature reviewed, it is seen that very few studies attempted harmonic analysis of stepped horn for UVAT. Moreover, a comparative evaluation of modal and harmonic analyses of a stepped horn considering the effect of different horn materials at varying displacements, forces, and pressures of recognized frequency is rarely reported. With this view, this paper focuses on a linear dynamical analysis, that is, frequency response analysis which is used to identify the system's reaction to excitement at particular frequencies (harmonic responses). The load supplied to the linear model is a constant sinusoidal load at a certain frequency with the harmonic response analysis.

6.2 DESIGN OF ULTRASONIC HORN

Reliance of most noteworthy conceivable vibrational amplitude ampleness is not straight-a-way on the attributes of basic elements of the sonotrode by use of which it is manufactured furthermore on the fundamental condition of the sonotrode. In regular concern, the sonotrode is produced using the material, for instance, titanium as well as aluminum amalgams, mild steel, and stainless steel.

An erroneous sonotrode configuration will have an adverse effect and will influence the whole metal-cutting activity. Few key factors have to be considered in designing UVT and can be referred to from Refs. [17,18]. The horn material should have a minimum energy loss. While assembling the horn with the transducer, the larger diameter of the horn should be on the side of the transducer and should have a range that is higher or equivalent to twice of radius, that is, the diameter of the transducer to avoid air contact.

The exhaustive study of literature underlined that the most extreme possible ultrasonic vibration amplitude is not only the function of properties of the material, in addition to the underlying state of the sonotrode, that is, the enhancement or intensification of the vibration energy relies upon the state of the horn. The respective form of energy stays steady throughout the sonotrode span demonstrating that as the cross-segment decreases, so does the energy density.

Cyclic loading is one of the significant specifications that ought to be taken into consideration for choosing the components of design. All these respective metal components will be exposed to exceptionally high cyclic stacking, and this can start fatigue. Therefore, this essential condition cannot be exempted. It is observed that titanium has the finest acoustical characteristics of any metal and its appreciable weariness strength allows it to tolerate fatigue rates at various incremental intensities. It likewise has high hardness when contrasted with different materials, which assists it with supporting wear conditions.

However, the horns comprised of titanium are ordinarily more costly than the rest of the materials because of the greater expense of the material. The aluminum heat-treated combination has incredible acoustical properties and is utilized to make horns not needing more vibrational amplitude and strength.[19,20] The most well-known and easy-to-make states of horn are stepped, tapered, exponential, cylindrical stepped, and catenoidal ventured shapes. The machinability improvement in UVAT measure has an immediate connection to the legitimate design of an ultrasonic horn, which guarantees the simple conveyance of vibration to the cutting tool and coming about increment in the rate of unwanted material from a workpiece, that is, chips. Hence given this, the most well-known states of horn are cone-shaped, exponential, bezier type, stepped, and cylindrical shapes. However, the stepped-type horn is highlighted more in this chapter due to its simplicity in manufacturing.

Bezier, as well as stepped sonotrodes, practically consists of similar intensification aspects. Notwithstanding, to manufacture a bezier type of sonotrode, possibly it requires a copy-turning machine which cannot be accessed easily, and as a result stepped horn type is picked over the bezier type.

Stepped horns comprise two distinct segments, each having a uniform cross-sectional region. They have the most elevated enhancement factor of all horns because of the unexpected change in the cross-sectional region at the nodal plane. This paper focuses on the design and analysis of stepped horns. As per the acoustic wave equation of stepped horn, motion of a plane wave traveling through a stepped horn is given by eqn. 6.1.[21]

$$\xi = \left[A\cos\frac{\omega x}{c} + B\sin\frac{\omega x}{c} \right] \cos \omega t \tag{6.1}$$

where ω = angular velocity, ξ = displacement. The above form of equation is simplified and can be written as eqn. 6.2.

$$\xi = \xi m \cos\frac{\omega x}{c} \cos \omega t \tag{6.2}$$

where ξm = maximum displacement at $x = 0$. In the case of a conventional balanced transducer, the nodal point is situated at the center of the piezoelectric parts. This highlights the most extreme stress that happens at the nodal point which can prompt an expansion in temperature and annihilation of the piezoelectric segment. Hence, the overall resonant length of the stepped horn can be referred from eqn. 6.3.[22,23]

$$L = \frac{c}{2f} \qquad (6.3)$$

where c = speed of sound traveling through horn material and it is given by,

$$c = \sqrt{\frac{E}{\rho}}$$

where E = Young's modulus of the material (N/m^2), δ = density of the material (Kg/m^3), and λ = wavelength. Furthermore, the amplitude of the stepped horn can be estimated as follows by the use of eqn. 6.4.

$$K = \frac{\xi_2}{\xi_1} = \left(\frac{d_1}{d_2}\right)^2 \qquad (6.4)$$

where K = overall amplitude gain, and ξ_1 and ξ_2 = displacements of larger and smaller ends of stepped horn, respectively ($\xi_1 < \xi_2$). The exhaustive study of the literature concludes that the best outcomes are achievable with stepped horn and therefore this research article aims toward an in-depth study of stepped horn. The nodal point is the point where the amplitude is zero.

Thus, by putting the $\xi = 0$ and $x = X_n$ in eqn 6.2, as for the stepped horn, $n = 2$; hence, $X_2 = L$ where "L" is the length of the horn, as shown in Figure 6.2.

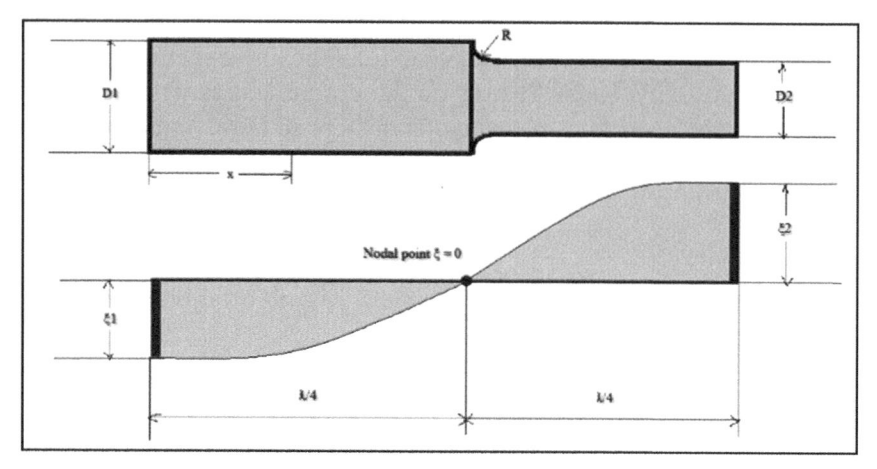

FIGURE 6.2 Nodal point location for stepped horn.[13]

Source: Adapted from Ref. [13]

6.3 FINITE ELEMENT ANALYSIS OF A STEPPED HORN

The design of the horn is a confounded interaction and it includes the choice of the right material. Inappropriately tuned horns can harm the entire assembly including the ultrasonic frequency generator. In concern of the same, finite element analysis (FEM) can be used as an insightful strategy to foresee, analyze, and simplify the physical issues.

In this method, the finite element or the mesh size is the fundamental unit of examining the actual problem incomprehensibly. The guidelines of FEM portray that the precision of a solution during the overall operation will be more if the mesh size is limitlessly small. The key factors that need to be considered for modeling are element type, properties of materials, and more importantly the boundary conditions.

6.3.1 GEOMETRICAL MODELING

For the investigation and assembling, the stepped horn is chosen because of its maximum intensification of amplitude at output. In this work, the stepped horn is designed using CATIA V5 R18 as shown in Figure 6.5. The resonant length of the horn is 126 mm with 40 and 20 mm as output and input diameters, respectively. The same is shown in Figure 6.3.

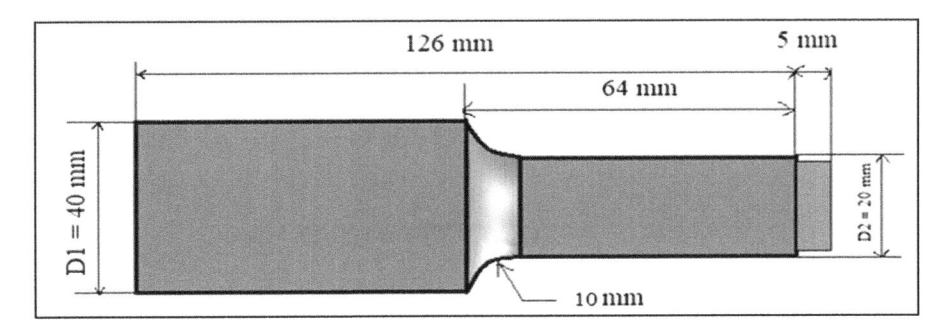

FIGURE 6.3 3D model of stepped horn.

The different materials are used to compare the analysis results of deformation, von Mises stresses, and equivalent elastic strains. These results are used to compare the different materials for the designed horn and hence the best and cost-effective material selection can be done, before actually manufacturing the horn.

Initially, the first step involves the designing of the horn and tool insert. The horn is designed according to the dimensions scale 1:1. Later, the tool insert is downloaded from the tool insert library of SECO tools. The tool insert utilized for our research purpose is CNMG120408-MF5 TH1000. The next step involves assembling the tool insert and horn designed using M5 bolt. The tool insert is assembled at the bottom tip of the horn. Then the assembly is analyzed for a harmonic response using the ANSYS mechanical 2021 R1.

For the harmonic response analysis, the first step involves dragging the harmonic response analysis system into the ANSYS workbench. The next step involves defining the material properties of different materials from the engineering data library. For the analysis of the assembly, we have defined the properties of four materials, namely, mild steel, aluminum alloy, titanium, and stainless steel. After inputting the engineering material data, the geometry is imported into the space claim geometry for verification. Once the geometry is verified for dimensional accuracy, the next step is to open the ANSYS mechanical model. In the mechanical model, the first step includes the assignment of materials. The tool insert is assigned with the material, that is, tungsten carbide substrate. The horn geometry is assigned with the four materials, one by one, and analyzed differently. The next step involves meshing the whole geometry. The element mesh size is 1 mm and the algorithm method for meshing used is solid 92 tetrahedron block Lanczos for all the four different material analyses and then the boundary conditions are applied. Lastly, the analysis is computed to find the minimum and maximum values of deformation, equivalent stresses (von Mises), equivalent elastic strain (von Mises), and maximum principal elastic strain in the horn and tool insert assembled geometry.

6.3.2 DESIGN PARAMETERS OF STEPPED HORN

The dynamic analysis is been carried out with the ANSYS® workbench as it is perhaps the most commanding and user-friendly tool. In this research work, for manufacturing of ultrasonic horn, four different materials have been considered, that is, aluminum, titanium, mild steel, and stainless steel as horn materials. The speed of sound principally relies on Young's modulus and the density of the respective specimen. The more the stiffer material, the quicker the sound wave flows through.

The thicker the medium, the slower the sound wave spread, since additional thickness implies more mass per volume, and therefore more inertia. Hence, it is a bit drowsy in the penetration. Most usually, the horns are fabricated from the material, for example, aluminum alloys, mild steel, stainless steel, and titanium alloys, as shown in Table 6.1.

TABLE 6.1 Mechanical Properties of Materials for Designing of Ultrasonic Horn.[24,25]

Name of material	Young's modulus E (GPa)	Density ρ = kg/m³	Bulk modulus K (GPa)	Shear modulus G (GPa)	Poisson's ratio (γ)	Velocity of sound C (m/s)
Aluminum alloy	70	2710	69.60	26.69	0.33	5082
Mild steel	210	7850	169.3	73.66	0.36	5172
Stainless steel	193	8000	169.3	73.66	0.36	4912
Titanium alloy	110	4700	114.29	35.29	0.33	4840

6.4 SELECTION OF ELEMENT TYPE

Selection of appropriate element types as per the material and designing of UVT is made to ensure guaranteed insightful rightness. The UVT is prevalently partitioned into metallic materials. Component choice fluctuates due to contrasting features. Considering the unique curved surface of the vibrating framework linear Lanczos solver with 22,864 nodes, 83,680 elements with a mesh size of each element as 1 mm is considered with four cores.

6.4.1 GENERATION OF MESH

As indicated by the fundamental standards of finite element investigation, more precise mesh component sizes give more exact outcomes of the analysis. If the mesh component estimates are limitlessly small, the model will move toward the ideal arrangement. Nonetheless, this is just a hypothesis. In the analysis cycle, when elements are excessively little, the meshing of elements will produce a large number of elements and notes with freedom for the model all in all. The meshed geometry of the stepped horn is shown in Figure 6.4.

This increments computational power, bringing about a model that is either excessively tedious to tackle or likely blunders in qualities. Accordingly, suitable cross-section element size (number of components/elements) is a factor that ought to be considered in a FEM investigation.

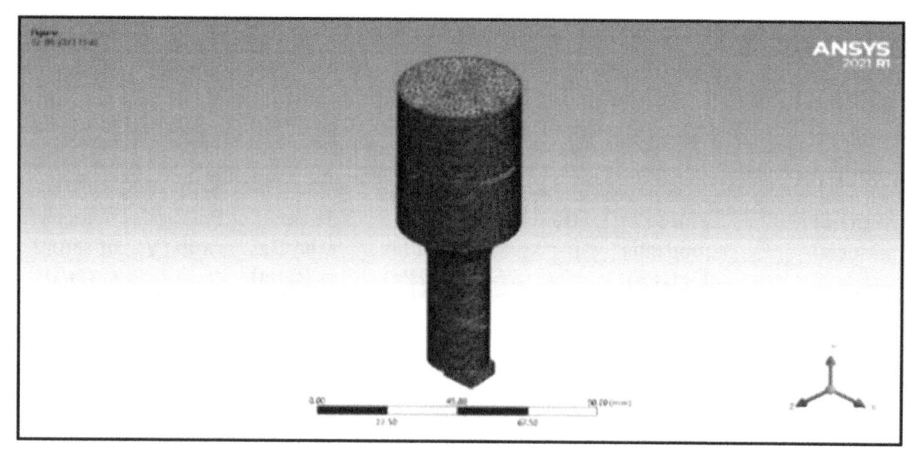

FIGURE 6.4 Meshed geometry of stepped horn.

6.4.2 BOUNDARY CONDITIONS

With defined boundary conditions, the harmonic analysis is carried out by systematically highlighting the additional input harmonic loads in terms of forces, pressure as well as displacement of the known frequency of the assembly which is necessary to be aware of the response of the assembly with respect to sinusoidal loads. The harmonic analysis gives the output results as stresses, strain, and displacement at each DOF of the structure or the horn assembly. The boundary conditions involve giving fixed support to the top geometry of the horn and the other boundary condition is given as a displacement of 20 micrometers (0.02 mm) to the extreme bottom geometry of the horn as shown in Figure 6.5. Then, the final step includes defining the analysis settings for performing the harmonic response. In the analysis settings, the frequency spacing is linear, the frequency range is 18 kHz–20 kHz (minimum to maximum), the solution method is full, and solution intervals are kept as ten.

6.5 RESULTS AND DISCUSSION

6.5.1 MODAL ANALYSIS

Modal analysis is carried out to calculate the natural frequency of the stepped horn. In the analysis settings, the limit for frequency is given as 0–25,000 Hz. The solver is direct and mode shapes are kept as 15 to find the three different mode shape phases—that is, torsional (twisting), longitudinal, and bending. The results for total deformation, equivalent elastic strain, and von Mises stress for three different phases—torsional (twisting), longitudinal, and bending are computed to find the maximum and minimum values.

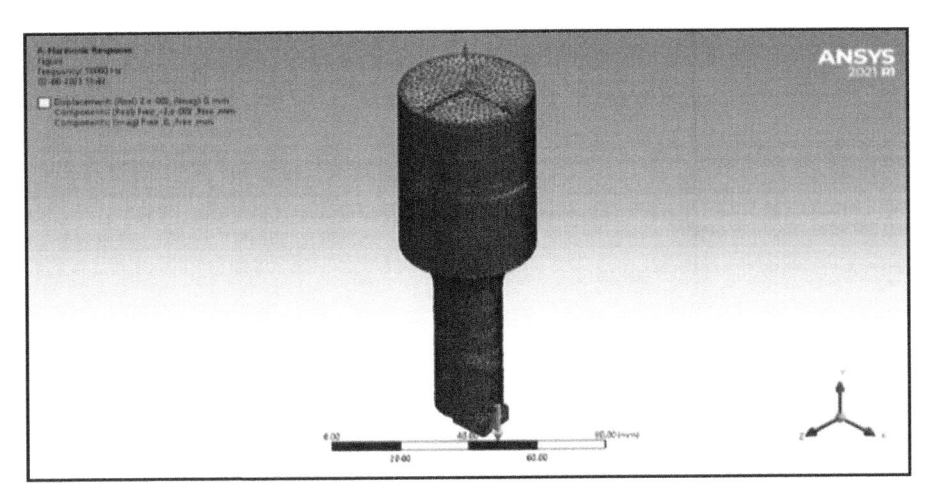

FIGURE 6.5 Boundary conditions for a stepped horn with tool insert.

This is hence compared to develop a standard and hence reach different conclusions. The comparative study of modal response for total deformation (longitudinal) for the materials, namely, aluminium alloy, mild steel, stainless steel, and titanium alloy is shown in Figure 6.6(a)–(d), respectively.

The detailed results of modal analysis for longitudinal total deformation for the randomly selected aluminum alloy and titanium alloy materials are mentioned in Tables 6.2 and 6.3, respectively. Furthermore, the analysis with bending and torsional deformation was carried out for mild steel as well as for stainless steel. Therefore, from the comparative analysis, it can be seen that the range of von Mises stresses for titanium alloy is found lowest as compared to other materials taken into consideration and it ranges from 3.28E+05 (max) to 209.71 (min) with a total deformation (longitudinal) as 130.75 mm. Hence, it can be concluded that titanium alloy is satisfying the condition for optimal design.

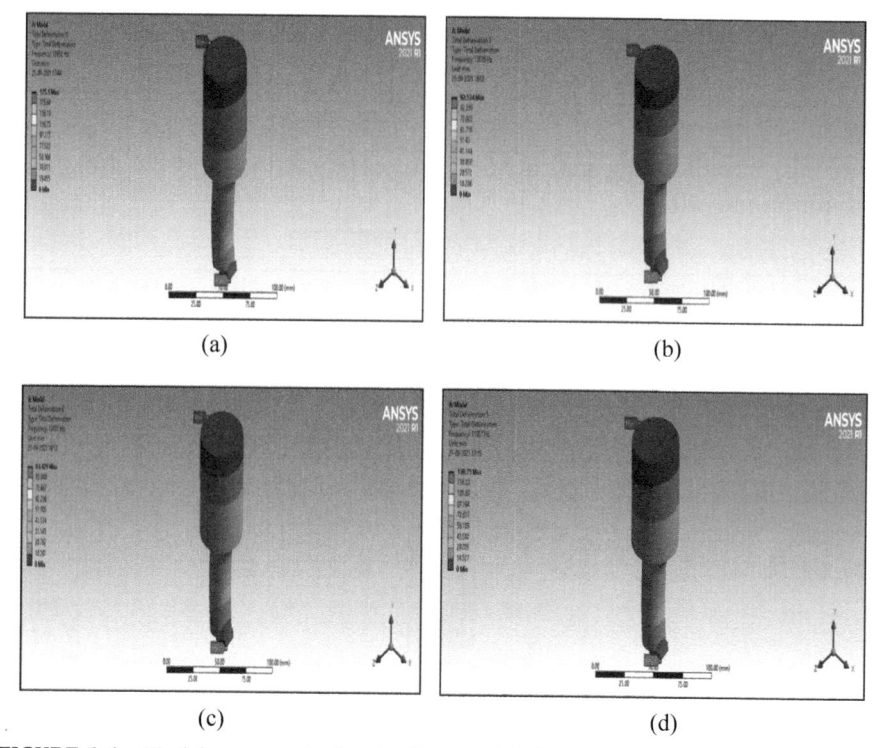

FIGURE 6.6 Modal response for longitudinal total deformation of (a) aluminum alloy, (b) mild steel, (c) stainless steel; and (d) titanium alloy.

TABLE 6.2 Modal Analysis Results for Aluminium Alloy.

Frequency (Hz)	Modal shapes	Deformation max (mm)	von Mises stress max (MPa)	von Mises stress min (MPa)	Equiv. strain max (mm/mm)	Equiv. strain min (mm/mm)
9842.6	Torsional	228.4	6.1958e+005	79.142	6.1403	1.2708e-003
12,632	Longitudinal	175.1	4.4442e+005	173.15	3.8084	3.5436e-003
2,2673	Bending	153.86	2.1164e+006	166.02	6.6208	2.6085e-003

6.5.2 HARMONIC ANALYSIS

Harmonic analysis is utilized to figure out how the design will respond when stacked at a specific frequency. Enforced displacements, forces, and pressures of recognized frequency are regarded as input specifications. Whereas harmonic displacement, current flow through the system, strains, stresses, and other output requirements are examined.

TABLE 6.3 Modal Analysis Results for Titanium Alloy.

Frequency (Hz)	Modal shapes	Deformation max (mm)	von Mises stress max (MPa)	von Mises stress min (MPa)	Equiv. strain max (mm/mm)	Equiv. strain min (mm/mm)
9023	Torsional	168.68	5.9061e+005	92.167	4.9227	1.119e-003
11,957	Longitudinal	130.75	3.2752e+005	209.71	2.2752	2.4263e-003
20,712	Bending	121.7	1.6e+006	272.83	5.2174	3.1922e-003

The response of an ultrasonic horn to an applied load with a sinusoidal time history is computed using harmonic response analysis, which means this analysis is performed to assess the reaction of a stepped ultrasonic horn at a certain frequency under a systematic (harmonic) load.

The ultrasonic horn has its natural frequency which has been calculated using modal analysis, and the response will be on the larger side when the harmonic load frequency matches with the natural frequency. The comparative study of harmonic response for total deformation (longitudinal) for the materials, namely, aluminium alloy, mild steel, stainless steel, and titanium alloy is shown in Figure 6.7(a)–(d), respectively.

The values of total deformation using harmonic analysis obtained from numerical simulations are 0.17054, 0.022588, 0.12588, and 0.082177 mm for aluminum alloy, titanium alloy, mild steel, and stainless steel, respectively.

Thus, based on the harmonic response analysis, it is observed that titanium alloys show the maximum strength followed by stainless steel, mild steel, and then aluminum alloy, respectively. It is also observed that aluminum alloy was showing a bit higher deformation, but the rate of cooling of aluminum alloy as compared to mild steel and stainless steel was found considerable.

6.5.2.1 FREQUENCY RESPONSE FOR TOTAL DEFORMATION

The detailed results of frequency responses for total deformation for aluminum alloy and titanium alloy are taken into consideration, which is mentioned in Table 6.4.

From the obtained results, it can be proposed that the maximum amplitude of vibration for the aluminum alloy, mild steel, stainless steel, and titanium alloy are at the frequency values of 19,400, 19,600, 19,599, and 18,200 Hz, respectively.

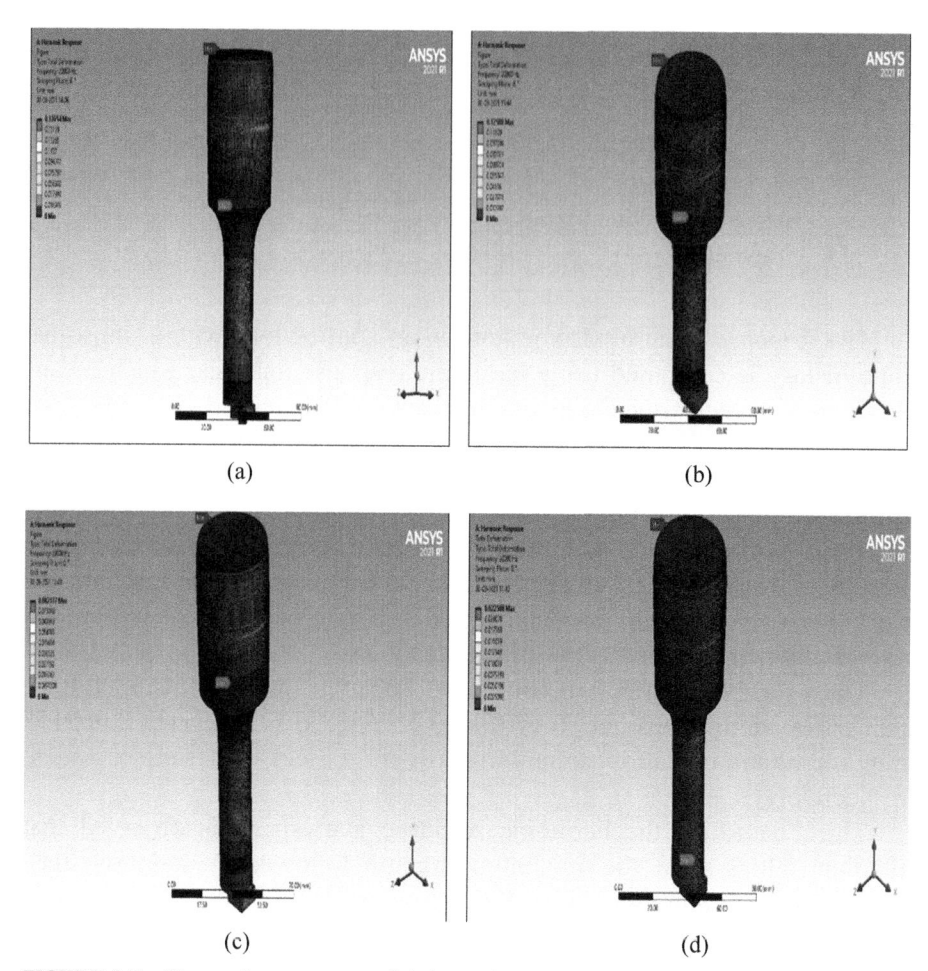

FIGURE 6.7 Harmonic response total deformation (longitudinal) of (a) aluminum alloy, (b) mild steel, (c) stainless steel, and (d) titanium alloy.

6.5.2.2 HARMONIC RESPONSE FOR EQUIVALENT STRESS (VON MISES)

Furthermore, the comparative study of harmonic response for equivalent stress (von Mises) for materials like aluminum alloy, mild steel, stainless steel, and titanium alloy is shown in Figure 6.8(a)a–(d), respectively. From the obtained outcomes, it can be concluded that the maximum value of equivalent stress for aluminum alloy, mild steel, stainless steel, and titanium alloy are 1134.6, 1587, 1034, and 195.06 MPa, respectively.

Comparative Evaluation of Modal and Harmonic Analysis

TABLE 6.4 Frequency Response for Total Deformation (Aluminium Alloy and Titanium Alloy).

Frequency (Hz)	Phase angle (degree) (for Al alloy)	Amplitude (mm) (for Al alloy)	Phase angle (degree) (for Sn alloy)	Amplitude (mm) (for Sn alloy)
18,200	180	1.8747e-004	180	4.0185e-004
18,400	180	1.7084e-004	180	2.8476e-004
18,600	180	1.7401e-004	180	2.2678e-004
18,800	180	1.9135e-004	180	1.915e-004
19,000	180	2.2691e-004	180	1.6745e-004
19,200	180	2.9786e-004	180	1.4984e-004
19,400	180	4.7193e-004	180	1.3627e-004
19,600	180	1.3889e-003	180	1.2542e-004
19,800	0	1.219e-003	180	1.1649e-004
20,000	0	4.0195e-004	180	1.0898e-004

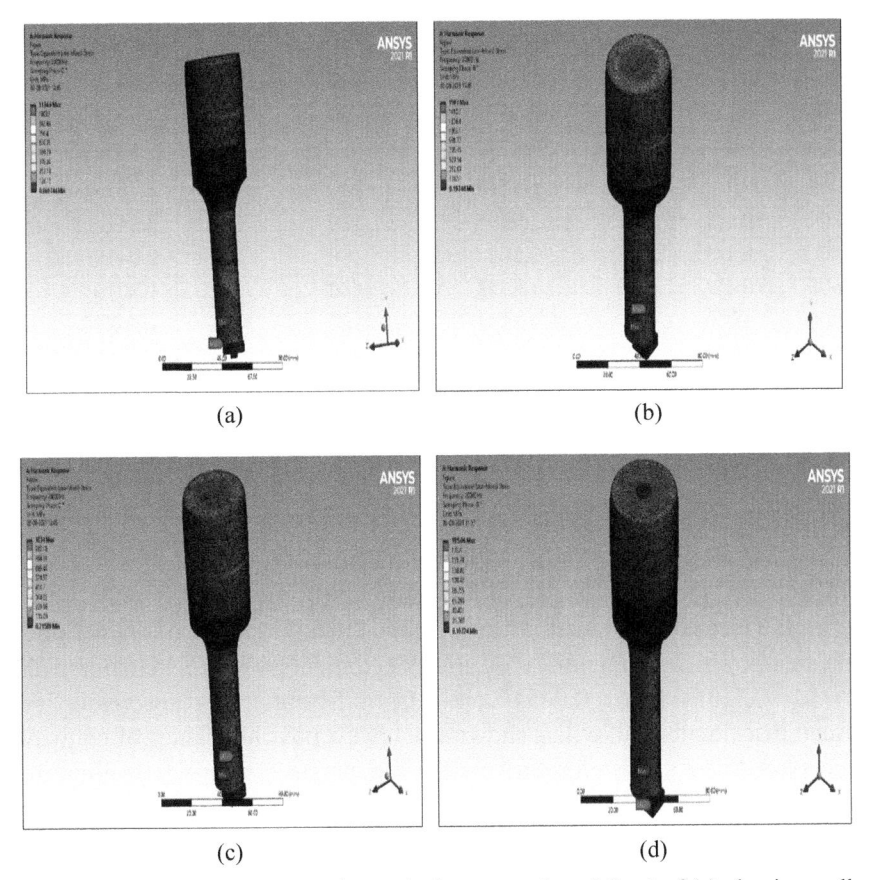

FIGURE 6.8 Harmonic response for equivalent stress (von Mises) of (a) aluminum alloy, (b) mild steel, (c) stainless steel, and (d) titanium alloy.

6.5.2.3 FREQUENCY RESPONSE FOR EQUIVALENT STRESS (VON MISES)

The detailed results of frequency responses for equivalent stress (von Mises) for aluminum and titanium alloys are taken into consideration, which is mentioned in Table 6.5.

From the data obtained, it can be proposed that the maximum amplitude of vibration for the aluminum alloy, mild steel, stainless steel, and titanium alloy are at the frequency values of 19,400, 19,600, 19,599, and 18,200 Hz, respectively. The lowest amplitude for titanium alloy is found to be 1.4491 at a frequency value of 2000 Hz with phase angle of 180°.

TABLE 6.5 Frequency Response for Equivalent Stress (Aluminum alloy and Titanium Alloy).

Frequency (Hz)	Phase angle (degree) (for Al alloy)	Amplitude (mm) (for Al alloy)	Phase angle (degree) (for Sn alloy)	Amplitude (mm) (for Sn alloy)
18,200	0	0.69092	180	8.8979
18,400	0	0.79119	180	5.7348
18,600	0	0.92682	180	4.2283
18,800	0	1.1227	180	3.3453
19,000	0	1.4314	180	2.7639
19,200	0	1.9898	180	2.351
19,400	0	3.3047	180	2.0419
19,600	0	10.123	180	1.8013
19,800	180	9.1991	180	1.6081
20,000	180	3.128	180	1.4491

6.5.2.4 HARMONIC RESPONSE FOR EQUIVALENT STRAIN (VON MISES)

Additionally, the comparative study of harmonic response for equivalent strain (von Mises) for materials like aluminum alloy, mild steel, stainless steel, and titanium alloy is shown in Figure 6.9(a)–(d), respectively.

From the results obtained, it can be concluded that the maximum value of equivalent strain for aluminum alloy, mild steel, stainless steel, and titanium alloy are 0.010733, 0.0081961, 0.00547, and 0.0014751 mm/mm, respectively. It can be seen that titanium alloy has shown the lowest possible values of equivalent strain (von Mises). Therefore, as per the comparative analysis, titanium alloy gives the best possible outcomes so it can be considered for design purposes.

It is found that the major issue for the horn's life is its dynamic qualities in general. The results withdrawn from the numerical analysis state that the lowest deformation is found in titanium alloy. Stepped horn has the highest stress concentration when the horn diameter varies, and the value should be less than the material's fatigue strength or endurance limit for titanium alloy. According to the harmonic analysis, the equivalent stress (von Mises stress) is 195 MPa, which is less than the horn material's fatigue strength (160 MPa). Hence, in contrast to the traditional horn design, which is a single-body horn design, a stepped horn could be efficient and reliable to use in UVAT.

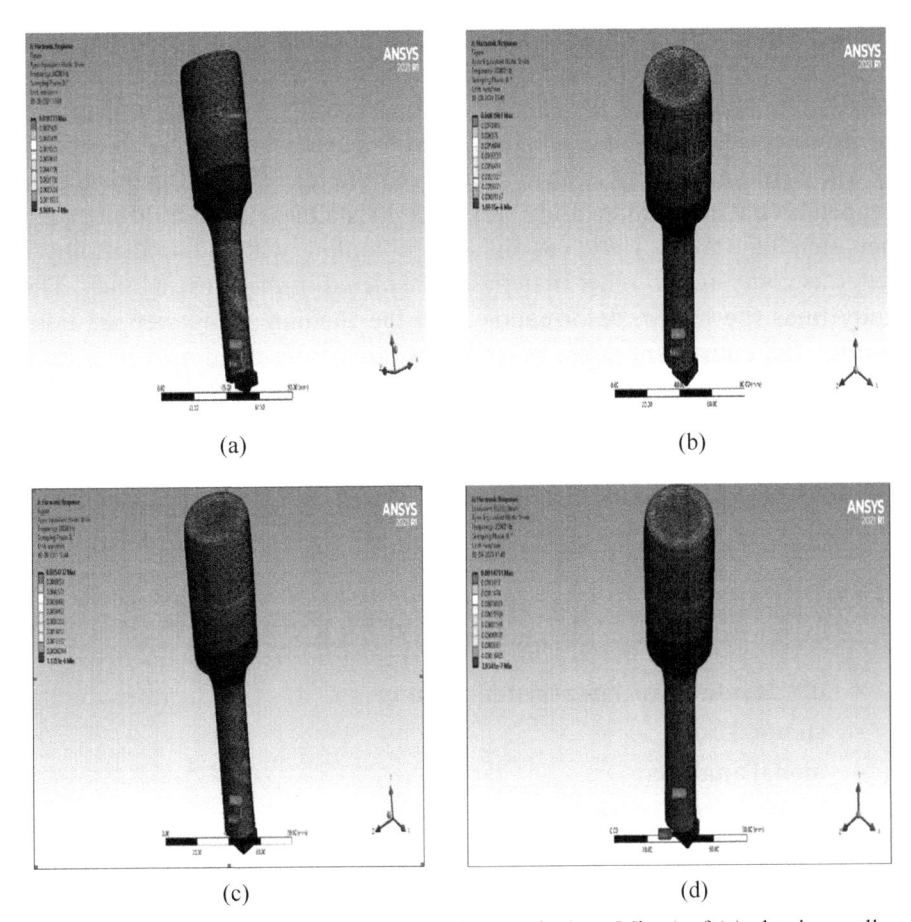

FIGURE 6.9 Harmonic response for equivalent strain (von Mises) of (a) aluminum alloy, (b) mild steel, (c) stainless steel, and (d) titanium alloy.

6.6 CONCLUSION

UVAT, wherein ultrasonic vibrations are superimposed on cutting tools, is increasingly used for sustainable machining of hard-to-cut materials. This paper discusses the design aspects and modal and harmonic analysis of a stepped horn that is used to transfer high-frequency vibrations from the source to cutting tools for different materials.

The harmonic analysis is carried out to estimate the working frequency and the vibration amplitude using FEA in ANSYS®. This study finds the lowest deformation under the stated conditions with the titanium alloy stepped horn followed by aluminum alloy, mild steel, and stainless-steel stepped horns. Modal analysis is carried out to obtain the natural frequency of the stepped horn. Harmonic analysis is carried out to understand the steady-state behavior of a stepped horn considering the effect of displacements, forces, and pressures of recognized frequency.

The harmonic response analysis showed the maximum strength for the stepped horn with titanium alloys followed by stainless steel, mild steel, and then aluminum alloy. However, the rate of cooling with aluminum alloy is better as compared to other materials considered in the present study. This study finds the lowest deformation with the titanium alloy stepped horn. Further, the equivalent stress (von Mises stress) observed as 195 MPa is less than the horn material's fatigue strength (160 MPa) for the geometry of the horn considered in the present study. However, aluminum alloy could be a better option for the stepped horn for UVAT from an economical and manufacturing perspective.

KEYWORDS

- ultrasonic vibration-assisted turning
- stepped horn
- modal analysis
- harmonic response
- FEA

REFERENCES

1. Madhavulu, G.; Ahmed, B. Hot Machining Process for Improving Metal Removal Rates in Turning Operation. *J. Mater. Process Technol.* **1994,** *44,* 199–206.
2. Kurada, S.; Bradley, C. A Review of Machine Vision Sensors for Tool Condition Monitoring. *Comput. Ind.* **1997,** *34* (1), 55–72.

3. Chinchanikar, S.; Choudhury, S. K. Effect of Work Material Hardness and Cutting Parameters on Performance of Coated Carbide Tool When Turning Hardened Steel, An Optimization Approach. *Measurement*, **2013**, *46*, 1572–1584.
4. Brehl, D. E.; Dow, T. A. Review of Vibration-Assisted Machining. *Precis. Eng.* **2008**, *32*, 153–172.
5. Marce, K.; Marek, Z.; Jozef, P. In *Investigation of Ultrasonic Assisted Milling of Aluminium Alloy AlMg4.5Mn*, 24th DAAAM International Symposium on Intelligent Manufacturing and Automation, 2013.
6. Vivekananda, K.; Arka, G. N.; Sahoo, S. K. Design and Analysis of Ultrasonic Vibratory Tool (UVT) using FEM, and Experimental study on Ultrasonic Vibration-assisted Turning (UAT). *Procedia Eng.* **2014**, *97*, 1178–1186.
7. Vivekananda, K.; Arka, G. N.; Sahoo, S. K. Finite Element Analysis and Process Parameters Optimization of Ultrasonic Vibration-Assisted Turning (UVT). *Procedia Mater. Sci.* **2014**, *6*, 1906–1914.
8. Sherrit, S.; Askins, S. A.; Gradziol, M.; Dolgin, B. P.; Chang, X. B. Z.; Bar-Cohen, Y. In *Novel Horn Designs for Ultrasonic Cleaning, Welding, Soldering, Cutting and Drilling*, Proceedings of the SPIE Smart Structures Conference 2002, vol. 34; p 4701.
9. Zou, P.; Xu, Y.; He, Y.; Chen, M.; Wu, H. Experimental Investigation of Ultrasonic Vibration Assisted Turning of 304 Austenitic Stainless Steel. *Shock Vibration.* **2015**, 1–19.
10. Rajput, P.; Siddiquee, A. N.; Sharma, R. Experimental Study of Ultrasonically Assisted Turning of AISI 52100 based on Taguchi Method. *Int. J. Recent Res. Asp.* **2015**, *2*, 42–47.
11. Ghule, G.; Ambhore, N.; Chinchanikar, S. In *Tool Condition Monitoring Using Vibration Signals During Hard Turning: A Review*, International Conference on Advances in Thermal Systems, Materials and Design Engineering, ATSMDE, 2017.
12. Ghule, G. S.; Sanap, S. Ultrasonic Vibrations Assisted Turning (UAT): A Review. *Adv. Eng. Design, Lect. Notes Mech. Eng.* **2021**, 275–285.
13. Ghule, G. S.; Sanap, S. A Bibliometric Analysis of Ultrasonic Vibration Assisted Turning. *Library Philosophy Practice*, Libraries at University of Nebraska-Lincoln, **2021**, 4988.
14. Amini, S.; Soleimanimehr, H.; Nategha, M. J.; Abudollahb, A.; Sadeghi, M. H. FEM Analysis of Ultrasonic-Vibration-Assisted Turning and the Vibratory Tool. *J. Mater. Process. Technol.* **2008**, *201*, 43–47.
15. Sharma, V.; Pandey, P. M. Experimental Investigations and Statistical Modeling of Surface Roughness During Ultrasonic-Assisted Turning with Self-Lubricating Cutting Inserts. *J. Process Mech. Eng.* **2017**, 1–14.
16. Kukkala, V.; Sahoo, S. K. Finite Element Analysis of Ultrasonic Vibratory Tool and Experimental Study in Ultrasonic Vibration-Assisted Turning (UVT). *Int. J. Eng. Sci. Invent. (IJESI)*, **2014**, 73–77.
17. Neppiras, E. In *The Pre-stressed Piezoelectric Sandwich Transducer*, Ultrasonics International Conference, 1973, pp 295–302.
18. Zhong, Z. W.; Lin, G. Ultrasonically Assisted Turning of Aluminum-Based Metal Matrix Composite Reinforced with SiC Particles. *Int. J. Adv. Manuf. Technol.* **2006**, *27*, 1027–1081.
19. Radmanovic, M.; Mancic, D. *Design and Modelling of the Power Ultrasonic Transducers*; Faculty of Electronics in NIS: Serbia, 2004.
20. Abdullah, M. S.; Pak, A. An Approach to Design a High Power Piezoelectric Ultrasonic Transducer. *J. Electroceramics*, **2009**, *22*, 369–382.

21. Airao, J.; Khanna, N.; Roy, A.; Hegab, H. Comprehensive Experimental Analysis and Sustainability Assessment of Machining Nimonic 90 using Ultrasonic-Assisted Turning Facility. *Int. J. Adv. Manuf. Technol.* **2020,** *109,* 1447–1462.
22. Kandi, R.; Sahoo, S. K. In *Design and Modeling of a Flexible Acoustic Horn for Ultrasonic Vibration Assisted Turning*, Proceedings of International Conference on Advances in Dynamics, Vibration and Control, National Institute of Technology: Durgapur, 2016; pp 197–201.
23. Sharma, V.; Pandey, P. M. Recent Advances in Ultrasonically Assisted Turning: A Step Towards Sustainability. *Int. J. Cogent Eng.* **2016,** *3* (1).
24. Florez García, L. C.; Gonzalez Rojas, H. A.; Sanchez Egea, A. J. Estimation of Specific Cutting Energy in an S235 Alloy for Multi-Directional Ultrasonic Vibration-Assisted Machining Using the Finite Element Method. *Materials* **2020,** *13,* 567.
25. Kandi, R.; Sahoo, S. K.; Sahoo, A. K. Ultrasonic Vibration-Assisted Turning of Titanium Alloy Ti–6Al–4V: Numerical and Experimental Investigations. *J. Braz. Soc. Mech. Sci. Eng.* **2020,** *42,* 399.

CHAPTER 7

Design and Analysis of Wing for Unmanned Aerial Vehicle

PRANIT DHOLE, RATNAKAR GHORPADE, and CHETAN PATIL

School of Mechanical Engineering, MIT World Peace University, Pune, India

ABSTRACT

Recent advancements in computer simulations, composite materials, and fabrication technology have led to the design of unmanned aerial vehicles with improved performance and structural stability without having adverse effects on weight. Wing design is one of the key components in the design of unmanned aerial vehicles. Wing design is a critical task having close relation with airfoil selection, winglets, and structural members. This paper describes the process of design and analysis of wings for unmanned aerial vehicles right from the selection of airfoils, a comparative study of NACA 4412 and Clark Y airfoils by computational fluid dynamics analysis, the addition of winglets and their effect on flight performance to material selection based on CAE structural analysis. A comparative study of various performance factors like lift force, drag force, and lift-to-drag ratio is done for NACA 4412 and Clark Y airfoils, and a better airfoil is selected. CAE simulation gives a good idea of structural behavior before the actual prototype is built. Commercially available polyurethane foam is used as skin material, which is given the shape of NACA 4412 by a hot wire cutter. Aluminum alloy is used for spar of structure, which has the advantages of being lightweight and having a high strength structure for stable flight performance. CAD modeling of the wing is done using Solidworks, and the Solidworks simulation workbench is used in order to carry out CAE and CFD simulations. This research concludes that the NACA 4412 airfoil has better performance than the Clark Y airfoil at 35 m/s speed, and the addition of a winglet has the advantages of increased lift and a reduction in wingtip vortices. Aluminum alloy 1060 for spar and polyurethane foam material for wing skin are used for high strength and weight reduction of the wing. The prototype of a UAV is manufactured using selected configurations, materials, and electronics for flight control. The actual manufacturing of the UAV prototype is explained in this study, which helps to understand the stages of UAV manufacturing.

7.1 INTRODUCTION

Unmanned aerial vehicle research has become important in various domains including private and commercial organizations, government agencies, military applications, and R&D universities. Commercial companies are developing unmanned aerial vehicles to perform specified tasks like photography, videography, food and drug delivery, etc. The military has its own applications of UAVs like reconnaissance operations, target and decoy, combat, R&D platforms, early warning systems, etc. Research in all domains serves a common purpose, which is to use these UAVs to their maximum limit in civilian as well as military purposes. Basically, unmanned aerial vehicles are of two types, one is a traditional fixed-wing plan-like structure, while the other is a multirotor drone. Each of these comes with benefits and consequences. The traditional fixed-wing designs have the capacity of high-altitude flights with a long range, but they cannot hover at one place and need a runway to operate. Multirotor drones are capable of taking off vertically without the need of long runways, but due to their multirotor configuration, their power consumption is high, which limits flight time. In this research, we are discussing about fixed-wing UAVs having advantages like high payload capacity, long range, and fuel efficiency.[1]

The unmanned aerial vehicles are a complicated technology having significant importance of all parts in order to operate smoothly. Wings are one of the parts of UAVs, which have a very high impact on performance. The basic forces acting on wings are lift, drag, weight, and thrust. The design of a wing starts from the selection of a proper airfoil which has the desired performance characteristics. CFD analysis of airfoils gives us a better understanding of airfoil characteristics in specified conditions and parameters.[2] Also, CAE analysis using proper materials for structural members ensures a foolproof design of the UAV wing.

7.2 METHODOLOGY

In this study, the aim and objectives are to design a wing for the micro-UAV category having a weight-carrying capacity of up to 2 kgs and a flight altitude of up to 200 ft., which is followed by the selection of an airfoil. NACA 4412 and Clark Y airfoils are studied as they are widely used in aircraft wing design. CAD models for NACA 4412 and Clark Y airfoils are prepared for CFD simulation using Solidworks CAD modeling software. CFD simulations for both airfoils are carried out in a virtual wing tunnel using the Solidworks Simulation toolbox. Each airfoil's flap is set at various angles of attack from 0° to 35°. By analyzing the results of CFD simulations, the optimum range for flap operation is found. Lift-to-drag ratios for NACA 4412 and Clark Y airfoils are compared, and a better airfoil is selected. At the end of this study, the addition of winglets is discussed with CFD simulation results as a future scope. At the end of this study, the actual UAV Wing prototype is manufactured using selected configurations and electronics. The following methodology is used in this research as shown in Figure 7.1.

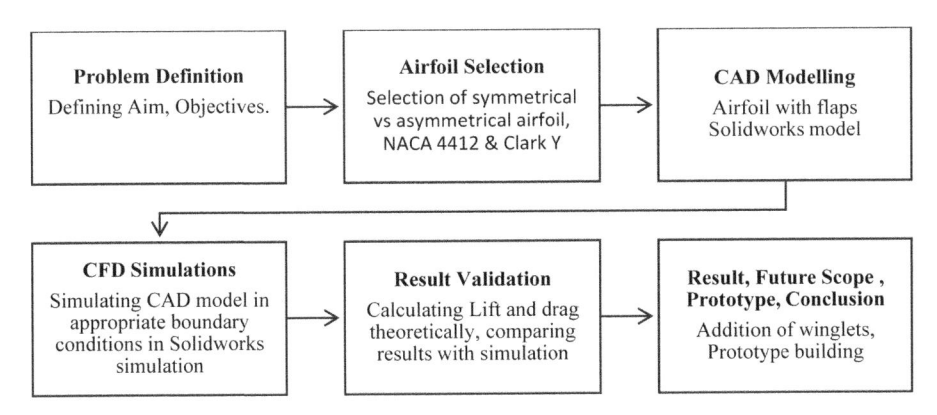

FIGURE 7.1 Methodology for the design and analysis of UAV wings.

7.3 AIRFOIL SELECTION AND SIMULATION

Airfoil is a streamlined body that plays a very important role in the performance of unmanned aerial vehicles. There are multiple factors to be considered while selecting an airfoil profile such as high lift, low drag, wing chord, and aspect ratio of the wing. There are different types of airfoil configurations like symmetric, asymmetric, flat bottom, and under-cambered, each having its own characteristics. In this research, two asymmetrical airfoils, NACA 4412 and Clark Y, are used because of their wide use in aircraft wing design. Also, asymmetrical airfoils are highly efficient and capable of generating lift even at a 0° angle of attack (α) of the wing.[4]

The basic terminology of an airfoil is as follows:

1. Chord – It is the distance between the leading edge and the trailing edge.
2. Angle of attack – It is the angle between the chord and the reference horizontal line.
3. Camber – It is the curvature of the airfoil.
4. Upper surface – Low-pressure surface area
5. Lower surface – High-pressure surface area

Considering UAVs in the micro-UAV category, the following parameters of wings are considered in this study:

1. Chord length (L) = 0.08 m
2. Wing span (L) = 0.4 m
3. Velocity (V) = 35 m/s
4. α varies from 0°–35°
5. Root chord = 0.08 m, Tip chord = 0.064 m
6. Average chord = 0.074 m
7. Area of wing = 0.028 m²
8. Air density = 1.225 kg/m³

Bernoulli's principle is responsible for the creation of lift forces on a wing. When air flows around airfoils, there is a point where the velocity of the fluid is close to zero called the stagnation point. Fluid flowing above the stagnation point, that is, over the top surface of airfoil, has a lower pressure zone, and the lower surface area has a high pressure zone, which creates a net vertical lift force due to the pressure difference.

The process of the simulation of fluid involves four different steps such as gathering design inputs and gathering 2D profiles for airfoils, preworking, meshing, and solver. In the first stage, NACA 4412 and Clark Y airfoil profiles are imported into the Solidworks environment for CAD modeling, followed by airfoil modeling by using Solidworks CAD modeling software using NACA 4412 and Clark Y airfoil profiles as shown in Figure 7.2.

Solidworks simulation automeshing is used for meshing the wing CAD model for simulation as shown in Figure 7.3.

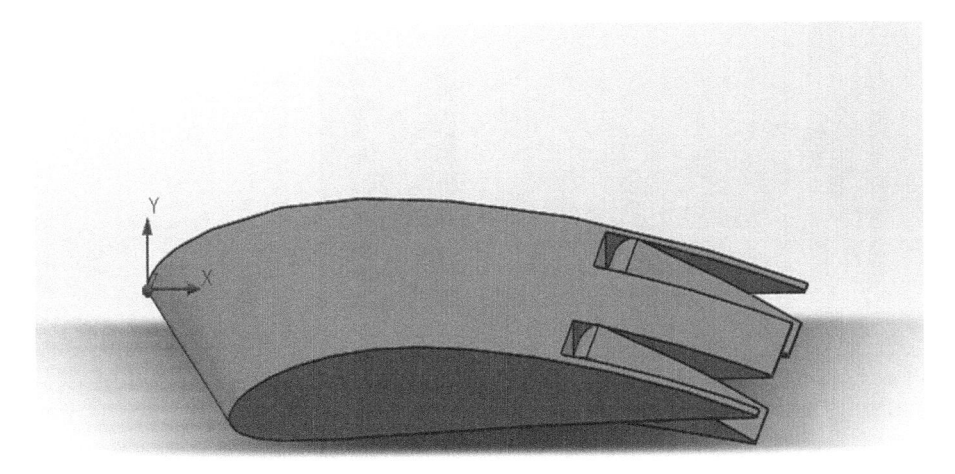

Design and Analysis of Wing for Unmanned Aerial Vehicle 119

FIGURE 7.2 CAD model of the wing.

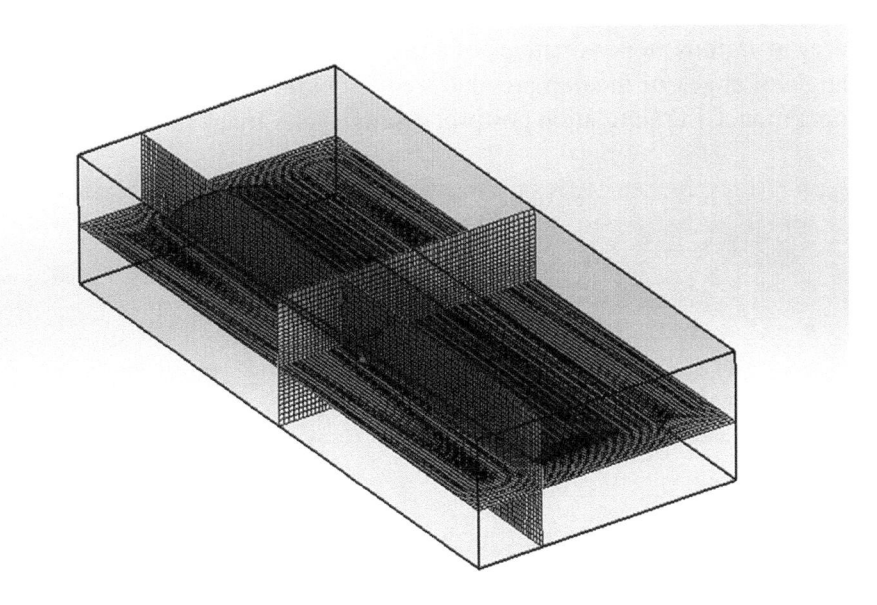

FIGURE 7.3 Meshing configuration.

The airfoil model is set into a virtual wing tunnel for CFD simulation as shown in Figure 7.4a. All parameters for the simulation environment including fluid as air, flow type as external, flow velocity as 35 m/s, and other characteristics are set as shown in Figure 7.4b.

(a) (b)

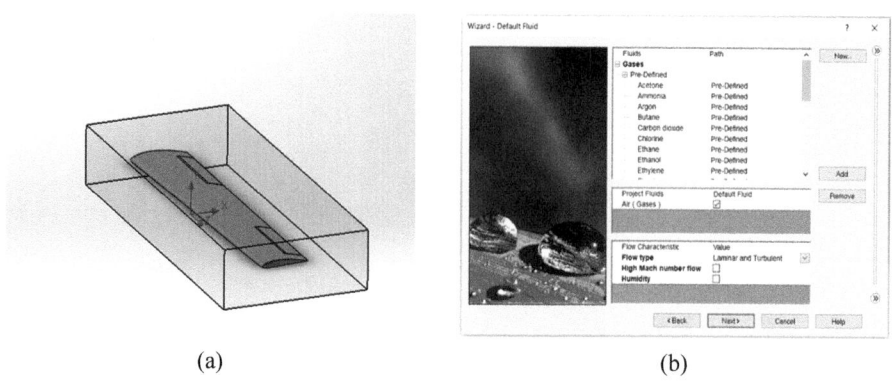

FIGURE 7.4 a. Virtual wing tunnel. b. Demonstrating the physics.

In the postprocessing stage, CFD results are analyzed. The important parameters for analysis are lift force, drag force, pressure distribution and velocity distributions around the airfoil, coefficient of lift, and coefficient of drag at various angles of attack of flaps. Graphical results for lift and drag vs angle of attack of the flap are analyzed to obtain the optimum range of flap operation. CFD simulation postprocessing results are as shown in Figure 7.5.

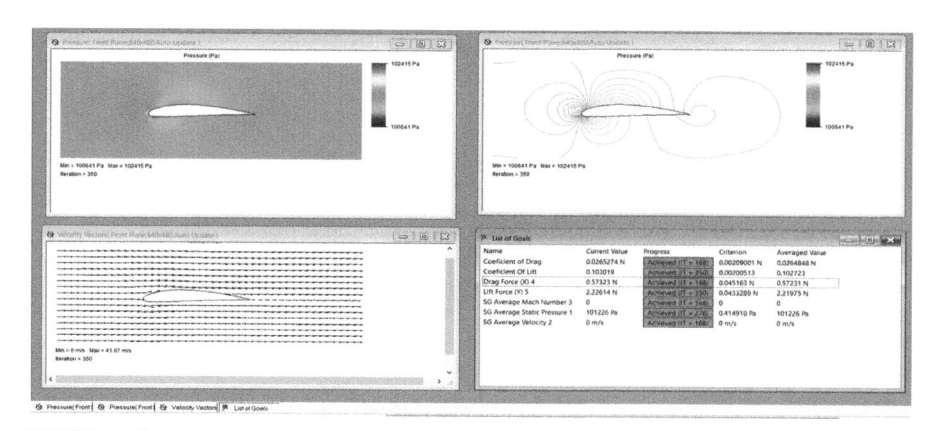

FIGURE 7.5 CFD postprocessing.

7.4 CFD SIMULATIONS

For this study, two asymmetric airfoils are considered: NACA 4412 and Clark Y. Asymmetric airfoils are used because of their characteristics of generating lift even at a zero-degree angle of attack.

7.4.1 AEROFOIL 1—NACA 4412

NACA 4412 CFD Simulations are carried out at various angles of attack of flaps. The NACA 4412 wing model is set at various angles of attack of flaps such as 0, 8, 15, 25, and 35° AOA as shown in Figure 7.6a. Also, a virtual wing tunnel with aerofoil placement is shown in Figure 7.6b.

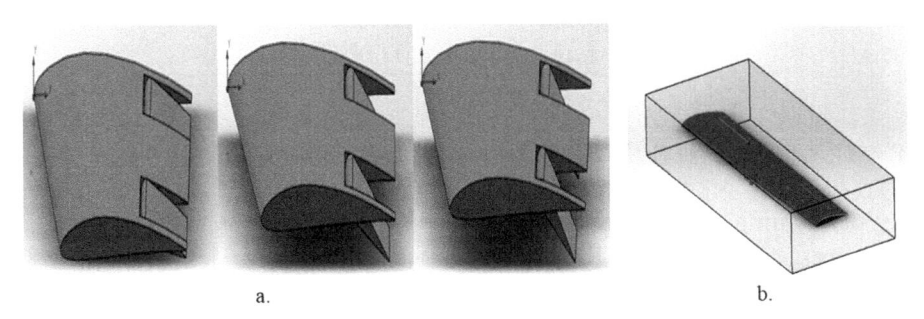

FIGURE 7.6 a. CAD model set at various AOA of flaps, b. Virtual wing tunnel.

CAD modeling is performed by using Solidworks, and an aerofoil CAD model is set into a Solidworks CFD environment with different boundary conditions such as external flow type, air as fluid at room temperature, and traveling at 35 m/s over aerofoil.

The results of the CFD simulation for NACA are shown in Table 7.1.

TABLE 7.1 NACA 4412 CFD Simulation at Various Angles of Attack of Flaps.

Angle of attack (α)°	Lift force (F_L) N	Drag force (F_D) N	Lift-to-drag ratio (L/D)
0 AOA	2.219747584	0.572310337	3.8785
8 AOA	3.401726171	0.744741445	4.5679
15 AOA	4.943257736	1.076776625	4.591
25 AOA	7.409100621	1.705407723	4.3444
30 AOA	7.563717554	1.740433201	4.3459
35 AOA	9.2794224951	2.231545766	4.15

7.4.2 AEROFOIL 2—CLARK Y

Clark Y CFD simulations are carried out at various angles of attack of flaps. The Clark Y wing model is set at various angles of attack of flaps such as 0, 8, 15, 25, and 35° AOA. The results of the CFD simulation at different AOA of flaps are shown in Table 7.2.

TABLE 7.2 Performance Parameters of Clark Y.

Angle of attack (α)°	Lift force (F_L) N	Drag force (F_D) N	Lift-to-drag ratio (L/D)
0 AOA	1.6588043427	0.57903575463	2.8647
8 AOA	2.868472135	0.7682922745	3.7335
15 AOA	4.5711083073	1.0910560178	4.1896
25 AOA	6.1702914680	1.6474459995	3.7453
30 AOA	7.0366753276	1.8398192954	3.8246
35 AOA	7.053802649	2.2628314129	3.1172

7.4.3 COMPARATIVE STUDY BETWEEN NACA 4412 AND CLARK Y

By performing CFD analysis on both airfoils, i.e., NACA 4412 and Clark Y, values of lift and drag force at various angles of attack of flaps and comparative graphs of lift force for both airfoils are as shown in Figure 7.7.

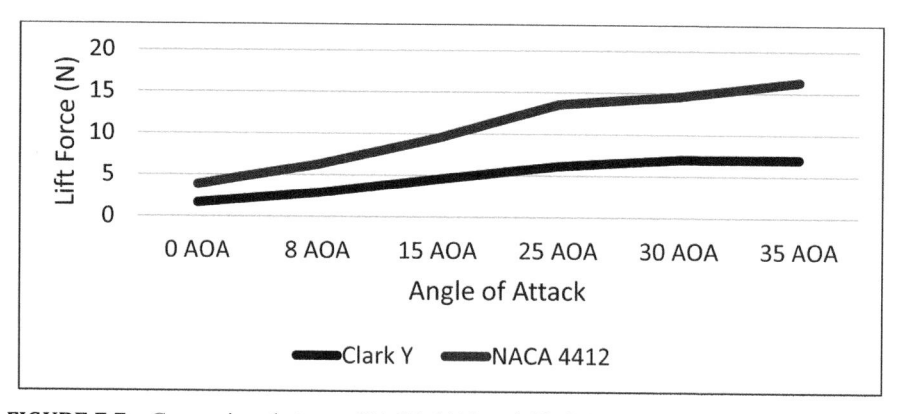

FIGURE 7.7 Comparison between NACA 4412 and Clark Y.

From the above Figure 7.7, the NACA 4412 airfoil has a higher lift force than the Clark Y airfoil at all angles of attack. Hence, the NACA 4412 airfoil is selected for further study of the winglet effect on the wing and also for structural analysis of the wing.

Theoretical validation is done using the following formulas and parameters:

$$\text{Lift Force } (F_L): 0.5 \times \rho \times A \times V^2 \times C_l \tag{7.1}$$

$$\text{Drag Force } (F_D): 0.5 \times \rho \times A \times V^2 \times C_d \tag{7.2}$$

where C_l = lift coefficient and C_d = drag coefficient are found using Solidworks simulation by setting

$$C_l = \frac{F_l}{0.5 \times \rho \times A \times V^2} \quad (7.3)$$

$$C_d = \frac{F_d}{0.5 \times \rho \times A \times V^2} \quad (7.4)$$

By simulation, we got values of C_l and C_d as shown in Table 7.3.

TABLE 7.3 C_l vs C_d at Various AOA.

Angle of attack (α)	C_l	C_d
0 AOA	0.10272329	0.10272329
8 AOA	0.158066901	0.158066901
15 AOA	0.228759208	0.228759208
25 AOA	0.342871055	0.342871055
30 AOA	0.350026265	0.080542052
35 AOA	0.429423967	0.429423967

By considering the above values, Table 7.4 is prepared, comparing the values of lift and drag force obtained from Solidworks simulation and theoretical calculations.

TABLE 7.4 Comparison between Theoretical and Simulation Values.

Angle of attack (α)°	Solidworks simulation values F_l (N)	F_d (N)	Lift Drag	Theoretical values F_l (N)	F_d (N)
0 AOA	2.219747584	0.572310337	3.8785	2.164279	0.557299
8 AOA	3.401726171	0.744741445	4.5679	3.333542	0.728312
15 AOA	4.943257736	1.076776625	4.5907	4.752095	1.040437
25 AOA	7.409100621	1.705407723	4.3444	7.192261	1.665321
30 AOA	7.563717554	1.740433201	4.3459	7.359028	1.690595
35 AOA	9.2794224951	2.231545766	4.1582	8.998089	2.653426

By evaluating all values of lift and drag forces at various angles of attack, as we increase the flap angle beyond 15°, the lift-to-drag ratio starts decreasing. From Table 7.4, it is concluded that the optimum range of flap angle should be in between 0 and 15°.

7.5 CAE SIMULATION OF WING STRUCTURE

The methodology used while performing structural analysis of the wing structure is as shown in Figure 7.8.

Problem Identification
- Defining Objectives for structural analysis study.
- Defining Procedure, Approach for Structural study.

Pre Processing
- CAD Modelling of Parts- Ribs, Spar, stringers.
- Material Assignment.
- Setting Contact constraints.

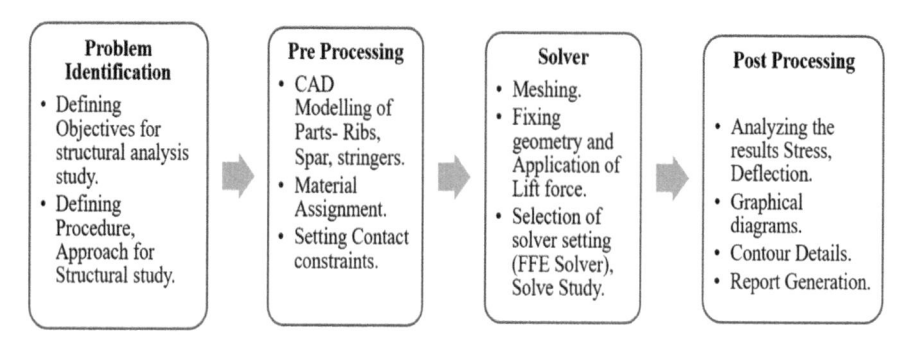

Solver
- Meshing.
- Fixing geometry and Application of Lift force.
- Selection of solver setting (FFE Solver), Solve Study.

Post Processing
- Analyzing the results Stress, Deflection.
- Graphical diagrams.
- Contour Details.
- Report Generation.

FIGURE 7.8 Methodology for structural analysis.

In the preprocessing stage, structural members like spar, stringer, and skin are modeled in Solidworks with material assigned to each part and then imported into the simulation environment, with contact set as bonded. Contact sets of structural members are shown in Figure 7.9.

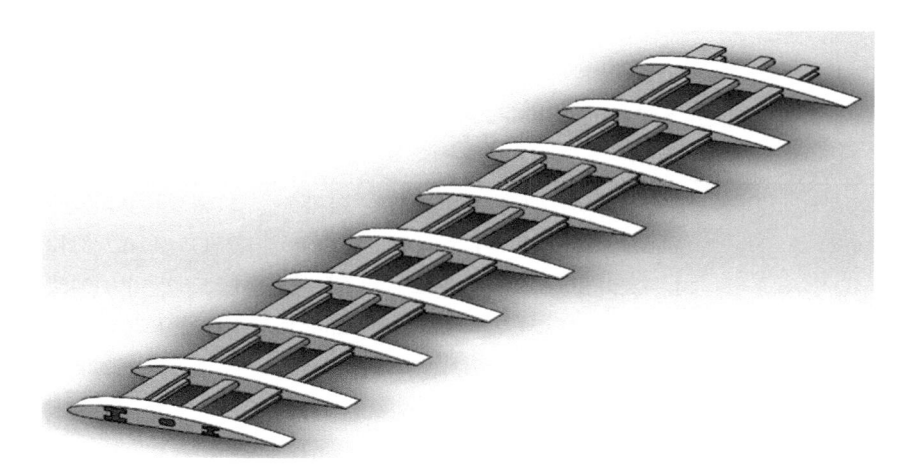

FIGURE 7.9 CAE preprocessing contact set.

The next stage of simulation is meshing and solver. Automesh is used, and it gives more accurate results with a minimum number of nodes. The FFE solver is used in this study. FFE is an iterative solver, which is more efficient than other Solidworks solvers.[7] Solidworks automesh technique is used for meshing. The detailed meshing parameters are as shown in Table 7.5.

TABLE 7.5 Meshing Details.

Element size	4 mm	Global contact	Bonded
Mesh quality	High	Analysis type	Static

Meshing configurations are shown in Figure 7.10.

FIGURE 7.10 CAE meshing configuration.

In the postprocessing stage, simulation results are analyzed by stress, strain, and defection diagrams. Postprocessing results are shown in Figure 7.11.

For this study, a material combination of aluminum 1060 alloy and commercially used polyurethane foam materials is used for spars and skin, respectively. By performing structural analysis with materials, results are shown in Table 7.6.

CAE structural results are well within the limits of maximum stress and deflection; hence, a material combination is selected for the manufacturing prototype of the UAV.

7.6 CFD ANALYSIS BY ADDITION OF WINGLET

When an aerofoil is traveling through air, there is a pressure difference between the upper and lower surfaces of the aerofoil. The wing has less pressure at the upper surface and high pressure at the lower surface. Air has a natural tendency of flowing from high pressure to low pressure through a path of less resistance; this flow causes vortice generation over the tips. This vortex results in whirlpools of air called vortex. Winglets are curved extensions of wingtips that improve an aircraft's fuel efficiency and range. Winglets are designed as small airfoils, which reduce the aerodynamic drag.

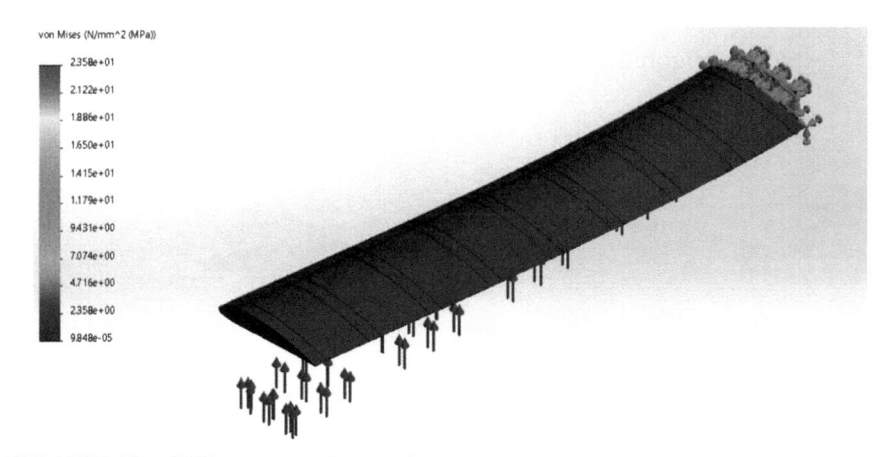

FIGURE 7.11 CAE postprocessing results.

TABLE 7.6 Structural Analysis Results.

Materials	Fixture and force conditions	Stress (σ) (Von Mises)	Deflection (δ)
Aluminum alloy 1060–Spar PVC plastic–Ribs PU Foam–Skin	- Fixed at the fuselage end. - 10 N force applied in the vertical direction.	**Min** 9.848e-05N/mm^2 (MPa) Node: 7172 **Max**- 2.358e+01N/mm^2 (MPa) Node: 23955	**Min** 0 mm Node:11012 **Max**- 3.890e+00mm Node: 41871

The CAD model of the winglet is shown in Figure 7.12.

A comparative study of lift force with the addition of winglets is shown in Table 7.7.

The results of lift force with and without winglets are shown in Figure 7.13.

7.7 PROTOTYPE BUILDING

By analyzing CFD and CAE simulation results, the actual prototype, consisting of selected airfoils and material combinations, is built using the proper selection of electronics. The prototype building is divided into three major sections as follows:

1. NACA 4412 airfoil-shaped wing manufacturing with selected material for structural members
2. Manufacturing UAV fuselage bodies and landing gear
3. Manufacturing elevators and rudders with the proper selection of electronics for flight controls

FIGURE 7.12 Winglet for the wing.

TABLE 7.7 Comparative Study for Winglet.

Angle of attack (α)°	Lift force without winglet (F_L) N	Lift force with winglet (F_{LW}) N	% Increase
0 AOA	2.219747584	2.7586933319	24.27%
8 AOA	3.401726171	5.7761425191	69.80%
15 AOA	4.943257736	7.1172669289	44%
25 AOA	7.409100621	8.8725960593	19.75%
30 AOA	7.563717554	10.222124582	35.14%
35 AOA	9.2794224951	9.4322349479	1.64%

7.7.1 PROTOTYPE MANUFACTURING

In stage 1 of the prototype building, a hot wire cutter is used to cut the NACA 4412 airfoil shape for the wing followed by the insertion of structural members inside the wing structure. In stage 2, the fuselage is built using lasercut aeroply sheets. Lasercut aeroply sheets are shown in Figure 7.14.

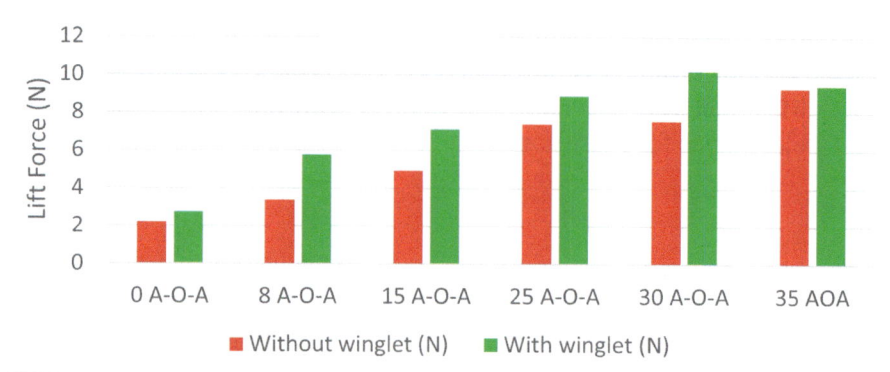

FIGURE 7.13 Without winglets vs with winglets.

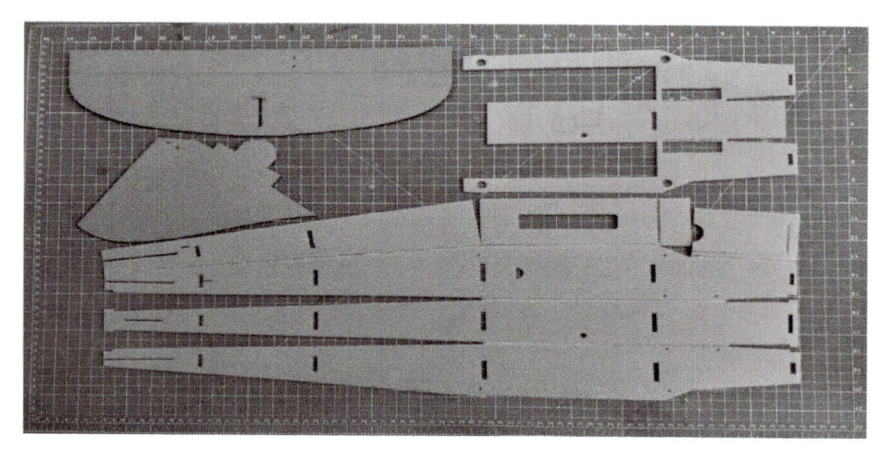

FIGURE 7.14 Lasercut fuselage aeroply sheet.

7.7.2 ELECTRONIC SELECTION

7.7.2.1 SELECTION OF ELECTRONICS IS IMPORTANT IN STAGE 3, ELECTRONIC SELECTION FOR MICRO-UAV

1. 2 Nos. (4 Nos. for EPP Aileron Wing) 9 gm Servos
2. 1 No. 30Amp Brushless ESC (Speed controller)
3. 1 No. 2822/14 1400 KV Brushless Motor
4. 1 No. 8 × 6 Propellor

5. 15 cm Servo Extension Cable
6. 1200 mAh battery
7. 6 channel transmitter receivers

7.7.2.2 ELECTRONIC CONNECTIONS

The electronic connections of components are shown in Figure 7.15.

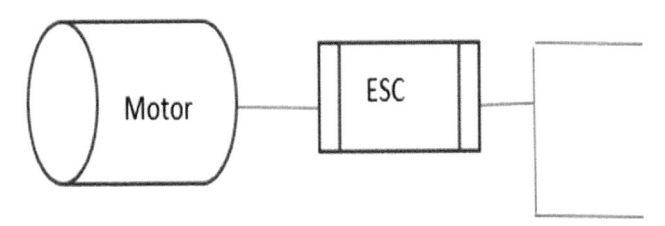

FIGURE 7.15 Electronic connections.

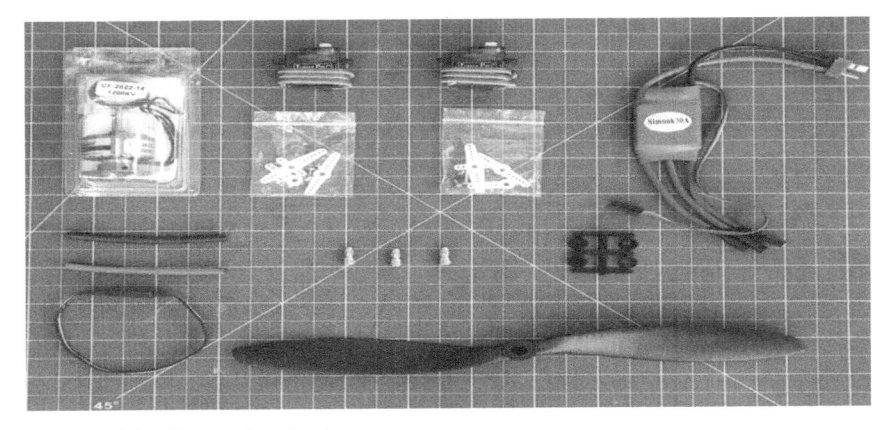

FIGURE 7.16 Electronic selection.

1400 KV BLDC motor, 2 nos. 9 G servos, ESC, propeller, and wire extensions are shown in Figure 7.16.

7.7.3 FINAL PROTOTYPE OF UAV

By combining all three stages of manufacturing, the UAV prototype is assembled. The wing structure is assembled with fuselage using rubber bands on both sides. The vertical rudder and horizontal elevators are fixed to the fuselage body using a hot glue gun. Landing gears of steel rods are manufactured in the required shape and attached to the fuselage body for a smooth run-on runway. The propeller is attached to the BLDC motor using a nut assembly, and rubber wheels are connected to the landing gear assembly. After the assembly of all parts, the final prototype of the UAV is as shown in Figure 7.17.

FIGURE 7.17 UAV wing prototype.

7.8 CONCLUSIONS

Based on a study done using CAD, CFD, and CEA techniques for the design and analysis of the wing of a UAV, it is concluded that–

1. The NACA 4412 airfoil has a higher lift force than the Clark Y airfoil at 35 m/s speed at various angles of attack of flaps.
2. The addition of a winglet at the tip of a wing increases lift and reduces vortices at the tip of the wing.
3. The combination of aluminum alloy 1060 and polyurethane foam material is better to avoid stress concentration and excessive deflection of the wing than traditional balsa wood material.
4. Proper selection of electronics, materials, and manufacturing methods has a huge impact on UAV wing operations.

KEYWORDS

- **unmanned aerial vehicle**
- **airfoil selection**
- **CAD**
- **CAE**
- **CFD**
- **winglet**

REFERENCES

1. Marques, P. Emerging Technologies in UAV Aerodynamics. *Int. J. Unmanned, Syst. Eng. (IJUSEng)*, **2013,** *1* (S1), 3–4.
2. Karthik, M. A.; Adithya, H. V. et al. Analysis and Selection of Airfoil sections for Low-Speed UAV's. *Int. J. Latest Eng. Res. Appl. (IJLERA)*, **2018,** *03* (05), 40–49.
3. Ockfen, A. E.; Matveev, K. I. Aerodynamic Characteristics of NACA 4412 Airfoil Section with Flap in Extreme Ground Effect. *Int. J. Nav. Archit. Ocean Eng.* **2009,** *1* (1), 1–12.
4. Długosz, A.; Klimek, W., In *The Optimal Design of UAV Wing Structure*, AIP Conference Proceedings 2018, 1922, p 120009.
5. Grodzki, W.; Łukaszewicz, A. Design and Manufacture of Unmanned Aerial Vehicles (UAV) Wing Structure using Composite Materials. *Article in Materialwissenschaft und Werkstofftechnik*, 2015.
6. Patil, C. K.; Ghorpade, R. Finite Element and Experimental Vibration Analysis of High-Pressure Fuel Injection Pipe for 8-Cylinder V Diesel Engine. *J. Univ. Shanghai Sci. Technol.* **2021,** *23* (3), 224–233.
7. Chahl, J. Unmanned Aerial Systems (UAS) Research Opportunities. *Aerospace* **2015,** *2* (2), 189–202.
8. Dündar, Ö.; Bilici, M. Design and Performance Analyses of a Fixed Wing Battery VTOL UAV. *Eng. Sci. Technol. Int. J.* **2020,** *23* (5), 1182–1193.
9. Escobar, A. G.; Lopez-Botello, O. Conceptual Design of an Unmanned Fixed-Wing Aerial Vehicle Based on Alternative Energy. *Int. J. Aerosp. Eng.* **2019,** 8104927.
10. Frulla, G.; Cestino, E. Design, Manufacturing and Testing of a HALE-UAV Structural Demonstrator. *Compos. Struct.* **2008,** *83,* 143–153.
11. Chung, P.-H.; Ma, D.-M. Design, Manufacturing, and Flight Testing of an Experimental Flying Wing UAV. *Appl. Sci.* **2019,** *9* (15), 3043.
12. Kontogiannis, S. G. Design, Performance Evaluation and Optimization of a UAV. *Aerosp. Sci. Technol.* **2013,** *29* (1), 339–350.
13. Fang, W.; Ma, H.; Zhang, H. In *Structural Design and Analysis of a Quadrotor Fixed-Wing Hybrid UAV Wing*, 31st Congress of the International Council of the Aeronautical Sciences, 2018.

14. Grodzki1, W.; Łukaszewicz, A. Design and Manufacture of Umanned Aerial Vehicles (UAV) Wing Structure using Composite Materials. *Mater. Werkst.* **2015,** *46* (3).
15. Maghazei, O.; Netland, T. H. Drones in Manufacturing: Exploring Opportunities for Research and Practice. *J. Manuf. Technol. Manag.* 2019.
16. Romeo, G. Design Proposal and Wing Box Manufacturing of a Self-Launching Solar Powered Sailplane. *Tech. Solar* **1997,** *21* (4), 106–115.
17. Heliplat, F. G. In *Structural Analysis of High Altitude Very-Long Endurance Solar Powered Platform for Telecommunication and Earth Observation Applications*, 23rd Congress of the International Council of the Aeronautical Sciences (ICAS), Toronto, Sept 8–13, 2002; Paper 395-ICAS2002-1.3.3.
18. Romeo, G. Design of High Altitude Very-Long Endurance Solar Powered Platform for Earth Observation and Telecommunication Applications. *Aerotec Missili Spaz*, **1998,** *77* (3–4), 88–99.
19. Romeo, G. In *Numerical Analysis, Manufacturing and Testing of Advanced Composite Structures for a Solar Powered Airplane*, XV AIDAA Congress, Torino, Nov, 1999; vol. II, p. 1001–1012.
20. Ayuso Gonzalvo, A. Scaled Size Prototype Manufacturing of CFRP Structural Parts and Metallic Joints Report. HE-D63-A1B-CAS-RP-04, Technical report; 2002.
21. Denisov, V. E.; Bolsunovksy, A. L.; Buzoverya, N. P.; Gurevich, B. l.; Shkadov, L. M. Conceptual Design for Passenger Airplane of Very Large Passenger Capacity in Flying Wing Layout. ICAS-96-4.6.1.
22. Roman, D.; Allen, J. B.; Liebeck, R. H. Aerodynamic Design Challenges of the Blended-Wing-Body. Subsonic Transport, AIAA-2000-4335.
23. Libeck, R. H.; Page, M. A.; Rawdon, B. K. Blended Wing-Body Subsonic Commercial Transport. *AIAA*, 98-0438, 1998.
24. Biber, K.; Tilmann, C. P. *Supercritical Airfoil Design for Future Hale Concepts*; AIAA
25. 2003-1095, 41st Aerospace Sciences Meetings and Exhibit, 6–9 January, **2003**.6.
26. Chandler, P.; Pachter, M., In *Research Issues in Autonomous Control of Tactical UAVs*, Proceedings of the American Control Conference, Philadelphia, PA, 1998, pp 394–398.
27. Prasanth, R. K.; Boskovi, J. D.; Li, S.-M.; Mehra, R. In *Initial Study of Autonomous Trajectory Generation for Unmanned Aerial Vehicles*, Proceedings of the 40th IEEE Conference on Decision and Control, Orlando, FL, 2001; pp 640–645.
28. McLain, T.; Chandler, P.; Pachter, M. In *A Decomposition Strategy for Optimal Coordination of Unmanned Air Vehicles*, Proceedings of the American Control Conference, Chicago, IL, 2000, pp 369–373.
29. Twigg, S.; Calise, A.; Johnson, E. In *On-Line Trajectory Optimization for Autonomous Air Vehicles*, AIAA Guidance, Navigation, and Control Conference, Austin, TX, 2003, pp AIAA 2003–5522.
30. Bellingham, J.; Richards, A.; How, J. In *Receding Horizon Control of Autonomous Aerial Vehicles*, Proceedings of the American Control Conference, Anchorage, AK, 2002, pp 3741–3745.
31. Bortoff, S. A. In *Path Planning for UAVs*, Proceedings of the American Control Conference, Chicago, IL, 2000; pp 364–368.
32. Hallberg, E.; Kaminer, I.; Pascoal, A. In *Development of the Rapid Flight Test Prototyping System for Unmanned Air Vehicles*, Proceedings of the American Control Conference, Philadelphia, PA, 1998; pp 699–703.

CHAPTER 8

Estimation of Modal Loss Factor of Viscoelastic Material Using the Modal Strain Energy Method

GORAKH PAWAR[1], PRAVIN HUJARE[2], ANIL SAHASRABUDHE[3], and DEEPAK HUJARE[4]

[1]*University of Utah, Salt Lake City, USA*

[2]*Vishwakarma Institute of Information Technology, India*

[3]*All India Council for Technical Education (AICTE), India*

[4]*MIT World Peace University, India*

ABSTRACT

A viscoelastic material is characterized by both viscous and elastic properties. A purely elastic material is one in which there is perfect energy conversion, i.e., all the energy stored in a material during loading is regained when the load is removed. Elastic materials have an inphase stress–strain relationship. A purely viscous material does not recover any of the energy stored during loading after the load is removed. Viscoelastic materials have behaviors that fall between elastic and viscous extremes. The rate at which the material essentially dissipates energy in the form of heat through shear, the main driving mechanism of damping materials, defines the efficiency of the viscoelastic material. Because a viscoelastic material falls between elastic and viscous behavior, some of the energy is obtained when the load is removed, and some part is lost or dissipated in the form of thermal energy. The phase shift between the stress and strain maximums, which does not exceed 90°, is a measure of the material's damping performance. The greater the phase angle between stress and strain during the same cycle, the more effective the material is at eliminating unwanted vibrations. The governing equations representing viscoelastic damping analysis are solved using the finite element method based on a modal frequency response analysis solver.

8.1 INTRODUCTION

Viscoelastic materials are elastomeric materials whose long-chain molecules cause them to convert mechanical energy into heat when they are deformed. Viscoelastic materials are usually polymers, which allow for a wide range of compositions that lead to a wide variety of materials and behaviors.[1]

Therefore, viscoelastic damping materials can be developed and adapted to specific applications. By proper tailoring, polymeric materials can be manufactured to have a wide range of damping, strength, durability, creep resistance, thermal stability, and other desirable properties over a selected temperature and frequency range.

The properties of the viscoelastic materials depend on the frequency of operation and the temperature of the surrounding environment.[2] In the past, many scientists have investigated the properties and mechanical behavior of polymeric materials. Ross and Kerwin were among the first to develop an analytical method for three-layered damping treatments.[3,4]

Soni used isoparametric thin shell elements for the face sheets and solid elements for the viscoelastic core that are fully compatible. The frequency domain vibration analysis of viscoelastically damped sandwich structures was achieved to obtain damped resonance frequencies and modal loss factors from the direct solution of complex structural equations of motion.[5]

Moreira and Rodrigues focused on the finite element modeling of constrained layer damping treatments using viscoelastic materials. They used commercial FEA software to simulate the dynamic response of aluminum plates with CLD treatment.[6]

8.2 HYSTERESIS BEHAVIOR OF THE VISCOELASTIC MATERIAL

The stress–strain relation for a linear viscoelastic material under a repeated cyclic loading is an ellipse. A viscoelastic material follows different patterns during its deformation when a force is applied than when a force is released. The correlation between measured stress σ compared to measured strain ε for all instants of a cycle of oscillations is found by plotting as shown in

Estimation of Modal Loss Factor of Viscoelastic Material 135

Figure 8.1. Within the limit, the plots of measured stress σ versus measured strain ε are elliptical in shape and maintain that shape as amplitude increases.[1]

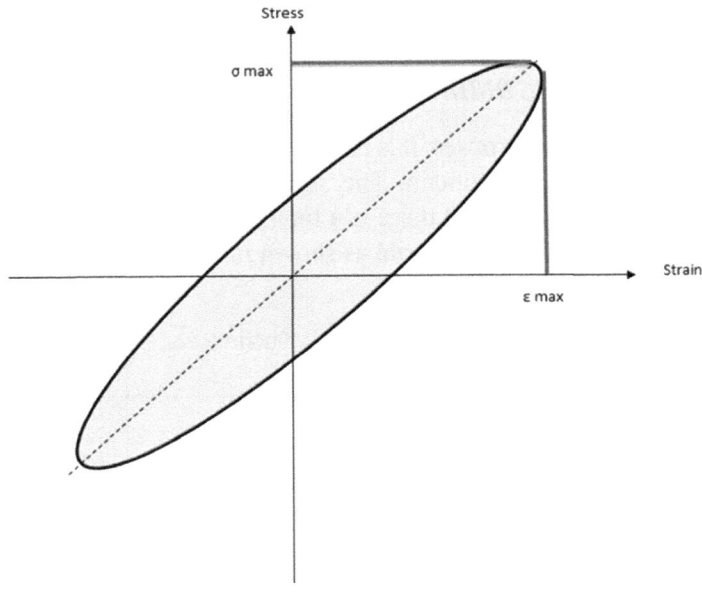

FIGURE 8.1 Ideal elliptical hysteresis loop.[1]
Source: Reprinted with permission from Ref. [1]

The maximum ellipse axis is the ratio of strength and aspect ratio (the ratio of the small axis to the large axis is the ratio of moisture). The elliptical loop equation has the following form in the form of a shear transformation:

$$\tau(t) = G_\beta(t) \pm G\eta_m \sqrt{\beta_0^2 + \beta(t)^2} \tag{8.1}$$

If the strain $\beta(t)$ varies harmonically with time during steady-state response, one can write $\beta(t) = \beta_0 \sin(\omega t)$, where ω is the frequency in radians per second, and insert this into eq 8.1 to give:

$$\tau(t) = G_\beta(t) \pm G\eta\beta_0 \sqrt{1 - \sin^2(\omega t)}$$
$$= G_\beta(t) \pm G\eta\beta_0 |\cos(\omega t)|$$
$$= G_\beta(t) \pm \frac{G\eta}{|\omega|} \frac{d\beta(t)}{dt} \tag{8.2}$$

where the + or − sign depends on whether cos(ωt) is positive or negative.

8.3 FREQUENCY DOMAIN BEHAVIOR OF VISCOELASTIC MATERIALS

The behavior of viscoelastic materials in the frequency domain is explained in this section.

8.3.1 VISCOELASTIC BEHAVIOR

For many engineering purposes, it is easy to learn how to vibrate viscoelastic materials in a frequency domain. The strain–time history and stress–time history are both harmonic, but there is a time or lag of phase between stress and strain for a viscoelastic material as shown in Figure 8.2.

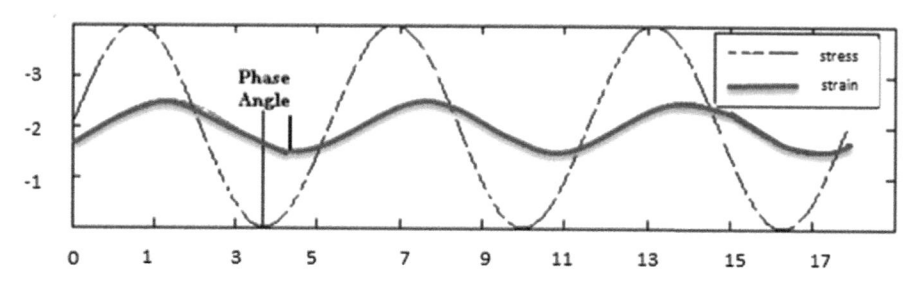

FIGURE 8.2 Harmonic excitation and response for a viscoelastic solid.[1]
Source: Reprinted with permission from Ref. [1]

The phase lag of viscoelastic material indicates that there is a velocity-dependent term in the stress–strain relationship, which, for $\tau(t) = \tau_0 \sin(\omega t)$ and $\beta = \beta_0 \sin(\omega t - \psi)$, is

$$\tau(t) = \tau_0 \sin(\omega t)$$
$$= \tau_0 \sin[(\omega t - \psi) + \psi]$$
$$= \tau_0 \sin(\omega t - \psi)\cos\psi + \tau_0 \cos(\omega t - \psi)\sin\psi$$
$$= \frac{\tau_0}{\beta_0}\cos\psi\,\beta(t) + \frac{\tau_0}{\beta_0|\omega|}\sin\psi\,\frac{d\beta(t)}{dt} \tag{8.3}$$

By writing that $G = (\tau_0/\beta_0)\cos(\psi)$ and $\eta = \tan(\psi)$, it gives

$$\tau = G\beta \pm \frac{G\eta}{|\omega|}\frac{d\beta}{dt} \tag{8.4}$$

Similarly, for extensional deformation, one can write

$$\tau = E\varepsilon \pm \frac{E\eta}{|\omega|}\frac{d\varepsilon}{dt} \qquad (8.5)$$

where ω is the frequency and $|\omega|$ is the numerical value of the frequency. G is the shear modulus, η is the loss factor, and E is the Young's modulus for extensional deformation.[1]

8.3.2 COMPLEX MODULUS MODEL

For a viscoelastic material, the modulus is represented by a complex quantity. This complex modulus describes the material's damping and stiffness properties as a function of temperature and frequency. The imaginary component E_2 narrates the material's viscous behavior and expresses the energy dissipative ability of the material. The real part of this complex term, E_1, relates to the elastic behavior of the material and describes its stiffness.

According to Hook's law, the complex modulus E^* is expressed as:

$$E^* = E_1 + E_2\, i = E_1\,(1 + i\eta) \qquad (8.6)$$

where $\eta = E_2/E_1 =$ loss factor. The loss factor (η) is a measure of the energy dissipation capacity of the material. The loss factor governs how much energy is dissipated. Viscoelastic materials are extensively used for passive damping in a variety of applications.

8.4 ANALYTICAL MODELS OF VISCOELASTIC MATERIALS

In this section, some parametric models often used to model the dynamic behavior of viscoelastic materials are described.

8.4.1 CLASSICAL MODELS

In the past, many simple viscoelastic behavior models were based on a combination of elastic and viscous elements. The elastic element can be modeled by a linear spring with a stiffness coefficient K, and the viscous element by a dashpot with a coefficient of viscosity C. Hence, viscoelastic models are combinations of linear springs and dashpots. Three basic classical models are described.[2,3]

8.4.2 THE MAXWELL MODEL

In the Maxwell model, the two elements, dashpot and spring, are combined in series. As shown in Figure 8.3, K is the spring constant and C is the viscosity.

FIGURE 8.3 The Maxwell model.

When a force F is applied to this model, the elongation is equal to the sum of the extensions in the elastic and viscous elements, as follows:

$$\Delta = \Delta e + \Delta v \qquad (8.7)$$

where Δ represents the total elongation of the Maxwell model, Δ_e the elongation of the spring element, and Δ_v the elongation of the viscous element. The differentiation of eq 8.7 with respect to time is given below:

$$\frac{d\Delta}{dt} = \frac{d\Delta_e}{dt} + \frac{d\Delta_v}{dt} \qquad (8.8)$$

As the force F is the same in the elastic and viscous elements, the differential equation between Δ and F is given as:

$$\frac{d\Delta}{dt} = \frac{1}{K}\frac{dF}{dt} + \frac{F}{C} \qquad (8.9)$$

The Maxwell model can be represented by the above equation. The response of Δ of the Maxwell model depends on the applied force F.

8.4.3 THE KELVIN-VOIGOT MODEL

In the Kelvin-Voigot model, the two elements, dashpot and spring, are combined in parallel. It is shown in Figure 8.4.

In this model, the sum of forces in the spring and in the dashpot is equal to the applied force F, that is,

$$F = Fe + Fv \qquad (8.10)$$

As the displacements of the spring and the dashpot are equal, the relation between the force and the displacement is given as:

$$F = K\Delta + C\frac{d\Delta}{dt} \tag{8.11}$$

The Kelvin-Voigot model can be represented by the above equation. This model is very good for modeling creep in materials but is much less accurate for modeling relaxation.

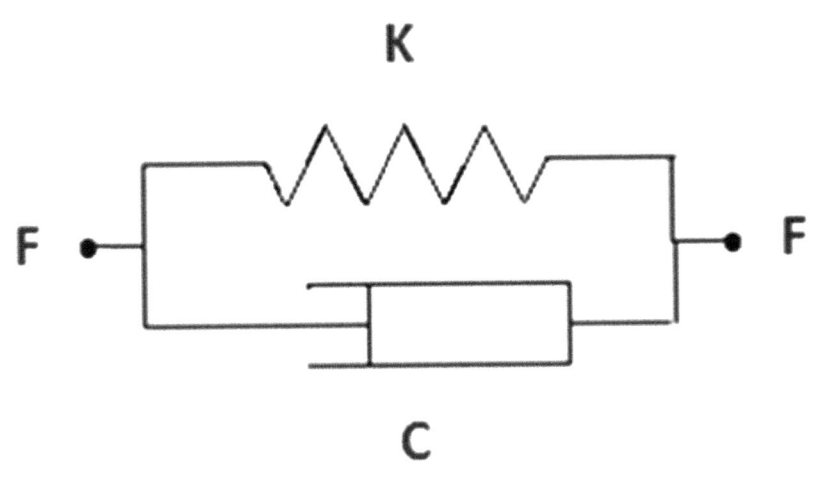

FIGURE 8.4 The Kelvin-Voigot model.

8.4.4 STANDARD LINEAR MODEL

A standard linear model is a combination of a Maxwell model and a linear spring connected in parallel. It is shown in Figure 8.5.

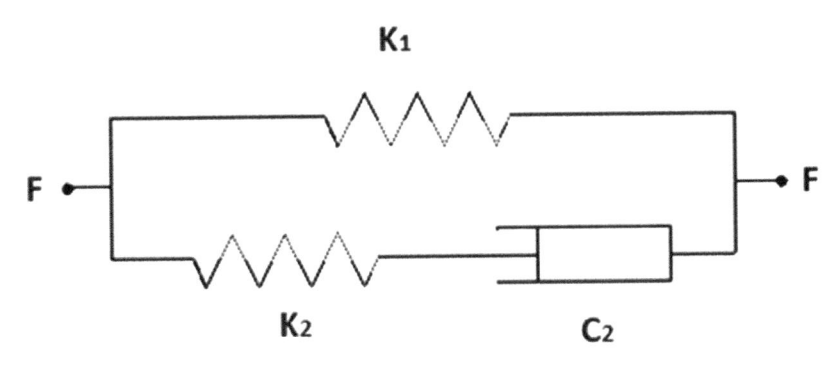

FIGURE 8.5 Standard linear model.

The equilibrium equation is as follows:

$$F = F_1 + F_2 \quad (8.12)$$

where F_1 is the force in the spring and F_2 is the force in the Maxwell model. Since the deformation in the linear spring and the Maxwell element are the same

$$\Delta = \frac{F_1}{K_1} \quad (8.13)$$

from the spring model and

$$\frac{d\Delta}{dt} = \frac{1}{K_2}\frac{dF_2}{dt} + \frac{F_2}{C_2} \quad (8.14)$$

$$= \frac{1}{K_2}\left(\frac{dF}{dt} - K_1\frac{d\Delta}{dt}\right) + \frac{1}{C_2}(F - K_1\Delta)$$

from the Maxwell model. By rearranging the terms,

$$\frac{dF}{dt} + \frac{K_2}{C_2}F = (K_1 + K_2)\frac{d\Delta}{dt} + \frac{K_1 K_2}{C_2}\Delta \quad (8.15)$$

The standard linear model can be represented by the above equation.

8.5 NUMERICAL MODELS OF VISCOELASTIC MATERIALS

The damping effect of the viscoelastic treatments is required to be predicted, and the damping treatments have been done appropriately in the design stage of the structure.[7] During this stage, the finite element method can provide a reliable tool to depict the structural response and analyze the effects of the treatment of design parameters on the structural dynamic performance. In general, there are three finite element methods available to resolve vibration problems of damped structures: complex eigenvalue, direct frequency response, and modal strain energy.

8.5.1 *COMPLEX EIGEN VALUE*

The discretized (i.e., finite element) version of the differential equation of motion of any structure is

$$[M]\{\ddot{x}\} + [C]\{\dot{x}\} + [K]\{x\} = \{F\} \quad (8.16)$$

Estimation of Modal Loss Factor of Viscoelastic Material

where

[M], [C], and *[K]* = physical coordinate mass, damping, and stiffness matrices.

$\{\ddot{x}\}$, $\{\dot{x}\}$, and $\{x\}$ = vectors of nodal accelerations, velocities, and displacements.

$\{F\}$ = vector of applied node loads.

If the structure contains a viscoelastic material, then the stiffness matrix [K] described in eq. 8.16 becomes complex.[8]

Hence, the differential equation of motion for a damped structure under free vibration contains a complex stiffness matrix. By suppressing the term, $[C\{\dot{x}\}]$ is

$$[M]\{\ddot{x}\} + [K]\{x\} = 0 \qquad (8.17)$$

The solution can be carried out in terms of damped normal modes.

$$\{x\} = \sum_{n=0}^{\infty} [X_n^*] e^{i\omega_n^* t} \qquad (8.18)$$

By substituting eq 8.18 in eq 8.17, one obtains the following eigenvalue problem

$$[K]\{X_n^*\} = \omega_n^{*2} [M]\{X_n^*\} \qquad (8.19)$$

where,

$$\omega_n^* = n^{th} \text{ complex eigenvalue}$$

$$\{X_n^*\} = n^{th} \text{ complex eigenvector}$$

Both the eigenvalue and eigenvectors will, in general, be complex in nature, but the complex eigenvalue method is, on the other hand, quite standard. However, there are two important limitations in the complex eigenvalue method. It is computationally expensive, and the properties of real viscoelastic materials are frequency-dependent instead of being constant for storage moduli, which results in the eigen-problem of eq 8.16 being nonlinear.

The complex eigenvalue method is numerically inefficient, and most commercial finite element software do not have the corresponding solver for the complex eigen-solution for a damped structure.

8.5.2 DIRECT FREQUENCY RESPONSE

If the applied load varies sinusoidally in time, energy dissipation in the structure can be calculated by handling the elastic constants of any or all the materials as complex quantities that are dependent on frequency and temperature.

These material properties are apparently available from sinusoidal tests. If the structure is linear, its response will be sinusoidal at the driving frequency, and the steady-state equations of motion will have the form:

$$\left[-M\omega^2 + K_1(\omega) + iK_2(\omega)\right]X(\omega) = F(\omega) \qquad (8.20)$$

where

$K_1(\omega)$, $K_2(\omega)$ = stiffness matrices calculated using the real and imaginary parts of material properties, respectively.

ω = radian frequency of excitation

$F(\omega)$, $X(\omega)$ = complex amplitude vectors of applied node loads and responses, respectively.

$$i = \sqrt{-1}$$

It should be clear that material constants are really complex quantities only in the sense that complex arithmic is used as an easy way to keep track of relative phases under sinusoidal excitation.

There are several limitations in the direct frequency response method. It is computationally affluent because a general sinusoidal solution wants the displacement impedance matrix [the bracketed quantity in eq 8.20] be recalculated, decomposed, and stored at each of several frequencies. In addition, this approach does not give information of direct use to a designer in improving the performance of a candidate structure.

8.5.3 MODAL STRAIN ENERGY METHOD

Modal strain energy method is one of the most economical ways to deal with the complex modulus of the damping material. In this approach, it is assumed that the damped structure has the same natural frequencies and the mode shapes as the undamped structure. It is flexible, accurate, and useful in many design analyses.

In this method, it is assumed that the damped structure can be designated in terms of the real normal modes of the associated undamped system if appropriate damping terms are included in the uncoupled modal equations of motion.[9]

$$\ddot{\alpha}_r + \eta^{(r)} \omega_r \dot{\alpha}_r + \omega_r^2 \alpha_r = f_r(t) \qquad (8.21)$$

$$x = \sum \Phi^{(r)} \alpha_r(t) \quad r = 1, 2, 3, \ldots \quad (8.22)$$

where

α_r = rth modal coordinate.
ω_r = natural radian frequency of the rth mode.
$\Phi^{(r)}$ = r^{th} mode shape vector of the associated undamped system.
$\eta^{(r)}$ = loss factor of rth mode.

It is implied that the physical coordinate damping matrix C of eq 8.16 need not be explicitly calculated but that it can be diagonalized, at least approximately, by the same real modal matrix that diagonalizes K and M.

The modal loss factors are calculated by using the undamped mode shapes and the material loss factor for each material. This general approach was first suggested by Ungar and Kerwin[10] in 1962. Its application by finite element methods was suggested by Johnson et al.[11]

The modal strain energy (MSE) method is simple and can be used in the analysis of viscoelastic damping treatments where the damping effectiveness is assessed by the energy balance between the strain energy of the entire structure and the strain energy of the viscoelastic layer. The normal modal analysis of the undamped system is performed to calculate the strain energy ratio. According to this approach the modal loss factor of the sandwich structure can be estimated as:

$$\eta^r = \eta_v^r \frac{U_v^r}{U_{Total}^r} \quad (8.23)$$

where η^r is the modal loss factor of the sandwich CLD structure for the rth mode, η_v^r is the material loss factor of the viscoelastic material for the rth mode, U_v^r is the strain energy stored in the viscoelastic core for the rth mode, and U_{Total}^r is the total strain energy for the rth mode.

In the modal strain energy approach, the structure is first considered undamped and modeled using the real part of the complex modulus as the modulus of the damping layer. The real eigenvectors of each mode are found from finite element analysis and strain energies in all layers of the structure are computed.

8.6 NUMERICAL ANALYSIS OF VISCOELASTIC MATERIALS

To deal with the complex modulus of the viscoelastic damping material, several different techniques have been developed, of which the modal strain energy method has become a commonly used approach.[12] The calculation of modal strain energy fits quite naturally within finite element methods and is a standard option available in the commercial software NASTRAN.

The damping performance is presented in terms of modal loss factor. The modal loss factor is found by the MSE method. The damping performance of a constrained layer-damped (CLD) beam is investigated by analyzing the numerical model. For analyzing the damping performance of the CLD beam, the first four bending modes are considered. The mode shapes corresponding to the bending modes of the nitrile CLD beams are shown in Figure 8.6.

Modeshape of bending mode 1

Modeshape of bending mode 2

Modeshape of bending mode 3

Modeshape of bending mode 4

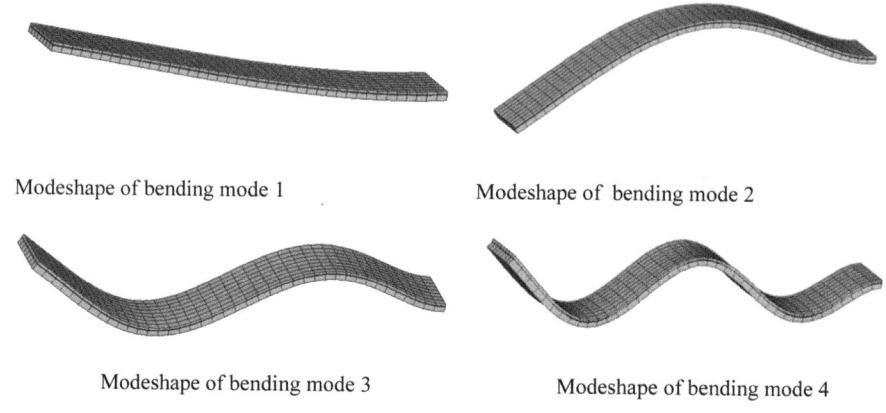

FIGURE 8.6 Modal frequency response analysis result of the nitrile CLD beam.

The strain energy fraction corresponding to each bending mode frequency is obtained from f06 file, as an output of the NASTRAN solver.[13] The modal loss factor is the product of the strain energy fraction, and the material loss factor is obtained by using eq 8.23. The frequencies corresponding to bending mode and modal loss factor of nitrile, butyl, urethane, and SBR viscoelastic material CLD beams are shown in Table 8.1.

8.7 CONCLUSIONS

The dissipative energy of the structure is computed proportional to the material loss factor and the strain energy in the damping layer, and the modal loss factor is obtained by calculating the ratio of the dissipative energy to the total strain energy. The numerical results are obtained in terms of modal loss factor (η) and modal frequency (Hz) for CLD beams at the first four bending modes.

Estimation of Modal Loss Factor of Viscoelastic Material

TABLE 8.1 Damping Performance of CLD Beams.

Mode no.	Nitrile CLD beam Frequency (Hz)	Modal loss factor (η)	Butyl CLD beam Frequency (Hz)	Modal loss factor (η)	Urethane CLD beam Frequency (Hz)	Modal loss factor (η)	SBR CLD beam Frequency (Hz)	Modal loss factor (η)
1	31.08	0.261	31.10	0.255	31.88	0.315	30.86	0.299
2	174.00	0.115	175.30	0.114	170.00	0.127	172.90	0.202
3	437.30	0.079	442.20	0.078	420.00	0.096	434.60	0.141
4	779.00	0.078	788.40	0.105	747.90	0.124	774.20	0.135

ACKNOWLEDGMENT

Dr. Pranab Saha, Principal Consultant from Kolano and Saha Engineers, Inc., USA, guided us to write this book chapter.

KEYWORDS

- **viscoelastic material**
- **modal loss factor**
- **complex modulus**
- **Maxwell model**
- **strain energy method (SEM)**

REFERENCES

1. Nashif, A. D., Jones, D. I. G, Henderson J. P. *Vibration Damping*; John Wiley and Sons: New York; 1985.
2. Ferry, J. D. *Viscoelastic Properties of Polymers*, 3rd ed.; John Wiley and Sons, 1980.
3. Sun, C. T., Lu, Y. P. *Vibration Damping of Structural Elements*; Prentice Hall: New Jersey, 1995.
4. Kerwin, E. M. Jr., Damping of Flexural Waves by a Constrained Viscoelastic Layer. *J. Sound Vib.* **1959**, *31*, 952–62.
5. Soni, M. L. Finite Element Analysis of Viscoelastically Damped Sandwich Structures. *Shock Vib. Bull.* **1981**, *51* (2), 97–109.
6. Moreira, R.; Rodrigues, J. D. Constrained Layer Damping Treatments: Finite Element Modeling. *J. Vib. Control* **2004**, *10*, 575–595.
7. Johnson, C. D. Design of Passive Damping Systems. *J. Vib. Acoust.* **1995**, *XX*, 117–124.
8. Kienholz, D. A.; Johnson, C. D. In *Prediction of Damping in Structure with Viscoelastic Materials*, Structural Dynamics and Material Conference, Atlanta, 1981.
9. Johnson, C. D.; Kienholz, D. A. Finite Element Prediction of Damping in Structures with Constrained Viscoelastic Layers. *AIAA J.* **1982**, *20* (9).
10. Ungar, E. E.; Kerwin, E. M. Jr. Loss Factors of Viscoelastic Systems in Terms of Energy Concepts. *J. Acoust. Soc. Am.* **1962**, *34* (7), 954–957.
11. Johnson, C. D.; Rogers, L. C.; Kienholz, D. A. Finite Element Prediction of Damping in Beams with Constrained Viscoelastic Layers. *Shock Vib. Bull.* **1981**, 78–81.
12. Koruk, H.; Kenan, Y. In *Assessment of Modal Strain Energy method: Advantages and Limitations*, Proceeding of the ASME 2012 Biennial conference on Engineering Systems Design and Analysis, France, July 2–4, 2012; ESDA2012.
13. NASTRAN 10.0 software user guide.

CHAPTER 9

Crop Yield Prediction and Leaf Disease Detection Using Machine Learning

SACHIN S. SAWANT, DYUTI BOBBY, ATHARVA DUSANE, and GAURAV DURGE

Department of Engineering, Sciences and Humanities (DESH), Vishwakarma Institute of Technology, Pune, Maharashtra, India

ABSTRACT

In the growing season, a farmer always must deal with two prominent issues: the first one is to make the correct choice of crops to grow, and the second is the detection of infestations. Indeed, crop yield is drastically affected by the above-mentioned factors. This issue can be efficiently addressed using machine learning techniques, which provide a helping hand in making the best decisions. In this project, a systematic procedure was adopted for learning to extract and synthesize the algorithms and features that have been used in the models. The Random Forest Regression (RF) algorithm is applied in yield prediction with an accuracy of 96.78%. Transfer learning for deep learning is used in the disease detection model with an accuracy of around 91.3%. This system was made with the inclusion of seasonal rainfall data in crop yield prediction. The proposed model predicts the production of a particular crop based on the data fed to it and classifies the disease of the leaf samples used.

9.1 INTRODUCTION

Even today, the survival of over 50% of India's population depends on agriculture, which undoubtedly serves as a pillar of the Indian economy. In fact, it is the mainstay of India's Gross Domestic Product (GDP), forming 20% of it in 2020–2021. Lately, frequent changes in the environmental conditions, accompanied by sudden climatic and weather changes, have posed a great threat for the healthy survival of agriculture and agriculture-based industries. In a country that is massively smitten by agriculture, technology for the advancement of crop turnout is an absolute necessity.[1–8]

In recent years, the role of machine learning (ML) as a support tool for crop yield prediction and leaf disease detection has been strongly highlighted, especially to revamp the yield rate of crops.[9] The tactics of crop choice are applied to boost crop production. The produce of crops might depend upon geographical circumstances of the region like river ground, hilly areas or the depth, and weather entities like temperature, cloud cover, rainfall, and humidity.[3,10,11] Its importance in futuristic prediction is supported by the dataset offered. Keeping this in mind, the project aims to accurately predict yield supported by specific input parameters. It additionally works on the intelligent classification of plants into affected and nonaffected leaves and, if pathologic, the identification of diseases affecting the plant.

Along with environmental circumstances and weather, the seasonal rain, which varies by location, is another key factor impacting agriculture in India. Moreover, district-specific options like soil type and alternative seasonal attributes play a particularly important role in the quality and number of crops.[3,12–19] The crop yield depends upon various parameters like country location, crop name, year, yield worth, average rain, pesticides, and average temperature. The proposed system includes parameters of the average seasonal rain, area of the agricultural farm wherever the crop is to be matured, district, and season for crop growth. The information used is from specific districts of the state of Maharashtra, for the years 1997–2014. The model used in this work is the Random Forest Regressor model. Random forest regression is a supervised learning model supported by the concept of ensemble learning. The ensemble learning technique combines the outputs of varied machine learning models with the outputs of varied decision tree models.

Agricultural depressions occur once the crops are undergoing some infestation, and farmers do not realize the infestation of plants at the time when it can be resolved easily. Once the disease is detected, the farmers are unaware of the true nature of the disease. Visual detection of leaf diseases is not a possibility.[20] Therefore, the examination of leaf disease detection in agriculture using ML may be an elementary subject of analysis, which reaps benefits within the observation of huge fields of yields. It can be used to establish manifestations of disease as they occur on plant leaves. Hyperspectral imaging and deep learning techniques have proven quite useful for plant disease recognition using image processing. The method described in the literature[21] tells us about the techniques considered for the detection of disease in a plant leaf through image processing. This work will be useful for researchers within the area of plant leaf disease recognition using deep learning techniques.[22-30]

Previous studies have also stated that models with more features do not always prove to be the best for yield prediction.[31] Dehzangi et al.[32] have developed an ensemble method that uses various classifiers, namely, Naive Bayes and Support Vector Machine (SVM), for the prediction of protein structural class. The autocorrelation-based feature extraction is used in the work to get better results. Kumar et al.[33] have described a machine learning model that solves the crop selection problem. Data mining process classification is predicting the target variable's value by creating a model based on some features of the categorical variable.[34-36] There is wide diversity in geographical characteristics and climatic conditions (rainfall, temperature, humidity, etc.) in India, which affects crop yield estimation.

In the present work, to predict more accurate crop production, the dataset has been compiled to include district-wise rainfall in the localized regions of Maharashtra. The rainfall parameter used depends on the season. Seasonal rainfall plays a vital role in determining the production in an area where maximum production occurs during the Kharif season, when rainfall is also at an all-year high in the Indian states. Due to this, for the first time, to the best of our knowledge, seasonal rainfall is considered in a model, instead of the average yearly rainfall, for a localized region. Another factor setting this project apart is the large dataset[37] used for the disease detection model, which provides some infestation markers for crops like potatoes, tomatoes, apples, maize, grapes, etc.

9.2 EXPERIMENTAL METHODS

9.2.1 ALGORITHM

Crop yield prediction:

1. Collect the datasets of rainfall[38] and crop production[39] in Maharashtra and then merge them with the help of algorithms or manually as the dataset used is small.
2. Data exploration: According to need, sorting the data and plotting the initially required plots.
3. Data cleaning: Filling the empty data with the average of those columns.

4. Data visualization: Plotting the density, histogram, scatter, pie, and bar plots based on the data with libraries such as seaborn and matplotlib.pyplot.
5. One-hot encoding: Converting the categorical data into numeric form and then a binary representation of that data.
6. Model building: Using a random forest regressor model to fit in our data and splitting it into two parts, i.e., Training (80%) and Testing (20%).
7. Prediction: Predicting the output based on the data fed.
8. Deploying this model on a website and taking input parameters to predict the results.

Crop disease detection:

1. Collect the datasets of leaves from Kaggle.[37]
2. Applying various deep learning algorithms to the data and taking transfer learning into account.
3. Increase the accuracy by modifying the parameters.
4. Deploying this model on the website and taking input images to detect different leaf diseases.

9.2.2 METHODOLOGY

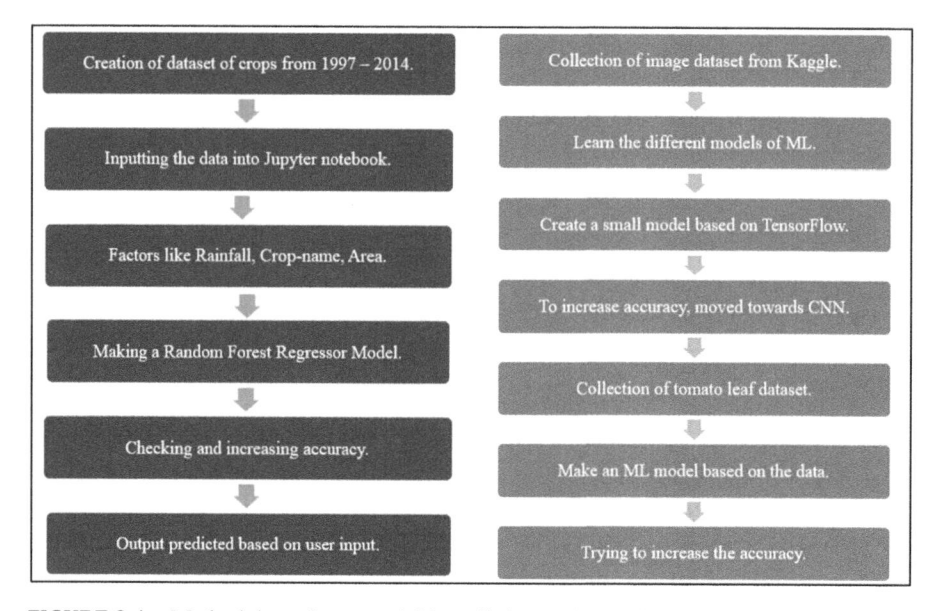

FIGURE 9.1 Methodology for crop yield prediction and crop disease detection.

9.2.3 DATASET

The final dataset used is created by incorporating the rainfall dataset[38] into the crop production statistics dataset.[39] The final crop yield prediction dataset values, which have been combined manually, are visible in Figure 9.2.

	State_Name	District_Name	Crop_Year	Season	Crop	Area	Production	Rainfall
0	Maharashtra	AHMEDNAGAR	1997	Autumn	Maize	1	1113.0	235.4
1	Maharashtra	AHMEDNAGAR	1997	Kharif	Arhar/Tur	17600	6300.0	406.5
2	Maharashtra	AHMEDNAGAR	1997	Kharif	Bajra	274100	152800.0	406.5
3	Maharashtra	AHMEDNAGAR	1997	Kharif	Gram	40800	18600.0	406.5
4	Maharashtra	AHMEDNAGAR	1997	Kharif	Jowar	900	1100.0	406.5
5	Maharashtra	AHMEDNAGAR	1997	Kharif	Maize	4400	4700.0	406.5
6	Maharashtra	AHMEDNAGAR	1997	Kharif	Moong(Green Gram)	10200	900.0	406.5
7	Maharashtra	AHMEDNAGAR	1997	Kharif	Pulses total	451	130.0	406.5
8	Maharashtra	AHMEDNAGAR	1997	Kharif	Ragi	2600	2100.0	406.5
9	Maharashtra	AHMEDNAGAR	1997	Kharif	Rice	5900	7200.0	406.5

FIGURE 9.2 Final crop yield dataset imported into the prediction model.

The "New Plant Diseases Dataset" used for leaf disease detection was obtained from Kaggle references.[37] It comprises a huge library of augmented images of healthy and diseased crop leaves; a few species from them are shown in Figure 9.3, in which 14 species of leaves are categorized into 38 different classes. The total dataset was divided into an 80/20 ratio of training and validation sets.

9.3 RESULTS AND DISCUSSION

9.3.1 VISUALIZATION

A few data visualization techniques are shown below to determine the impact factor of each variable and identify the relationships between different independent and dependent variables.

Figure 9.5 provides a validation of the concept that seasonal production is influenced by rainfall. Comparing the two graphs, it is evident that production is the highest in the Kharif season when the average rainfall is also the highest. Production is the lowest in Autumn when the average rainfall is the lowest.

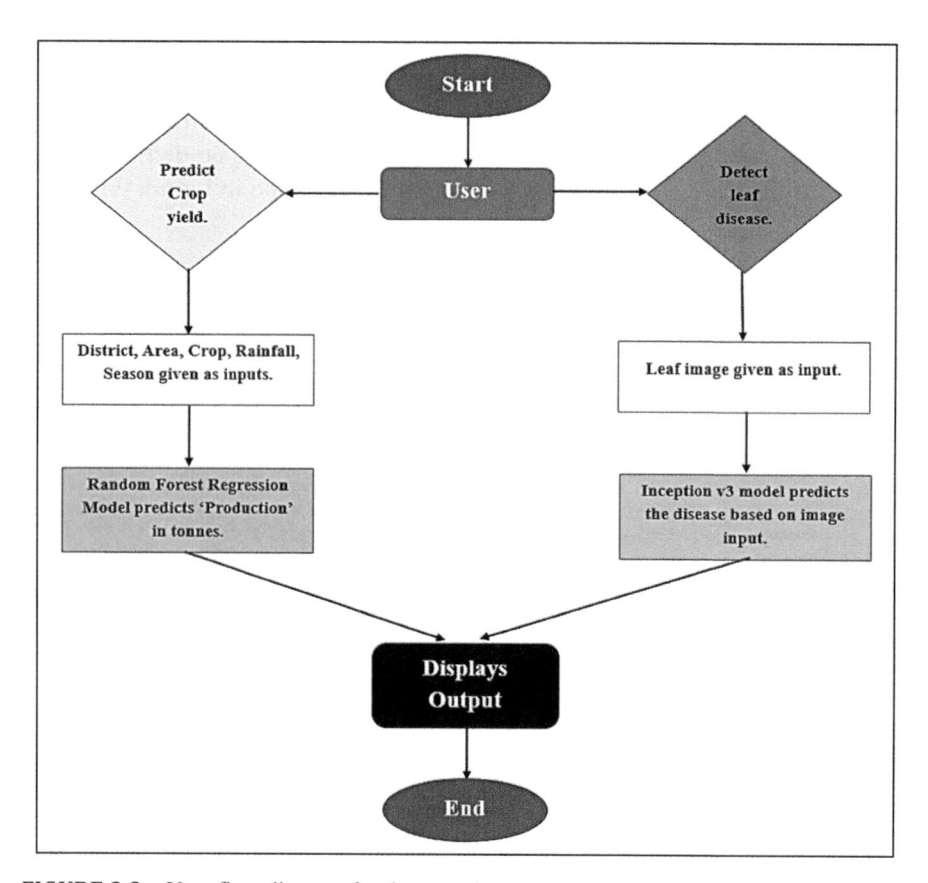

FIGURE 9.3 User flow diagram for the complete system.

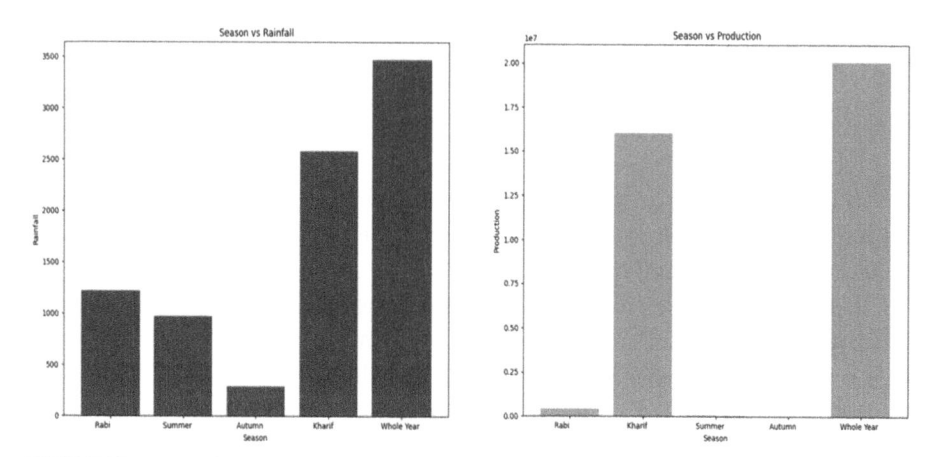

FIGURE 9.4 Maximum rainfall in Kharif leading to maximum production.

Crop Yield Prediction and Leaf Disease Detection Using Machine Learning 153

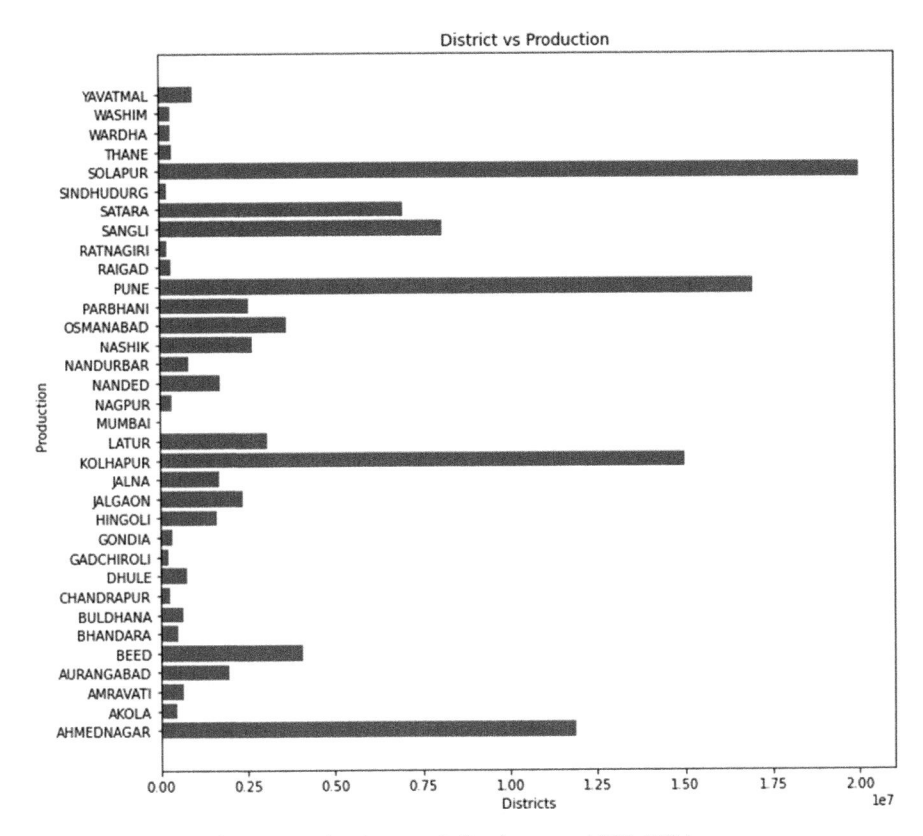

FIGURE 9.5 Districts vs. production graph for the years 1997–2014.

Figure 9.6 depicts the production characteristics for the districts in Maharashtra over the year span ranging from 1997 to 2014. From the plot, it is evident that the maximum production, combined over the years, has been achieved by Solapur district, followed by Pune and Kolhapur districts. The lowest production, almost zero, is from the urban district of Mumbai. The districts of Ratnagiri and Gadchiroli also present reduced production in the timeline of data collection.

In Figure 9.7, maximum crop production is achieved in the years after 2012, mainly in the year 2014. The least production years are 2006 and 2007.

In Figure 9.8, the distribution of data according to seasons is plotted with the help of a pie chart. Kharif season data is in greater quantity in the dataset, which is 57.6%, and the data for Autumn is in the least quantity in the dataset, which is 0.1%. This is representative of a general scenario where production is highest in Kharif and lowest in Autumn.

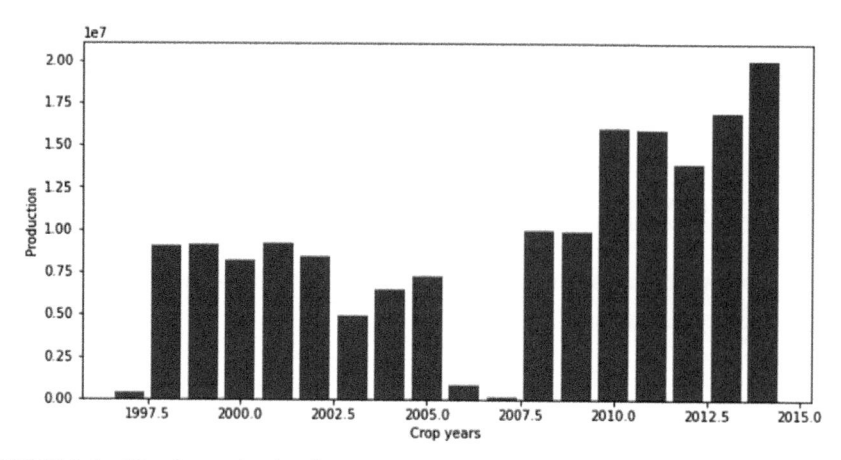

FIGURE 9.6 Plot for production in Maharashtra over the years 1997–2014.

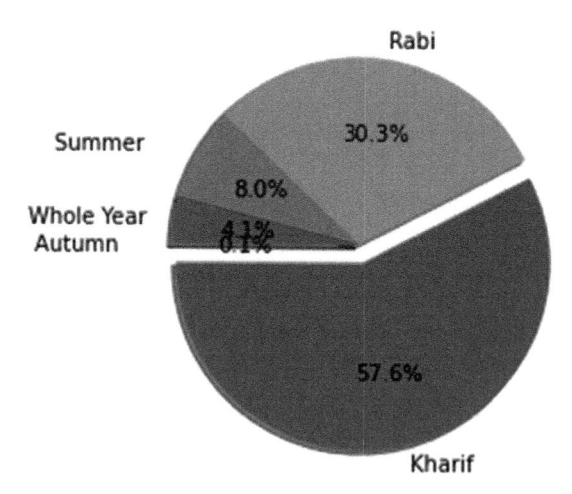

FIGURE 9.7 Dependence of production on seasonal characteristics.

Figure 9.9 shows the density plot for the amount of rainfall received. The density of rainfall in Maharashtra is higher for rainfall between 0 and 1000 mm.

9.3.2 MODELING

As shown in Figure 9.10, one-hot encoding is performed on the dataset to convert categorical data into numerical data. The factors "State Name,"

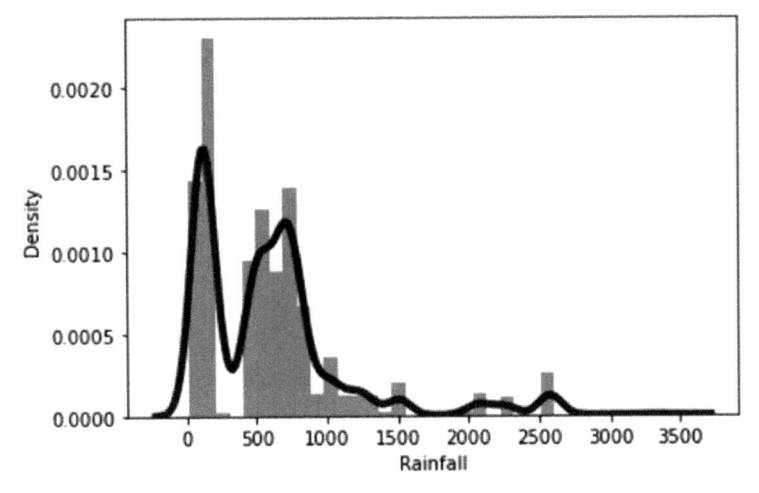

Crop Yield Prediction and Leaf Disease Detection Using Machine Learning 155

FIGURE 9.8 Frequency plot of the amount of rainfall within different ranges.

FIGURE 9.9 Dataset after one-hot encoding is performed.

```
In [28]: dist = input("Enter district: ")
         area = input("Enter Area of Farm in hectares: ")
         crop = input("Enter crop you want to grow: ")
         season = input("Enter season - Autumn,Kharif, Rabi,Summer or Whole year: ")
         rain = input("Enter current average rainfall in your area :")

         Enter district: PUNE
         Enter Area of Farm in hectares: 17200
         Enter crop you want to grow: Rice
         Enter season - Autumn,Kharif, Rabi,Summer or Whole year: Kharif
         Enter current average rainfall in your area :100

In [29]: X_new = pd.DataFrame()
         X_new["District_Name"] = [dist]
         X_new["Crop"] = [crop]
         X_new["Season"] = [season]
         new = pd.get_dummies(X_new)
         new["Area"] = [area]
         new["Rainfall"] = [rain]
```

FIGURE 9.10 User-entered data.

"Crop," and "Season," which have a massive impact on the predicted yield, are converted to binary numerical values to enable us to reach an accurate conclusion from our model. The one-hot-encoded dataset has a shape of 78 columns, a huge increase from the initial shape of 8 columns.

With the input of just a few relevant factors—the district in which the farm is located, the area of the farm in hectares, the crop to be grown on the farm, the season for the crop to be grown, and the average rainfall received in that season—within seconds the farm's yield for the season can be predicted. The input user interface is depicted in Figure 9.11, while the predicted production output in tonnes is shown in Figure 9.12.

Figure 9.13. depicts the explained variance score calculated for the model. Here, the score the model has is 0.9798, where 1.0 is the best score, thus supporting the calculation of accuracy for the crop yield prediction model. The R2 score for the model is also 0.9798, where 1.0 is the best score.

```
In [31]: y_new = model.predict(new)
         y_new
         print("The predicted crop yield is :",y_new[0])

         The predicted crop yield is : 14020.450999532557
```

FIGURE 9.11 Yield prediction with user input.

```
In [23]: import sklearn.metrics
         sklearn.metrics.explained_variance_score(y_test, y_predicted)
         #best value is 1.0, worst is 0
Out[23]: 0.979865779942532

In [24]: sklearn.metrics.r2_score(y_test, y_predicted, sample_weight=None, multioutput='uniform_averag
         #best score is 1.0, worst is 0
Out[24]: 0.9798638491158106

In [25]: from sklearn.metrics import mean_absolute_percentage_error
         mean_absolute_percentage_error(y_predicted,y_test)
Out[25]: 4.942591872193217
```

FIGURE 9.12 Validation results calculated for the Random Forest Regressor model.

The Absolute Percentage Error of the model stands at 4.94%, concluding that the model has highly accurate forecasting. Figure 9.14 shows the accuracy score of the model calculated with the cross-validation score.

The scatter plot shown in Figure 9.15 depicts the actual value of production with dots in blue and the predicted value with dots in orange. The overlapping of both shows that prediction accuracy is quite a good match with the calculated value of around 96.78%.

```
from sklearn.model_selection import KFold
cv = KFold(n_splits=10, random_state=1, shuffle=True)

from sklearn.model_selection import cross_val_score
scores = cross_val_score(model, X, y, cv=cv, n_jobs=-1)
scores

array([0.91936323, 0.98246796, 0.96393641, 0.96625176, 0.96519128,
       0.96824684, 0.98374461, 0.96620101, 0.97790801, 0.9850957 ])

print("%0.2f accuracy with a standard deviation of %0.2f" % (scores.mean()*100, scores.std()))
96.78 accuracy with a standard deviation of 0.02
```

FIGURE 9.13 Use of cross-validation scores to calculate accuracy.

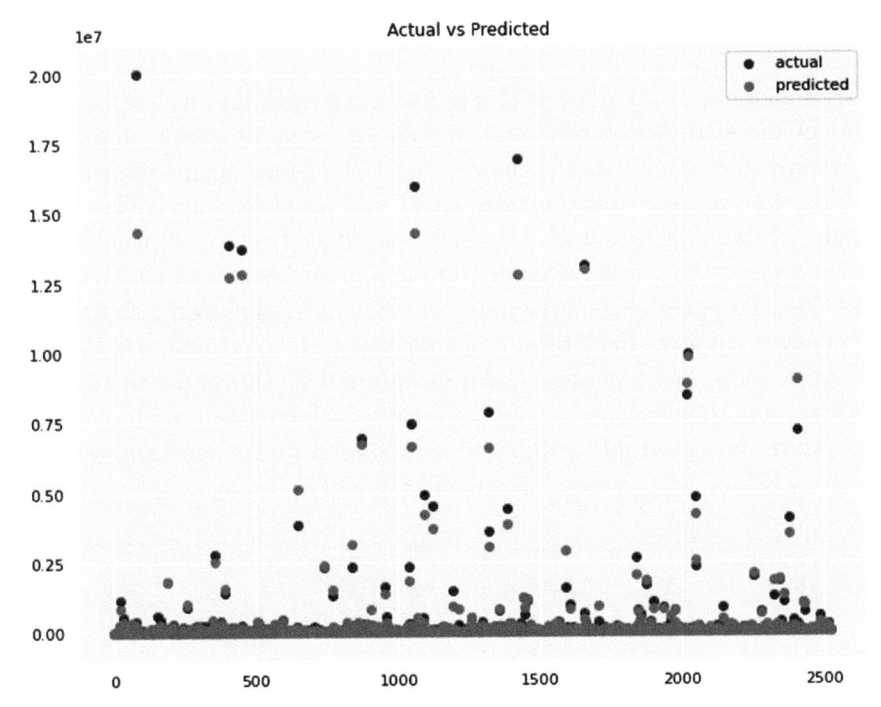

FIGURE 9.14 Actual vs. predicted output scatter plot for yield prediction.

A random forest represents an ensemble of decision trees. Essentially, this means that decision trees are constructed in a random manner. Each tree is created from a different sample of data rows, and at each juncture (node), a different set of features is selected for partitioning. Each of the trees predicts a different result from its set of observations. Random Forest Regression calculates the mean of these predictions to produce a single result. The ensemble of various trees thus increases the overall accuracy. Figure 9.14 shows the calculated accuracy of 96.78% by using the cross-validation score.

```
Epoch 95/100
100/100 [==============================] - 59s 588ms/step - loss: 2.3019 - accuracy: 0.9416 - val_loss: 3.5231 - val_accuracy: 0.9153
Epoch 96/100
100/100 [==============================] - 59s 591ms/step - loss: 1.7966 - accuracy: 0.9463 - val_loss: 2.6704 - val_accuracy: 0.9281
Epoch 97/100
100/100 [==============================] - 58s 582ms/step - loss: 1.9508 - accuracy: 0.9378 - val_loss: 2.7375 - val_accuracy: 0.9316
Epoch 98/100
100/100 [==============================] - 59s 590ms/step - loss: 2.4788 - accuracy: 0.9341 - val_loss: 2.9645 - val_accuracy: 0.9209
Epoch 99/100
100/100 [==============================] - 59s 591ms/step - loss: 2.0709 - accuracy: 0.9464 - val_loss: 3.0567 - val_accuracy: 0.9259
Epoch 100/100
100/100 [==============================] - 58s 583ms/step - loss: 1.5644 - accuracy: 0.9478 - val_loss: 3.1986 - val_accuracy: 0.9191
```

FIGURE 9.15 Epochs executed for the inception 3 model.

The increase in accuracy is due to the cross-validation technique, which involves the statistical partitioning of data into smaller subsets and is used to evaluate the model's performance. The K-fold method involves creating multiple (k) training–validation tests, which will lead to k events of testing and training of the model. K-fold also results in reduced computational time and variance. Here, the k-fold parameter for maximum accuracy is 10. The scatter plot of actual vs predicted output for yield prediction is depicted in Figure 9.15.

Figure 9.16 shows the total number of epochs executed, which is 100, with 100 steps each. The line plot shown in Figure 9.17 shows the training loss and testing (validation) loss observed during each epoch of model building. It is clearly observed that training and validation losses are decreasing and accuracy is increasing with each epoch of model building.

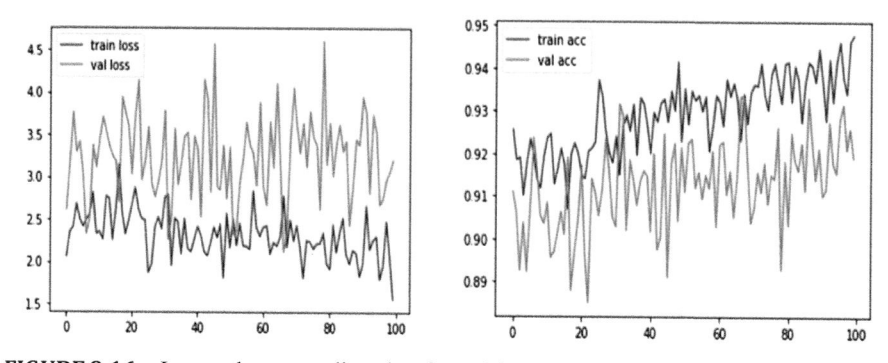

FIGURE 9.16 Loss and accuracy line plots for training and validation datasets for 100 epochs.

Crop Yield Prediction and Leaf Disease Detection Using Machine Learning 159

As shown in Figure 9.18, the percentage accuracy shown by our work is 91.3%, which is far higher than previous models as visualized in Table 9.1. An enhancement in accuracy can be attributed to the InceptionV3 model used.

```
[33] final_acc = model.evaluate(validation_set)[1]
     print(f"Final accuracy of model is = {final_acc*100}% ")

550/550 [==============================] - 92s 164ms/step - loss: 3.4997 - accuracy: 0.9130
Final accuracy of model is = 91.30434989929199%
```

FIGURE 9.17 Final accuracy of the inception model is calculated as 91.3%.

```
inception = InceptionV3(input_shape=IMAGE_SIZE + [3], weights='imagenet', include_top=False)

# don't train existing weights
for layer in inception.layers:
    layer.trainable = False
```

FIGURE 9.18 InceptionV3 model imported from Keras.

```
def predict_disease(path):
    img = image.load_img(path, target_size=(224,224))
    plt.imshow(img)
    i = image.img_to_array(img)
    im = preprocess_input(i)
    img = np.expand_dims(im, axis=0)
    pred = np.argmax(model.predict(img))
    print(f"The disease is {ref[pred]}")

predict_disease('/content/test/test/PotatoEarlyBlight4.JPG')
The disease is Potato___Early_blight
```

FIGURE 9.19 Predict leaf disease function.

The flexibility of transfer learning enables us to use pretrained models for feature extraction, which transforms the model completely. Inception networks are computationally more efficient in terms of the economic cost incurred as well as the number of parameters generated by the network. Figure 9.19 depicts the import of the InceptionV3 model, while the leaf disease detection function is depicted in Figure 9.20.

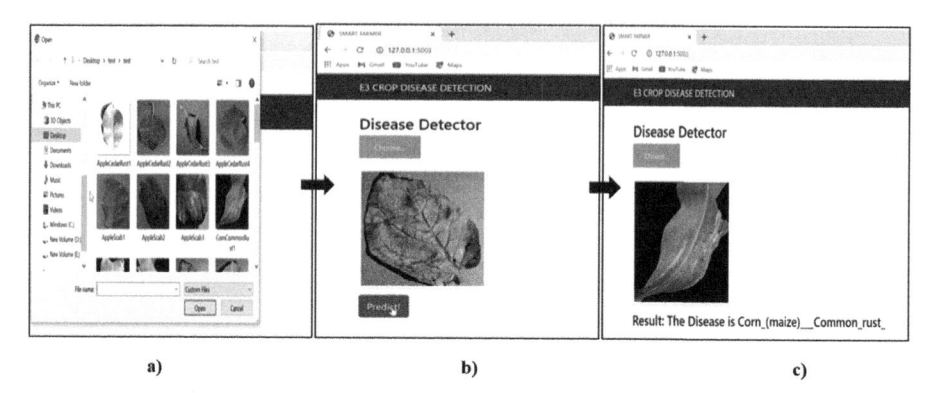

FIGURE 9.20 Website deployment of the leaf disease detector model: (a) Image input taken from the user, (b) Chosen image uploaded onto the website, (c) Disease is detected by the model.

When the model is deployed on a website we created, the consumer has an interactive user interface for easy disease detection as shown in Figure 9.20. The final model takes input from the user and predicts crop yield or detects leaf disease as per the request received. Upon filling in the required details, a near-accurate output is received.

Thus, in comparison with the earlier results reported in the literature,[40–43] the system proposed in this paper has achieved better accuracy, with 96.78% for crop yield prediction and 91.3% for leaf disease detection models.

The devised dataset is limited to Maharashtra to consider localized parameters which differ from district to district. This system can be expanded to inculcate all regions across India, which will greatly impact the accuracy of the model, considering the district-wise seasonal rainfall for all states. Disease detection can be integrated into this model to create an efficient structure for precision agriculture. Thus, this paper is a modern discussion on the techniques of random forest regression for crop yield estimation and InceptionV3 in infestation observation.

9.4 CONCLUSIONS

Machine learning is the future, and more people utilizing it for different prediction models can aid in global development. It has applications in every branch and is already being implemented commercially as it can solve issues that are tough and exhausting for humans. The predicted yield will help farmers reduce the probability of growing those crops that will go in vain. Facilities such as storage, transport, and transactions of the produce can be created by accurately predicting the yield. The timely detection of leaf infestations will lead to the healthy growth of plants and crops, benefiting farmers by saving plants and crops from getting affected. The increased accuracy of 96.78% in crop yield prediction and 91.3% in leaf disease detection, as compared to previous models, vindicates the use of the Random Forest Regressor and InceptionV3 models successfully.

KEYWORDS

- crop yield
- deep learning
- disease detection
- machine learning
- random forest regressor

REFERENCES

1. Reddy, D.; Kumar, M. In *Crop Yield Prediction using Machine Learning Algorithm*, 5th International Conference on Intelligent Computing and Control Systems (ICICCS), IEEE, 2021, 1466–1470.
2. Ghadge, R.; Kulkarni, J.; More, P.; Nene, S.; Priya, R. Prediction of Crop Yield using Machine Learning. *Int. Res. J. Eng. Technol.* **2018,** *5* (2), 2237–2239.
3. Alagurajan, M.; Vijayakumaran, C. ML Methods for Crop Yield Prediction and Estimation: An Exploration. *Int. J. Eng. Adv. Technol.* **2020,** *9* (3), 3506–3508.
4. Fuentes, A.; Yoon, S.; Kim, S.; Park, D. A Robust Deep-Learning-Based Detector for Real-Time Tomato Plant Diseases and Pests Recognition. *Sensors* **2017,** *17* (9), 1–21.
5. Bondre, D.; Mahagaonkar, S. Prediction of Crop Yield and Fertilizer Recommendation using Machine Learning Algorithms. *Int. J. Eng. Appl. Sci. Technol.* **2019,** *4* (5), 371–376.

6. Pandith, V.; Kour, H.; Singh, S.; Manhas, J.; Sharma, V. Performance Evaluation of Machine Learning Techniques for Mustard Crop Yield Prediction from Soil Analysis. *J. Sci. Res.* **2020**, *64* (2), 394–398.
7. Sagar, B.; Cauvery, N. Agriculture Data Analytics in Crop Yield Estimation: A Critical Review. *Indones. J. Electr. Eng. Comput. Sci.* **2018**, *12* (3), 1087–1093.
8. Manje, S.; Samanta, S.; Gupta, A.; Bhave, M., Crop Yield Prediction Using Machine Learning Techniques. *Int. J. Res. Eng. Appl. Manag.* **2021**, *7* (2), 27–32.
9. Palanivel, K.; Surianarayanan, C. An Approach for Prediction of Crop Yield using Machine Learning and Big Data Techniques. *Int. J. Comput. Eng. Technol.* **2019**, *10* (3), 110–118.
10. Bose, P.; Kasabov, N.; Bruzzone, L.; Hartono, R. Spiking Neural Networks for Crop Yield Estimation Based on Spatiotemporal Analysis of Image Time Series. *IEEE Trans. Geosci. Remote Sens.* **2016**, *54* (11), 6563–6573.
11. Keerthana, M.; Meghana, K.; Pravallika, S.; Kavitha, M., In *An Ensemble Algorithm for Crop Yield Prediction*, 3rd Third International Conference on Intelligent Communication Technologies and Virtual Mobile Networks (ICICV), IEEE, February, 2021. pp 963–970.
12. Kumar, Y.; Spandana, V.; Vaishnavi, V.; Neha, K.; Devi, V. In *Supervised Machine learning Approach for Crop Yield Prediction in Agriculture Sector*, 5th International Conference on Communication and Electronics Systems (ICCES), *IEEE*, 2020, pp 736–741.
13. Chandgude, A.; Harpale, N.; Jadhav, D.; Pawar, P.; Patil, S. A Review on Machine Learning Algorithm used for Crop Monitoring System in Agriculture. *Int. Res. J. Eng. Technol. (IRJET)* **2018**, *5* (04), 1468–1471.
14. Gandhi, N.; Armstrong, L.; Petkar, O.; Tripathy, A. In *Rice Crop Yield Prediction in India using Support Vector Machines*, 13th International Joint Conference on Computer Science and Software Engineering (JCSSE), *IEEE*, 2016, 1–5.
15. Priya, P.; Muthaiah, U.; Balamurugan, M. Predicting Yield of the Crop using Machine Learning Algorithm. *Int. J. Eng. Sci. Res. Technol.* **2018**, *7* (1), 1–7.
16. Bhatnagar, R.; Gohain, G. Crop Yield Estimation using Decision Trees and Random Forest Machine Learning Algorithms on Data From Terra (EOS AM-1) & Aqua (EOS PM-1) Satellite Data. In *Machine Learning and Data Mining in Aerospace Technology. Studies in Computational Intelligence;* Hassanien, A., Darwish, A., El-Askary, H., Eds.; Springer: Cham, 2020; Vol. 836, pp 107–124.
17. Raina, S.; Gupta, A. In *A Study on Various Techniques for Plant Leaf Disease Detection Using Leaf Image*, International Conference on Artificial Intelligence and Smart Systems (ICAIS), IEEE, 2021; pp 900–905.
18. Chlingaryan, A.; Sukkarieh, S.; Whelan, B. Machine Learning Approaches for Crop Yield Prediction and Nitrogen Status Estimation in Precision Agriculture: A Review. *Comput. Electron. Agric.* **2018**, *151*, 61–69.
19. Jeong, J.; Resop, J.; Mueller, N.; Fleisher, D.; Yun, K.; Butler, E.; Kim, S. Random Forests for Global and Regional Crop Yield Predictions. *PLoS One* **2016**, *11* (6), e0156571.
20. Goel, N.; Jain, D.; Sinha, A. In *Prediction Model for Automated Leaf Disease Detection & Analysis*, 8th International Advance Computing Conference (IACC), IEEE, 2018; pp 360–365.
21. Khirade, S.; Patil, A. In *Plant Disease Detection using Image Processing*, International Conference on Computing Communication Control and Automation, IEEE, 2015; pp 768–771.

22. Nazki, H.; Yoon, S.; Fuentes, A.; Park, D. Unsupervised Image Translation using Adversarial Networks for Improved Plant Disease Recognition. *Comput. Electron. Agric.* **2020**, *168*, 105117.
23. Chaitanya, P.; Yasudha, K. Image based Plant Disease Detection using Convolution Neural Networks Algorithm. *Int. J. Innovat. Sci. Res. Technol.* **2020**, *5* (5), 331–334.
24. Zhang, S.; Zhang, S.; Zhang, C.; Wang, X.; Shi, Y. Cucumber Leaf Disease Identification with Global Pooling Dilated Convolutional Neural Network. *Comput. Electron. Agric.* **2019**, *162*, 422–430.
25. Li, L.; Zhang, S.; Wang, B. Plant Disease Detection and Classification by Deep Learning—A Review. *IEEE Access* **2021**, *9*, 56683–56698.
26. Singh, U.; Chouhan, S.; Jain, S.; Jain, S. Multilayer Convolution Neural Network for the Classification of Mango Leaves Infected by Anthracnose Disease. *IEEE Access* **2019**, *7*, 43721–43729.
27. Sharma, P.; Berwal, Y.; Ghai, W. Performance Analysis of Deep Learning CNN Models for Disease Detection in Plants using Image Segmentation. *Inf. Process. Agric.* **2020**, *7* (4), 566–574.
28. Fina, F.; Birch, P.; Young, R.; Obu, J.; Faithpraise, B.; Chatwin, C. Automatic Plant Pest Detection and Recognition Using K-Means Clustering Algorithm and Correspondence Filters. *Int. J. Adv. Biotechnol. Res.* **2013**, *4* (2), 189–199.
29. Chai, A.; Li, B.; Shi, Y.; Cen, Z.; Huang, H.; Liu, J. Recognition of Tomato Foliage Disease based on Computer Vision Technology. *Acta Horticulturae Sinica* **2010**, *37* (9), 1423–1430.
30. Barbedo, J.; Factors Influencing the use of Deep Learning for Plant Disease Recognition. *Biosyst. Eng.* **2018**, *172*, 84–91.
31. Klompenburg, T.; Kassahun, A.; Catal, C. Crop Yield Prediction using Machine Learning: A Systematic Literature Review. *Comput. Electron. Agric.* **2020**, *177* (10), 105709.
32. Dehzangi, A.; Paliwal, K.; Sharma, A.; Dehzangi, O.; Sattar, A. A Combination of Feature Extraction Methods with an Ensemble of Different Classifiers for Protein Structural Class Prediction Problem. *IEEE/ACM Trans. Comput. Biol. Bioinform.* **2013**, *10* (3), 64–575.
33. Kumar, R.; Singh, M.; Kumar, P.; Singh, J. In *Crop Selection Method to Maximize Crop Yield Rate using Machine Learning Technique*, International Conference on Smart Technologies and Management for Computing, Communication, Controls, Energy and Materials (ICSTM) IEEE 2015; pp 138–145.
34. Balakrishnan, N.; Muthukumarasamy, G. Crop Production-Ensemble Machine Learning Model for Prediction. *Int. J. Comput. Sci. Softw. Eng.* **2016**, *5* (7), 148–153.
35. Zhou, C.; Cule, B.; Goethals, B. Pattern based Sequence Classification. *IEEE Trans. Know. Data Eng.* **2015**, *28* (5), 1285–1298.
36. Tang, B.; He, H.; Baggenstoss, P.; Kay, S. A Bayesian Classification Approach using Class-Specific Features for Text Categorization. *IEEE Trans. Know. Data Eng.* **2016**, *28* (6), 1602–1606.
37. Disease detection dataset—New Plant Diseases Dataset: Samir Bhattarai [Online]. https://www.kaggle.com/vipoooool/new-plant-diseases-dataset
38. Rainfall dataset—Rainfall in India: Rajanand Ilangovan [Online]. https://www.kaggle.com/rajanand/rainfall-in-india
39. Crop production dataset—Crop Production in India: Abhinand [Online]. https://www.kaggle.com/abhinand05/crop-production-in-india

40. Moraye, K.; Pavate, A.; Nikam, S.; Thakkar, S. Crop Yield Prediction Using Random Forest Algorithm for Major Cities in Maharashtra State. *Int. J. Innov. Res. Comput. Sci. Technol.* **2021,** *9* (2), 40–44.
41. Agarwal, M.; Singh, A.; Arjaria, S.; Sinha, A.; Gupta, S. ToLeD: Tomato Leaf Disease Detection using Convolution Neural Network. *Procedia Comput. Sci.* **2020,** *167*, 293–301.
42. Kudagi, S.; Patil, S.; Bewoor, M. Sugarcane Crop Disease Prediction and Expected Yield Estimation using SVM. *Int. J. Adv. Sci. Technol.* **2020,** *29* (4), 2936–2945.
43. Suresh, A.; Manjunathan, N.; Rajesh, P.; Thangadurai, E. Crop Yield Prediction using Linear Support Vector Machine. *Eur. J. Mol. Clin. Med.* **2020,** *7* (6), 2189–2195.

PART III
Structural Engineering

CHAPTER 10

Buckling Analysis of Thin-Walled Cylinder under Axial Compression and Internal Hydrostatic Pressure Using Finite Element Model

SAURABH PATIL[1], ASHOK MACHE[2], and SHARDUL JOSHI[3]

[1]*Vishwakarma Institute of Information Technology, Pune, India*

[2]*Department of Mechanical Engineering, Vishwakarma Institute of Information Technology, Pune, India*

[3]*Department of Civil Engineering, Vishwakarma Institute of Information Technology, Pune, India*

ABSTRACT

The efficiency of thin-walled structures can be improved by accurately predicting critical design parameters. The study on the effects of hydrostatic load on the stability of thin-walled structures has been limited to external hydrostatic load. This paper presents the buckling analysis of thin-walled cylinders under axial compression and internal hydrostatic pressure using the finite element method. The numerical results in the study have shown that the internal hydrostatic pressure had an appreciable strengthening effect on the cylindrical shell. The paper also discusses the effect of internal hydrostatic pressure on the characteristic deformation modes of a thin-walled cylinder. In the design of lightweight structures, stiffening elements improve the structural efficiency. Lastly, this paper studies the effect of the number of stiffeners on the critical buckling load for axially compressed cylinders using a general-purpose finite element code.

Smart Innovations and Technological Advancements in Civil and Mechanical Engineering.
Satish Chinchanikar, Ashok Mache, Shardul Joshi, & Preeti Kulkarni (Eds.)
© 2025 Apple Academic Press, Inc. Co-publis hed with CRC Press (Taylor & Francis)

10.1 INTRODUCTION

Thin-walled elements are an integral part of aircrafts, rockets, petroleum storage, pressure vessels, and launch vehicles. In practical applications, these structures are subject to various loading conditions that can cause instability and failure. The efficiency of such thin-walled structures can be improved by accurately predicting critical design parameters.[1] In thin-walled elements, membrane stiffness is several orders greater than the bending stiffness. The consequence of this property is that a thin shell can absorb large amounts of strain energy without any significant deformation. When the strain energy accumulates in a portion of an object, it suddenly converts to kinetic energy resulting in rapid deformation. This process of a sudden change of potential energy to kinetic energy causing large deformation is termed buckling. The load at which this happens is called the critical buckling load. Buckling is mainly caused due to compressive forces in the form of axial compression, bending, external pressure, and torsion.

The buckling behavior of cylindrical shells under axial compression, bending, and torsion has recieved more attention for its immediate practical applications. In the field of pressure loads, the effect of uniform internal and external pressure is well-documented using experimental and theoretical methods.[1,2] The study on the effects of hydrostatic load on the stability of thin-walled structures has been limited to external hydrostatic load for its varied practical applications in marine engineering, leaving the effects of internal hydrostatic load much less explored. Since the effect of internal hydrostatic pressure is minimal for the small length of cylinders, any analysis to effectively observe any significant effects may need cylinders with a large size of orders. This makes it harder for any experimental efforts to study such effects. Hence, numerical approaches can be effectively used to perform studies that are harder to conduct through experimental approaches. This paper endeavors to understand the effects of internal hydrostatic pressure on the buckling of thin-walled cylinders under axial compression.

The cutout is an important feature in the practical application of thin-walled cylinders. Large tank, silo, or pressure vessel designs can include holes in the cylindrical wall for inlets, outlets, manways, or windows. In aerospace applications, the cutouts on the shell body are required to feature windows, doors, electrical cables, or pipes. The inclusion of cutout imperfections has been shown to compromise the overall stability of the cylinder structure.[3] The first studies on the analysis of axially compressed cylindrical shells with cutouts using numerical approaches have found good agreement with the experimental results.[4,5] With the development of more sophisticated numerical methods, the stability analysis of thin-walled structures with intricate geometry became easier. In this paper, the model under study includes a circular cutout, $D/R = 0.3$, at the center of the cylinder length. The effect of cutout size under the combined loading, axial and hydrostatic, is also presented. Further, as a practical design application, the paper also presents studies on the effect of axial stiffeners on the model under considered loading conditions.

10.2 ANALYSIS MODEL

For this study, a thin cylinder model with a 2 mm thickness is considered. The length and diameter of the cylinder model are 5 m and 4 m, respectively. The material properties of a homogenous isotropic material, common to steel, are assigned to the model. The properties with higher elasticity were considered as they would retain initial buckling modes longer for clear observations. Details are mentioned in Table 10.1. A $d = 600$ mm flange hole cutout is made along the midway of the cylinder length. The cylinder is modeled in ABAQUS-CAE using a standard S4R element shell with linear order and reduced integration. The model mesh size was set to 50 mm throughout the shell. In the analysis model, the top edge was radially restrained, and the bottom plate was fully constrained.

TABLE 10.1 Material Properties.

E	193×10^3 MPa
v	0.33
ρ	7.85×10^{-9} kg/mm³

10.3 BUCKLING ANALYSIS

10.3.1 ANALYTICALLY ESTIMATED

The classical equation for the critical stress in axisymmetric deformation mode can be used to analytically predict the critical buckling load.[6] It describes that for an axially compressed open cylindrical shell of radius r, thickness t, Young's modulus E, and Poisson's ratio v, the critical stress is given:

$$\sigma_{cr} = \frac{E}{\sqrt{3(1-v^2)}} \cdot \frac{t}{r} \qquad (10.1)$$

$$P_{cr} = 2\pi t^2 \cdot \frac{E}{\sqrt{3(1-v^2)}} \qquad (10.2)$$

Multiplying the critical stress with the cross-sectional area, we can calculate the critical load. Using the above eq 10.2 and the given properties, the critical value for the considered model under an axial compression can be estimated to be 2.965193 × 106 N. Though the observed modes in the numerical study are largely asymmetric, the derived load can still be used to gauge the critical buckling load.

10.3.2 LINEAR BUCKLING ANALYSIS

Eigenvalue buckling analysis, or linear buckling analysis, is generally used to estimate the critical (bifurcation) load of slender structures. The linear buckling analysis can provide an upper limit for determining the critical buckling load. The more important function of the linear eigenvalue buckling problem is that the critical load is defined at the point where the model stiffness matrix in the governing equation becomes singular. This can be written in the form,

$$[K + \lambda_i[S]]\{\psi_i\} = \{0\} \qquad (10.3)$$

where λ_i is the load multiplier and ψ_i is the mode shape. So, the critical load of each mode can be calculated as

$$P_{cr} = \lambda_i \times Q \qquad (10.4)$$

where Q is the applied load.

The applied load value is kept at 1 N so that λ_i directly reflects the critical load.

Almost all buckling modes are restricted to the vicinity of the cutout, as shown in Figure 10.1. The linear buckling analysis fails to distinguish between stable local bucking and hence predicts low values of the critical load factor for initial mode results.

10.3.3 NONLINEAR BUCKLING ANALYSIS

The linear buckling analysis does not account for the effect of large deformation on the structure during loading. This may result in impractical failure modes and an overestimation of the critical load. Hence, nonlinear analysis is preferred when solution accuracy and postbuckling behavior are important aspects of the study. In a nonlinear static analysis, the load is applied gradually, and at the onset of buckling, the solution fails to converge due to a sudden loss of stiffness. This can be inferred in a force–displacement plot. Many different methods have been developed over the years to simulate nonlinear buckling problems using finite element codes. The widely used arc length method traces the global load–displacement response and determines the critical load at the point of negative stiffness.[6] However, it fails to capture the instability arising from the local buckling of the structure. The buckling problems with local buckling can be effectively analyzed using dynamic or artificial damping methods.

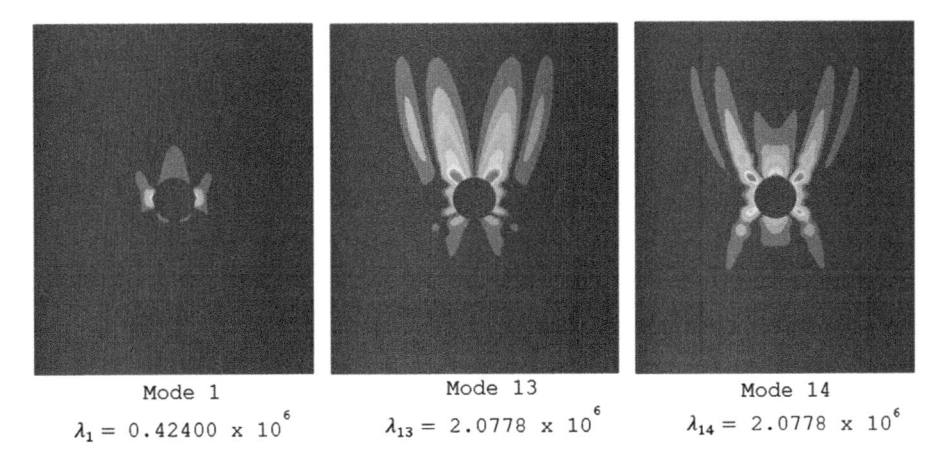

Mode 1
$\lambda_1 = 0.42400 \times 10^6$

Mode 13
$\lambda_{13} = 2.0778 \times 10^6$

Mode 14
$\lambda_{14} = 2.0778 \times 10^6$

FIGURE 10.1 Normalized displacement contour of linear buckling analysis.

The artificial damping method was used for nonlinear buckling analysis in this study. In this method, the strain energy from the local buckling is dissipated using artificial damping, which stabilizes the structure making the quasistatic process possible.[7] In Abaqus, automatic stabilization is included using the automatic addition of volume-proportional artificial damping. A viscous force of this form is added to the global equilibrium equation, as shown in eq 10.6,

$$F_v = cM * v \quad (10.5)$$
$$P - Q - Fv = 0 \quad (10.6)$$

where M* is an artificial mass matrix with unit density, c is the damping factor, v is the vector of nodal velocity, P is the total applied load, and Q is the internal force.[8] The damping factor was set to a low value of 1 E-09 with an activated nonlinear geometric effect (for large displacement) in the static general step. The reaction force and displacement time histories are recorded for a reference point tied to the top edge.

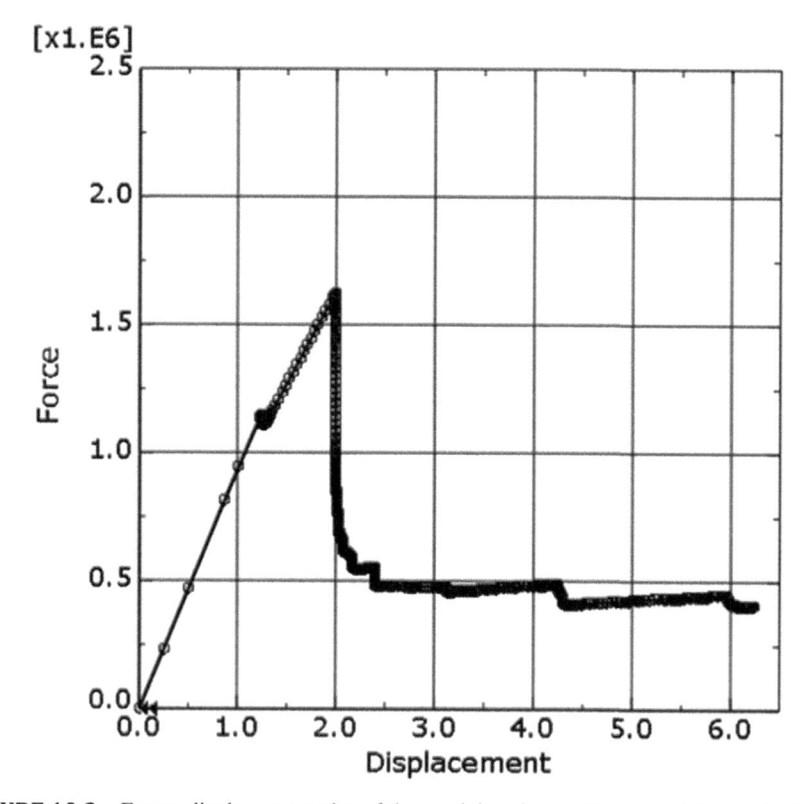

FIGURE 10.2 Force–displacement plot of the model under nonlinear bucking analysis.

The force and end-shortening results for the considered model are shown in Figure 10.2. Initially, localized stress develops on the edges of the cutout imperfection. The very first local buckling is observed when the area near the cutout deforms out of the plane. When the critical point is reached, the cutout becomes the source of successive buckling and travels around the cylinder. As the analysis progresses, the diamond-shaped buckling pattern merges and collapses into subsequent buckling modes. The critical load is obtained at a load value of 1.62714×10^6 N. Using the artificial damping method, the local deformation is well-observed in the end-shortening results.

10.4 EFFECT OF INTERNAL HYDROSTATIC PRESSURE

In combined loading under axial compression and internal pressure, the buckling load increases as the pressure preload infuses circumferential tensile stress that acts against the axial load to increase the buckling capacity. Moreover, the experimental studies suggest that internal pressure restores local buckling during loading, increasing the overall buckling capacity.[2] It is observed that in some cases, the initial pressure compensates for the loss in stiffness due to geometric imperfections. Though the internal hydrostatic pressure amplitude is lower, its effects are found to be significant in this study. For this study, the nonlinear analysis model is added with an additional internal hydrostatic-pressure load on the cylinder walls. The maximum pressure value is 0.04905 MPa at the bottom of the tank.

The comparison of end-shortening responses is shown in Figure 10.3. The plot compares the local snap-through and buckling responses of both cases. The critical load of the shell under hydrostatic load is ~1.2 times higher than the normal case. Also, from animation plots, it is seen that the effect of the preloaded stress is more prominent in delaying buckling than restoring local buckling.

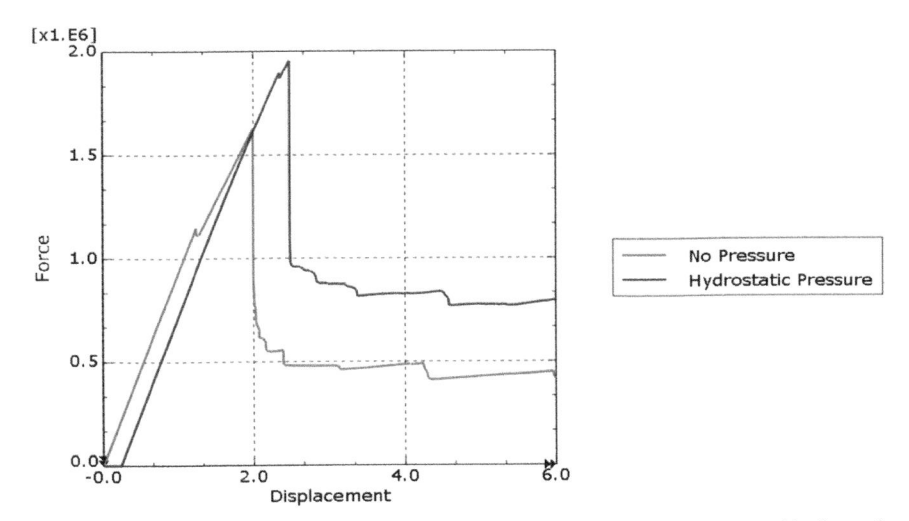

FIGURE 10.3 Comparison of force–displacement behavior under no-pressure and hydrostatic-pressure load.

The initial deformation modes in both cases are similar, but as the buckling progresses, the postbuckling modes under hydrostatic pressure modes translate up as shown in Figures 10.4 and 10.5. After the postprimary buckling modes, the subsequent modes merge and move back to the midplane of the shell. Though, throughout the postbuckling, the deformation modes are observed slightly above the no-pressure case.

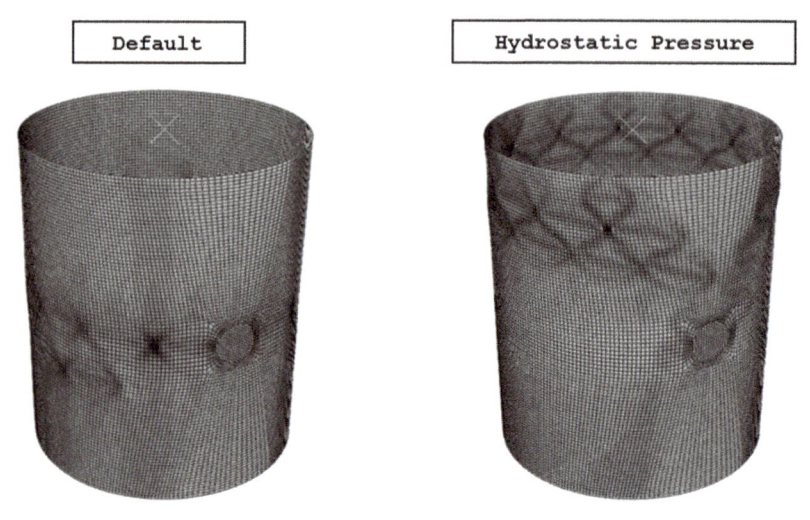

FIGURE 10.4 Comparison of buckling deformation of the model cylinder under no-pressure and hydrostatic-pressure load.

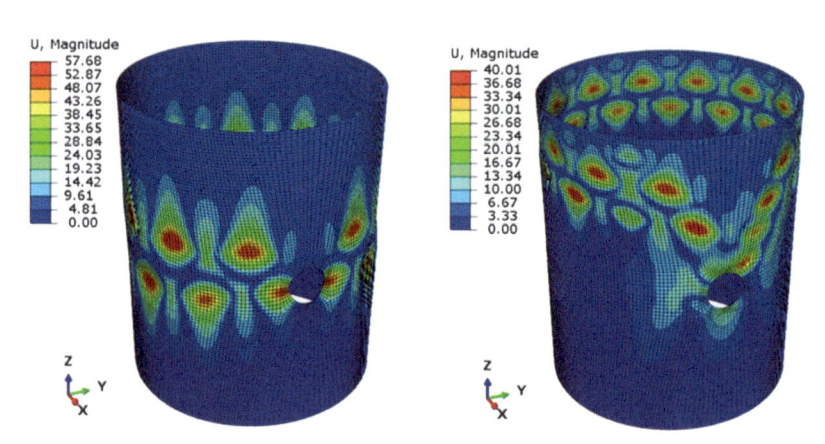

FIGURE 10.5 Comparison of displacement contour plots of buckling under no-pressure and hydrostatic-pressure load.

In both cases, the number of axial half-waves remains constant at a value of $m = 2$. In the case of hydrostatic pressure, the buckling mode is restricted to a small area. The displacement contour shows the typical diamond-shaped buckling pattern that forms in cylinders under axial compression. The flange imperfection distorts this, but in both cases, the pattern is well-preserved around the circumference.

The change in buckling load against different D/R ratios is illustrated in Figure 10.6. As the graph shows, the effect of hydrostatic pressure decreases with increasing the D/R ratio. At larger cutout sizes, the effect of hydrostatic pressure diminishes as the larger cutout size weakens the cylinder and significantly reduces the critical load. The variation of the critical load under normal and hydrostatic pressure with varying L/R ratios, for $D/R = 0.3$, was studied and is illustrated in Figure 10.7. Under normal axial compression, increasing length shows a steady decline in critical load. In the case of axial compression under internal hydrostatic load, the effect of the hydrostatic load has shown no apparent relationship with the L/R ratio. Since the effect of cutout imperfections adds to the buckling behavior, a more comprehensive study by isolating the effect of sole hydrostatic pressure is required.

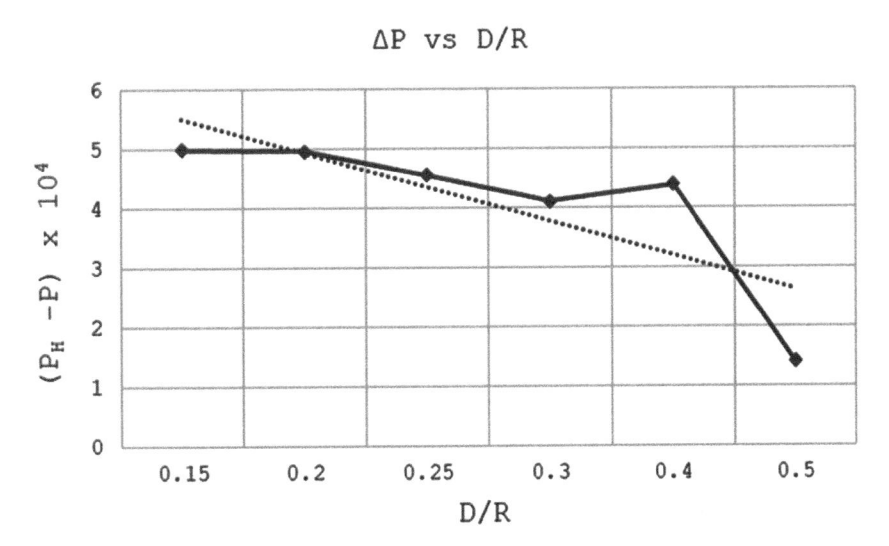

FIGURE 10.6 Critical load difference against different D/R ratios of the cylinder.

10.5 EFFECT OF AXIAL STIFFENERS

In the design of lightweight structures, stiffening elements improve the structural efficiency. Stiffened cylinders exhibit higher buckling loads and less imperfection sensitivity. This paper studies the effect of the number of equally spaced stiffeners on the critical buckling load for an axially compressed cylinder using FEA (Figure 10.8). A 2 mm thick and 500 mm wide integrally stiffened stiffener is added to the shell around the inner wall of the shell. Boundary conditions are similar to the previous model, and displacement-controlled static GNA analysis is set up. The number of stiffeners is varied from none to 100, and the critical buckling load is recorded for each iteration.

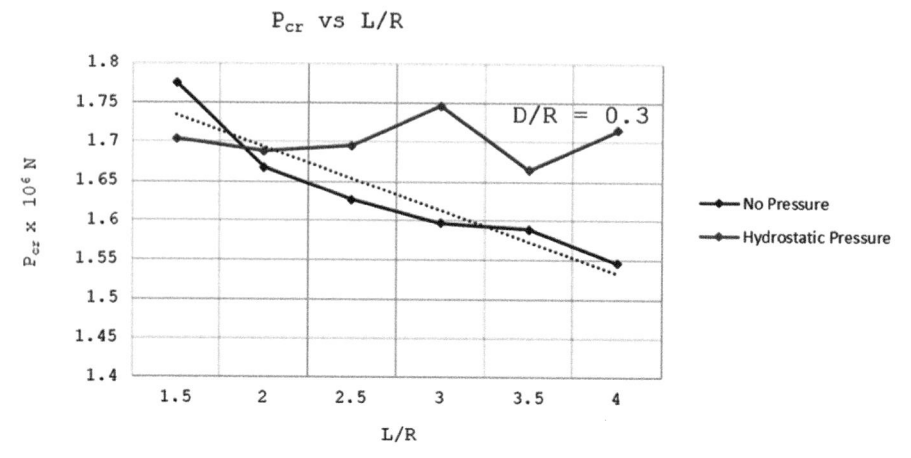

FIGURE 10.7 Critical buckling load against different L/R ratios of the cylinder.

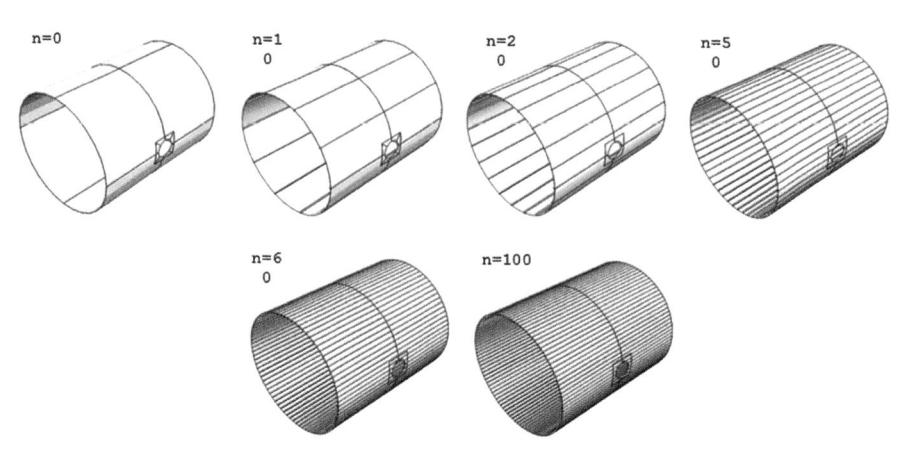

FIGURE 10.8 Model under study with an increasing number of stiffeners.

The plot in Figure 10.9 illustrates the effect of the number of stiffeners on the critical buckling load. From the plot, it was noted that the buckling resistance drastically increases after a certain number of stiffeners. Initially, from $n = 0$ to $n = 20$, the addition of stiffeners slightly increases the buckling resistance. At $n = 50$, the critical buckling load increases dramatically by 1.35 times, after which the buckling load linearly increases with a noticeably constant slope.

Effect of Number of Stiffners on Critical Buckling Load

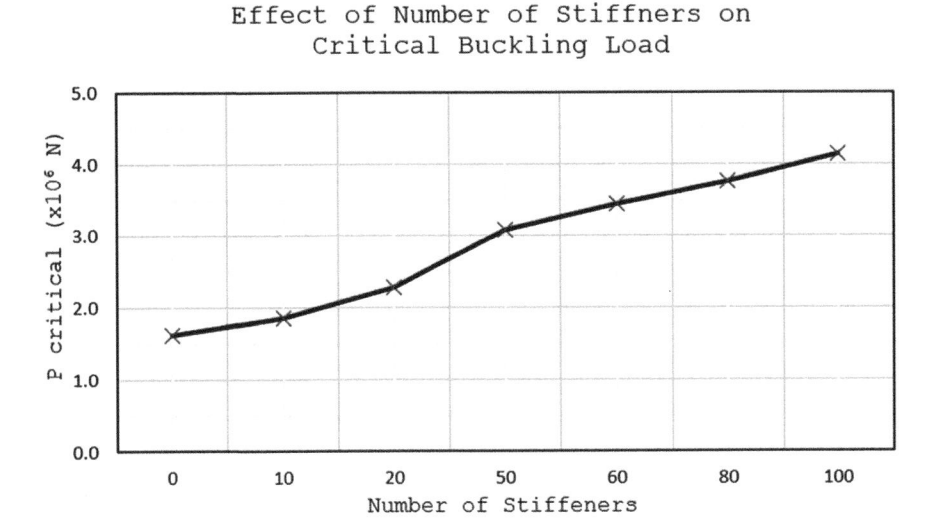

FIGURE 10.9 The effect of the number of stiffeners on the critical load.

Analyzing the buckling response, it was noted that the stiffer cylinder has a longer characteristic wavelength of buckling modes. This was also experimentally observed by Singer and Abramovich.[9] The studies have shown that this is due to increased axial bending stiffness and lower imperfection sensitivity.[10] The resistance to axial load by the stiffeners has resulted in an increased number of local bucklings before the shell reaches the critical point, as shown in Figure 10.10. The highly stiffened shell also exhibits a low loss of stiffness in the postbuckling modes.

10.6 CONCLUSIONS

The buckling analysis of a thin-walled cylinder with a hole under axial compression and internal hydrostatic pressure was studied using FEM. For this, nonlinear buckling analyses were conducted using the artificial damping method. The numerical results in the study have shown that the internal hydrostatic pressure had an appreciable strengthening effect on the cylindrical shell. Moreover, the effect of hydrostatic pressure has also affected the characteristic deformation modes of a thin-walled cylinder under axial compression. In the later part of the study, the effect of stiffeners on a thin-walled cylinder was presented. It was well-observed that the number of stiffeners has shown a linear increase in the buckling resistance of cylinders.

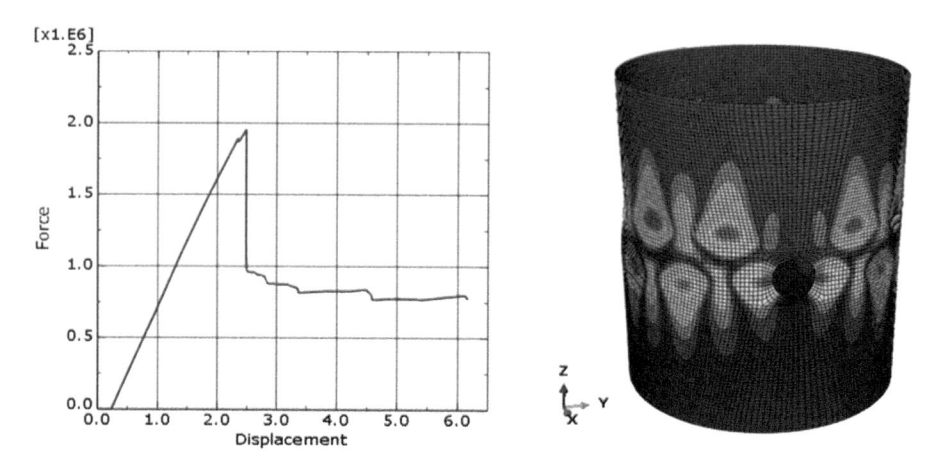

FIGURE 10.10 Results for no-stiffened ($n = 0$) iteration.

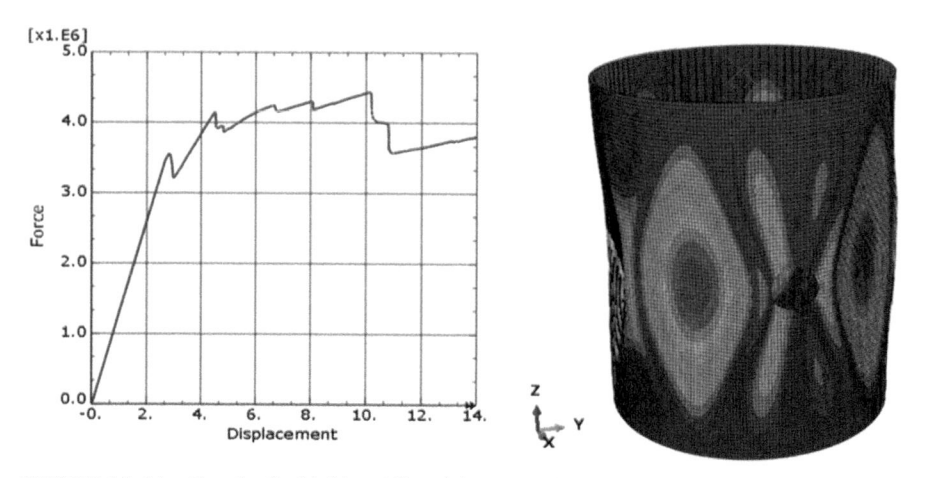

FIGURE 10.11 Results for highly stiffened ($n = 100$) iterations.

KEYWORDS

- **thin-walled element**
- **buckling**
- **axial compression**
- **hydrostatic pressure**
- **stiffeners**

REFERENCES

1. Thompson, L. E. Effects of Internal Pressure upon the Buckling of thin Circular Cylindrical Shells Under Axial Compression. Masters Theses, 1965; p 5231.
2. Buckling of Thin-Walled Circular Cylinders'; NASA/SP-8007-2020/REV 2/
3. Starnes, J. Effect of a Circular Hole on the Buckling of Cylindrical shells loaded by Axial Compression, *AIAA J.* **1972**, *10,* 1466–1472.
4. Almroth, B. O., Brogan, F.; Marlowe, M. B. Stability Analysis of Cylinders with Circular Cutouts. *AIAA J.* **1973**, *11,* 1582–1584.
5. Almroth, B. O.; Holmes, A. M. C, Buckling of Shells with Cutouts, Experiment and analysis. *Int. J. Solids Struc.* **1972**, *8,* 1057.
6. Kobayashi, T.; Mihara, Y.; Fujii, F. Path-tracing Analysis for Post-buckling Process of Elastic Cylindrical Shells Under Axial Compression. *Thin-Walled Struc.* **2012,** *61,* 180–187.
7. Kobayashi, T.; Mihara, Y. Artificial Damping Method for Local Instability Problems in Shell. *Shell Struc: Theory Applic.,* **2014,** 203–206.
8. Smith, M. *ABAQUS/Standard User's Manual*, Version 6.9; Dassault Systèmes Simulia Corp; 2009.
9. Singer, J.; Abramovich, H. Vibration Techniques for Definition of Practical Boundary Conditions in Stiffened Shells. *AIAA J.* **1979,** *17* (7), 762–769.
10. Bushnell, D. *Computerized Buckling Analysis of Shells*; ISBN 90-247-3099-6.

CHAPTER 11

Seismic Behavior of Buckling Restrained Brace Installed Steel Buildings

PRAJAKTA SHETE[1], SHWETA SAJJANSHETTY[2], and SUHASINI MADHEKAR[3]

[1]Department of Civil Engineering, College of Engineering Pune, Pune, India

[2]College of Engineering Pune, Pune, India

[3]Department of Civil Engineering, College of Engineering Pune, Pune, India

ABSTRACT

The design of earthquake-resistant structures with adequate ductility and stiffness is a fundamental requirement. During seismic events, structures are susceptible to undergo large inelastic displacements or collapse and therefore require special attention to limit the deformations and forces. The use of *Energy-Dissipating Devices* (EDDs) is an effective solution for structural vibration control. The devices are installed in structures at appropriate locations to dampen the input seismic *energy*. The present work investigates the seismic behavior of Steel Moment-Resisting Frame (SMRF) buildings with and without Buckling-Restrained Braces (BRBs). Symmetrical and unsymmetrical ten-story buildings located in earthquake zone V of India are designed as SMRF as per IS 800 (2007)[7] and IS 1893 (2016).[8] Three Indian earthquake ground motions are scaled to match with 5% damped target spectra developed according to IS 1893 for soil-type II using ETABS 2016. The scaled earthquake ground motions are used to carry out nonlinear time history analysis. The buildings are then installed with BRBs, and nonlinear time history analysis is performed. A comparison of different parameters describing the response of structures reveals that BRBs perform a vital role in mitigating the seismic response of the structures.

11.1 INTRODUCTION

During earthquake events, immense energy from the ground is input into the structure, which must be either absorbed or dissipated through various mechanisms. In India, the conventional approach for earthquake protection of structures is to increase the member sizes and thereby increase the mass and stiffness of the structure. However, this approach makes the members very bulky and strong, thus transferring the seismic energy to the joints, which are the weakest links of structures. Also, an increase in stiffness makes the structure more susceptible to future earthquake attacks. All types of structures have 5% inherent damping, as per IS 1893.[8] The addition of EDDs can substantially increase the damping ratio up to 20–30% and dissipate seismic energy effectively.[2] A concentrically braced frame system is an effective conventional lateral force-resisting system. These bracing elements offer lateral stiffness; however, these braces are expected to yield in tension and buckle in compression. This motivated the researchers to overcome the buckling of the conventional bracing system and to fully utilize the capacity of bracing members. To this end, the buckling-restrained braces are developed. Buckling-Restrained-Braced Frames (BRBFs) are one of the newer seismic force-resisting systems used in modern building designs. BRB is a displacement-dependent energy dissipator that dissipates energy while yielding. BRB is a fabricated assembly. Figure 11.1 shows the common BRB assembly. The most common BRBs consist of a steel core plate that is surrounded by a steel tube casing filled with grout or concrete. The core generally consists of a rectangular steel plate. The other core cross-sections, such as cruciform with outer squares and circular sections, can also be used, as shown in Figure 11.2. The core is axially decoupled from the fill and casing by various means that produce physical isolation or a gap. As the name states, the BRB assembly restrains core buckling under compressive loading and achieves a compressive yield strength that is approximately equal to its tensile yield strength. Therefore, the core area can be sized for design-level seismic loads based on the yield stress of the core, F_{ysc}, as opposed to braces in conventional CBFs, which are sized based on the critical buckling stress, F_{cr}, of the section.

The BRB core is manufactured with several distinct regions along its length that enable a stable cyclic response. Studies were conducted on BRBs to investigate the seismic response of the structures. Sabelli[11] performed an extensive analytical investigation of the seismic response of three and six-story concentrically braced steel frames using BRB and suggested effective means of proving BRB to overcome many of the potential problems associated with a special concentrically braced frame. An extensive numerical and experimental study has been carried out previously, which shows the capabilities of BRBs during seismic events.[3,4,9,10,12] The literature review revealed that the effect of symmetrical and unsymmetrical steel buildings installed with BRBs had not been studied so far. In the present work, the innovative work of comparing the seismic performance of SMRF buildings with and without BRBs for a symmetrical and unsymmetrical plan for Indian earthquake records has been carried out.

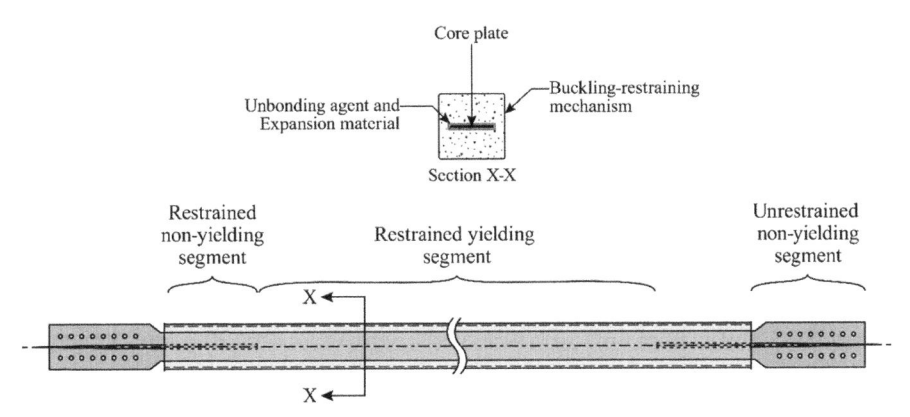

FIGURE 11.1 Common BRB assembly.

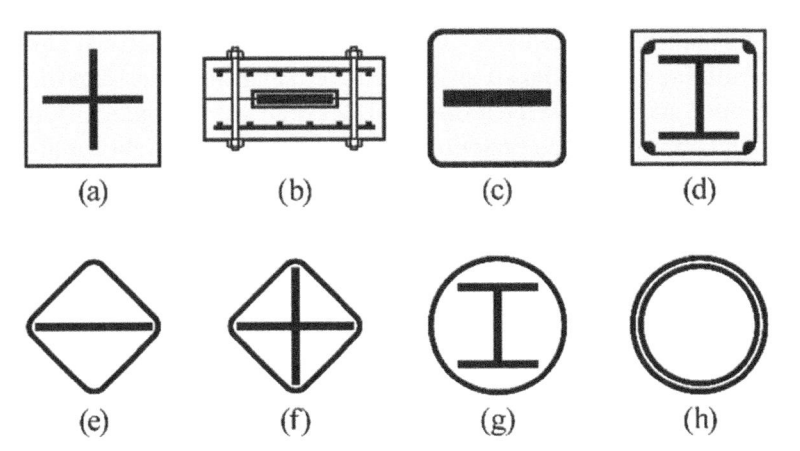

FIGURE 11.2 BRB cross-sections.

11.2 MODELING OF STEEL BUILDINGS

In the present study, ten-story symmetrical and unsymmetrical steel buildings are modeled and analyzed with and without BRBs in ETABS 2016. The lateral dynamic load is applied to the buildings using a nonlinear time history for three different Indian earthquake ground motions, which occurred in seismic zone V. As per the guidelines of ASCE, the time history records from zone V are scaled to match with 5% damped target response spectra of IS 1893[8] for the soil of type II. The maximum response among the three earthquakes is considered. Three different configurations of BRBs are considered for symmetrical buildings to study the optimized performance of SMRF and BRB. For unsymmetrical buildings, two different configurations are studied. BRBs substantially reduce the structural seismic response. From the analysis of the results, it is observed that C2 is the best configuration as it shows a substantial reduction of the response quantities. Installing BRBs in the building reduces the abrupt change in story drifts, thereby protecting the glass claddings. The use of BRBs for seismic response control proves to be very beneficial.

11.2.1 STRUCTURAL SYSTEM

Ten-story symmetrical and unsymmetrical buildings are modeled in ETABS 2016 to study the performance of buckling-restrained braces in structures subjected to earthquake ground motions. Figures 11.3 and 11.4 show the typical plans of symmetrical and unsymmetrical steel buildings, respectively. Figures 11.5 and 11.6 show the elevation of a ten-story building. Table 11.1 represents the details of ten-story symmetrical and unsymmetrical buildings. All the buildings are designed as per IS 800[7] using the limit state of design and the limit state of serviceability IS 1893, Part 1,[8] for the soil-type II in zone V and an importance factor of 1. The typical story height for all buildings is 3 m.

Figure 11.8 shows the three different configurations used to study the optimized performance of SMRF and BRB. They have been studied for ten-story symmetrical buildings. For unsymmetrical buildings, two different configurations, as shown in Figure 11.4, are studied. The red lines represent the location of BRB along the height of the building. The elevation of configurations C1, C2, and C3 with BRBs is represented in Figures 11.5–11.7, respectively.

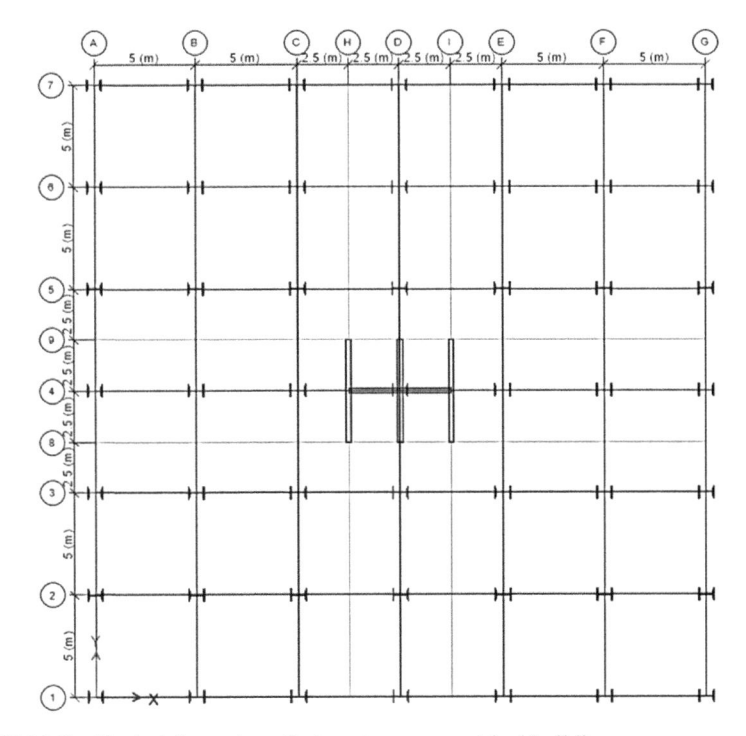

FIGURE 11.3 Typical floor plan of a ten-story symmetrical building.

TABLE 11.1 Particulars of a Ten-Story Model.

Story no.	Beam	Builtup column section details
1–3	ISMB 350	ISHB 450 + Plate width = 350 mm, thk = 16 mm
4–5	ISMB 350	ISHB 300 + Plate width = 350 mm, thk = 12 mm
6	ISMB 300	ISHB 300 + Plate width = 350 mm, thk = 12 mm
7–10	ISMB 300	ISHB 250 + Plate width = 350 mm, thk = 10 mm

11.2.2 NONLINEAR ANALYSIS

A nonlinear time-history analysis is performed in which a model that incorporates the nonlinear load-deformation capacities of individual members in the structure is subjected to earthquake ground motions to obtain forces and displacements. The basis, modeling approaches, and acceptance criteria of the nonlinear dynamic analysis are like those for the nonlinear static analysis. The main exception is that the design displacement is not established using

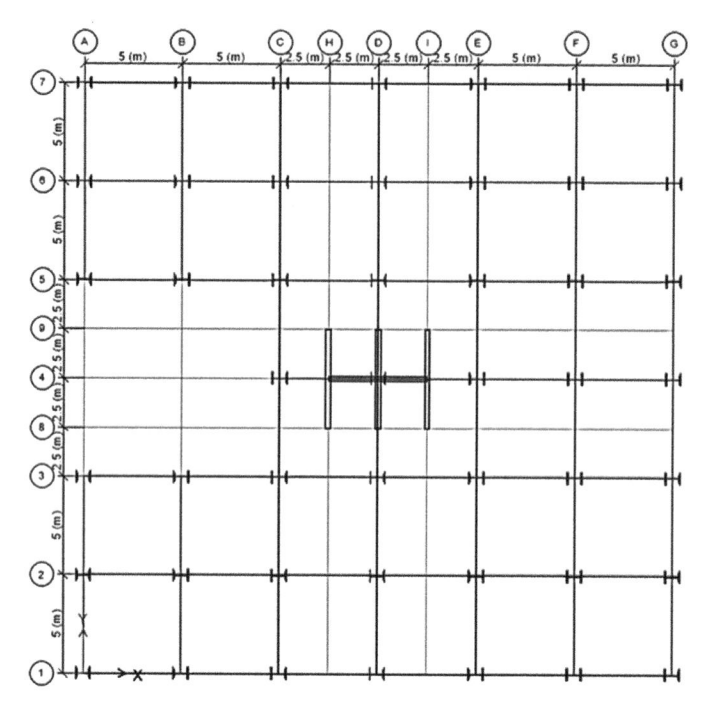

FIGURE 11.4 Typical floor plan of a ten-story unsymmetrical building.

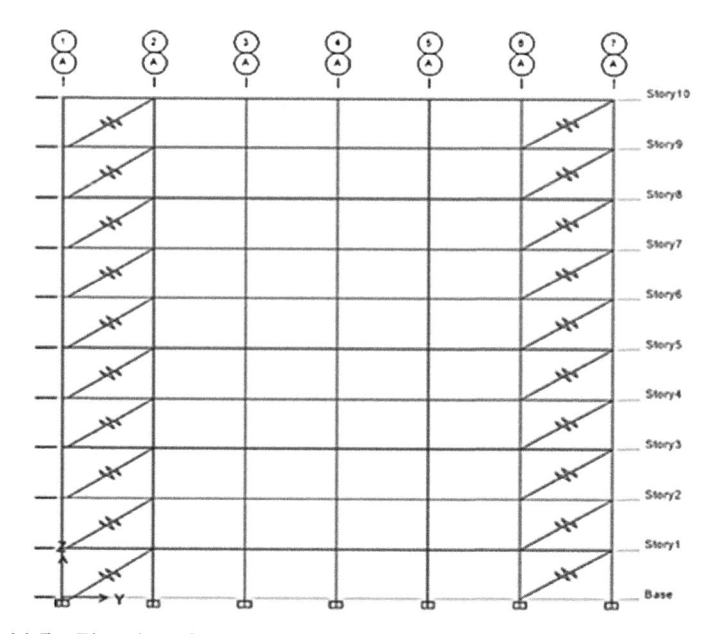

FIGURE 11.5 Elevation of a ten-story building with BRB (C1).

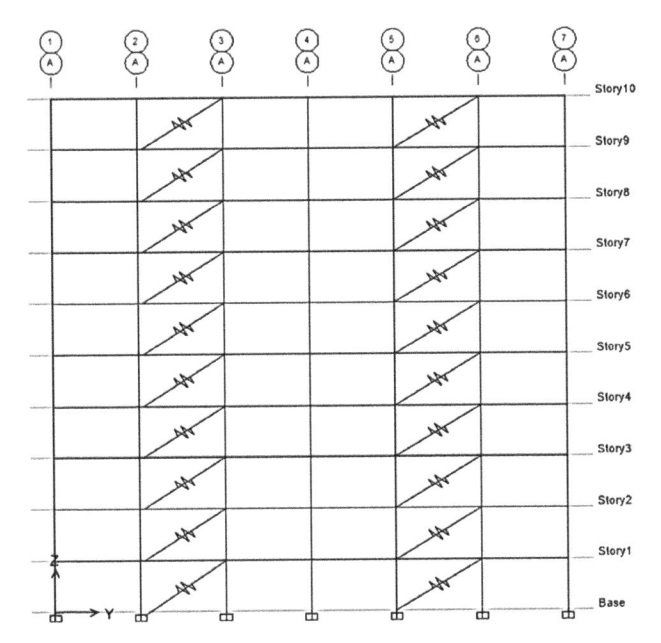

Seismic Behavior of Buckling Restrained Brace Installed Steel Buildings 187

FIGURE 11.6 Elevation of a ten-story building with BRB (C2).

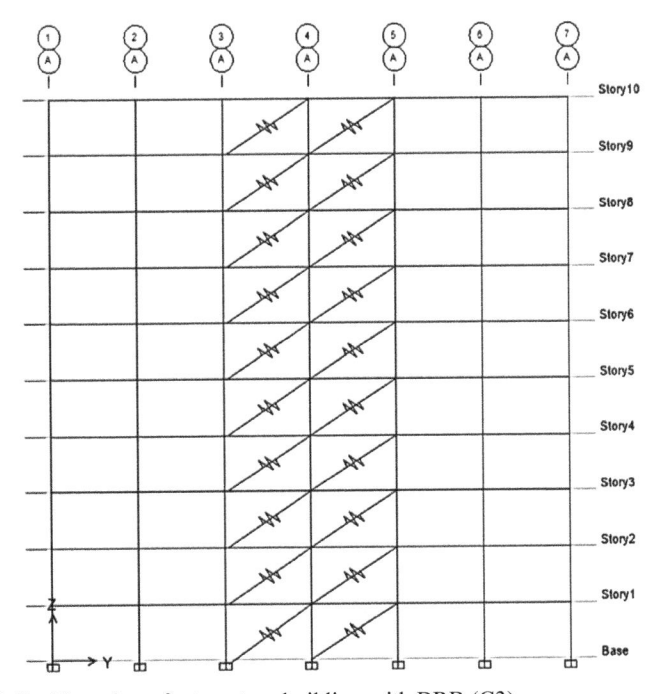

FIGURE 11.7 Elevation of a ten-story building with BRB (C3).

a target displacement but instead is determined directly through dynamic analysis using the ground motion time histories. As a calculated response can be highly sensitive to the characteristics of one ground motion, it is required that the analysis must be carried out with a minimum of three ground motions. As per ASCE/SEI 7_05,[1] if less than seven ground motions are used, the maximum demand in any member may be used. For the present study, three ground motions are used, and the one giving higher forces is used. The particulars of earthquake records are given in Table 11.2.

TABLE 11.2 Particulars of Earthquake Records.

Earthquake records	Earthquake recording station	Recording component	PGA (m/s^2)	PGV (m/s)	Date	Zone	Magnitude
Bhuj	Ahmedabad	N 78 E	1.08	0.113	26-Jan-06	V	7.7
Uttarkashi	Bhatwari	N 85 E	2.48	2.48	26-Oct-91	V	6.8
Dharmshala	Shahpur	N 75 E	2	0.059	26-Apr-86	V	5.7

Earthquake time-history records selected for nonlinear time-history analysis are Bhuj, Dharmshala, and Uttarkashi. These three time histories are then scaled to match with 5% damped target spectra developed according to IS: 1893 (Part I)[8] for soil-type II in zone V and with an importance factor of 1. The response reduction factor considered for the analysis and design is 5.

The scaling is done in such a way that the average spectral acceleration of all three records remains above the design target spectrum over the range of 0.2–1.5 times the fundamental period as specified by the ASCE/SEI 7_05[1] standard for nonlinear dynamic analysis. Figure 11.10 shows the earthquake ground motion for the Uttarkashi time history record and the matched scaled response spectra used for the analysis.

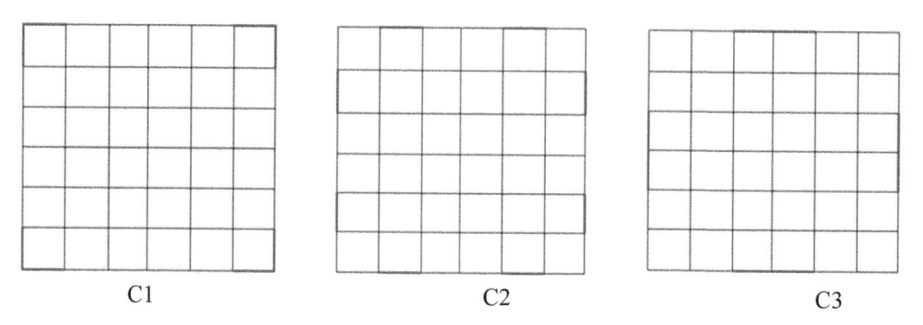

FIGURE 11.8 Configurations of BRB for a ten-story symmetrical building.

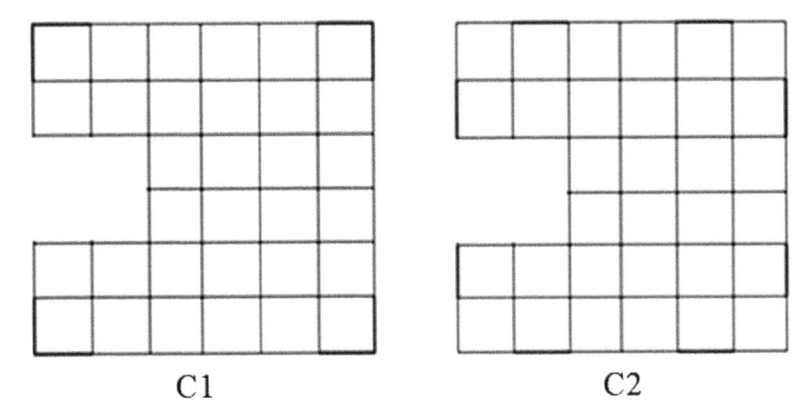

FIGURE 11.9 Configurations of BRB for a ten-story unsymmetrical building.

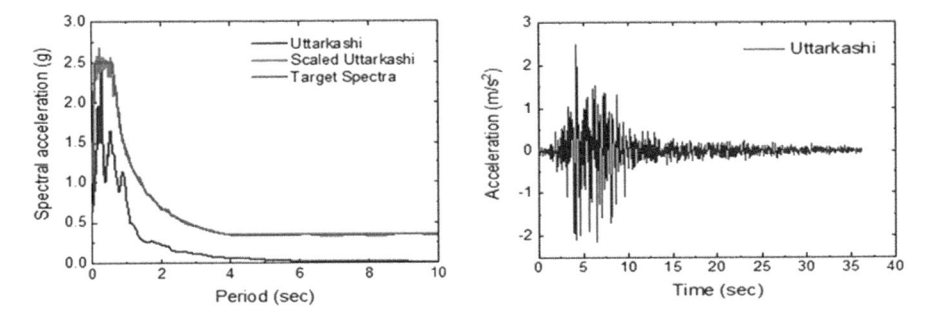

FIGURE 11.10 Earthquake ground motion and scaled response spectra of time history.

11.3 RESULTS AND DISCUSSION

Analytical results obtained from nonlinear time histories for various buildings are described here. Results show that BRBs have reduced the seismic response, i.e., base shear, story displacement, and story drift, of the structure effectively for symmetrical and unsymmetrical buildings for different configurations.

11.3.1 EFFECT ON BASE SHEAR

A considerable reduction is observed in base shear after installing BRB. The reduction in base shear is as shown in Table 11.3 with BRBs. Reduction in base shear also reduces story shear and hence the member forces.

TABLE 11.3 Base Shear (kN) for Steel Buildings.

Building	Type	Configuration	For earthquake records		
			Dharmshala	Uttarkashi	Bhuj
Ten-story symmetric building	SMRF		5527	21,582	12,118
	SMRF+BRB	C1	3790	15,154	8418
		C2	3705	14,819	8231
		C3	3744	14,971	8316
Ten-story unsymmetric building	SMRF		4096	19,355	11,300
	SMRF+BRB	C1	2521	11,745	7811
		C2	2341	11,488	6617
	SMRF+BRB	C1	24.27	74.70	43.82
		C2	23.93	73.69	43.21

11.3.2 EFFECT ON STORY DISPLACEMENT

During earthquakes, the lateral displacement of buildings is quite high. Installation of BRB in buildings can reduce the story displacement considerably. Table 11.4 below shows the reduction in story displacement for steel buildings with BRBs.

TABLE 11.4 Top Story Displacement of Steel Buildings.

Building	Type	Configuration	For earthquake records		
			Dharamshala	Uttarkashi	Bhuj
Ten-story symmetric building	SMRF		32.85	107.42	60.32
	SMRF+BRB	C1	23.75	80.94	44.45
		C2	23.45	79.95	43.89
		C3	22.87	78.06	42.84
Ten-story unsymmetric building	SMRF		41.71	123.30	73.98

Figures 11.11 and 11.12 represent the story displacement of the symmetrical and unsymmetrical steel buildings along the height of the structure for ten storys.

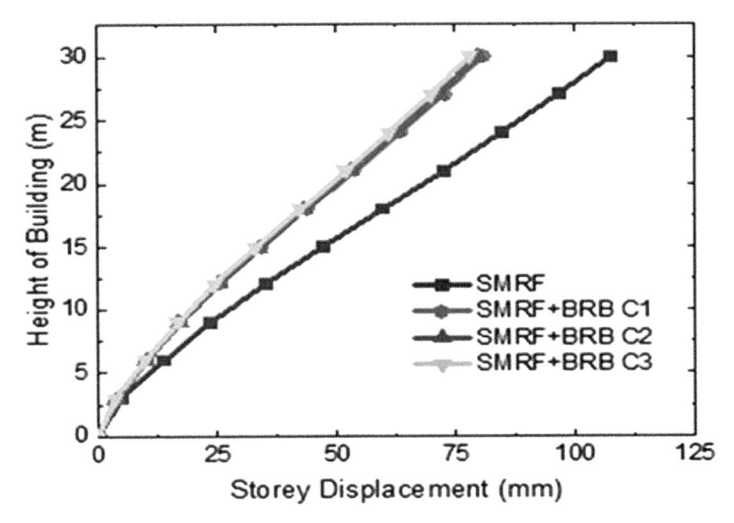

FIGURE 11.11 Story displacement vs. the height of the building with BRB for a symmetrical building.

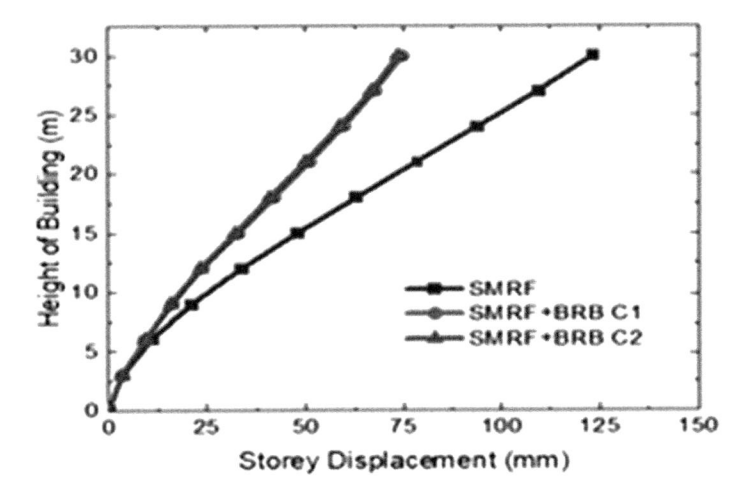

FIGURE 11.12 Story displacement vs. the height of the building with BRB for an unsymmetrical building.

11.3.3 EFFECT ON STORY DRIFT

During earthquakes, the lateral drift of the buildings is also quite high. BRBs play a very vital role when it comes to reducing the drift. In time history analysis, as the stiffness of the structure changes from time to time and if there is any soft-story formation, the interstory drift responses deteriorate such that the peak drift values are significantly worse than those of SMRFs. The graphs below show the reduction in story drift for symmetrical buildings and unsymmetrical buildings.

11.3.4 HYSTERESIS LOOP

The nonlinear behavior of a member is usually described by an idealized hysteresis loop. The energy dissipated by a BRB is the area inside a hysteresis loop. The ETABS 2016 developed hysteresis for BRB, which shows different phases of strain hardening. Figure 11.15 shows the typical hysteresis loop of BRB for symmetrical and unsymmetrical buildings.

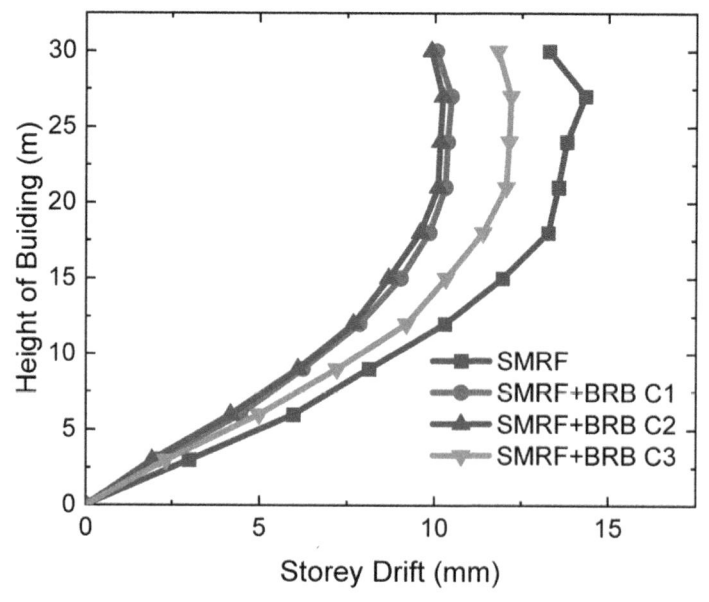

FIGURE 11.13 Story drift vs. the height of the building with BRB for a symmetrical building.

The use of BRB for improving the seismic performance of structures is very beneficial. It reduces the base shear, story displacements, and story drift of the structure. C2 is the best configuration as it shows the highest amount of reduction in base shear, story displacements, and story drifts. A total of 44% base shear reduction is observed for ten-story symmetrical buildings, whereas for unsymmetrical buildings, a 48% base shear reduction is observed.

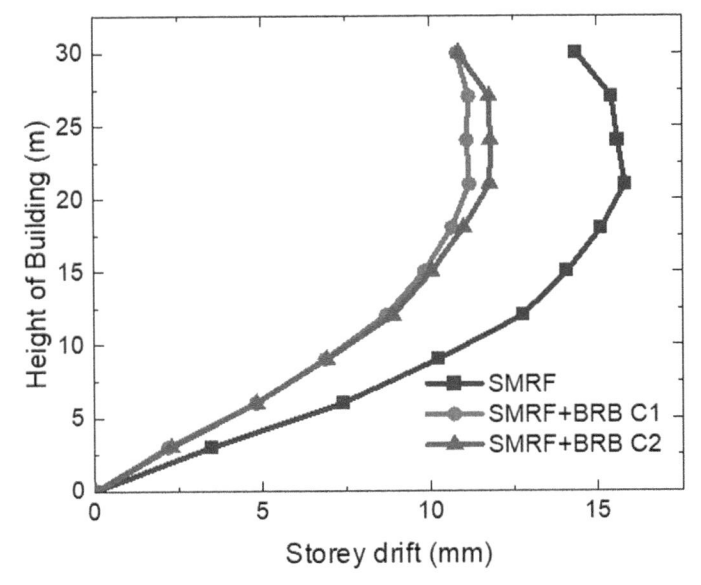

FIGURE 11.14 Story drift vs. the height of the building with BRB for an unsymmetrical building.

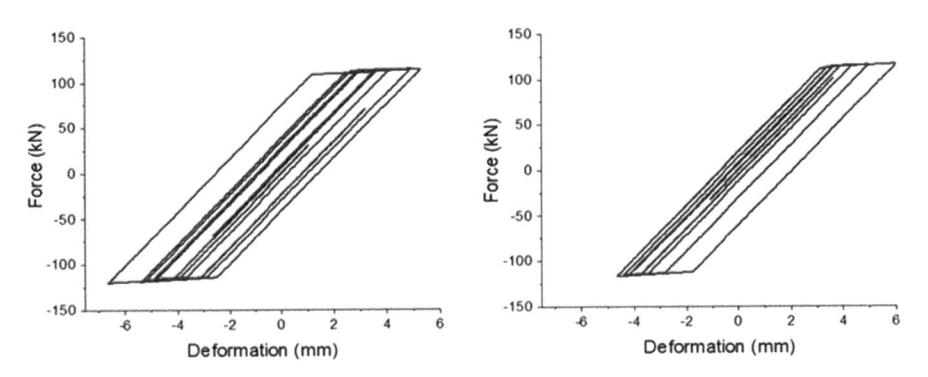

FIGURE 11.15 Hysteresis loop of a ten-story symmetrical and unsymmetrical building.

11.4 CONCLUSIONS

The BRB improves the seismic performance of the structure, which can be observed in the form of response reduction in terms of base shear, story displacements, and story drifts. The use of BRBs for seismic response control proves to be very beneficial, compared to other types of dampers, as BRBs are a cost-effective option and can be manufactured locally.

KEYWORDS

- **buckling-restrained braces**
- **nonlinear time history analysis**
- **moment-resisting frames**
- **seismic response control**

REFERENCES

1. ASCE/SEI 7-10. *Minimum Design Loads for Building and Other Structures*; American Society of Civil Engineers: Reston, VI; 2006.
2. Bartera, F.; Giacchetti, R. Steel Dissipating Braces for Upgrading Existing Building Frames. *J. Constr. Steel Res.* **2004,** *60* (3–5), 751–769.
3. Black, C. J.; Makris, N.; Aiken, I. D. Component Testing, Seismic Evaluation and Characterization of Buckling-Restrained Braces. *J. Struc. Eng.* **2004,** *130* (6), 880–894.
4. Fahnestock, L. A.; Ricles, J. M.; Sause, R. Experimental Evaluation of a Large-Scale Buckling-Restrained Braced Frame. *J. Struc. Eng.* **2007,** *133* (9), 1205–1214.
5. Kersting, R. A.; Lopez, W. A.; Fahnestock, L. A. *NEHRP Seismic Design Technical Brief No. 11: Seismic Design of Steel Buckling-Restrained Braced Frames*; 2015.
6. Hussain, S.; Benschoten, P.; Satari, A.; Lin, S. In *Buckling Restrained Braced Frame (BRBF) Structures: Analysis, Design, and Approvals Issues*, Proceedings of the 75th SEAOC Annual Convention; Long Beach, CA, United States, 2006.
7. IS 800. General Construction in Steel—Code of Practice; 2007.
8. IS 1893 (Part I). Criteria for Earthquake Resistant Design of Structures; 2016.
9. Kim, D. H.; Lee, C. H.; Ju, Y. K.; Kim, S. D. Subassemblage Test of Buckling-Restrained Braces with H-shaped Steel Core. Struct. Des. Tall Spec. Build. **2015,** *24* (4), 243–256.
10. Mirtaheri, M.; Gheidi, A.; Zandi, A. P.; Alanjari, P.; Samani, H. R. Experimental Optimization Studies on Steel Core Lengths in Buckling Restrained Braces. *J. Constr. Steel Res.* **2011,** *67* (8), 1244–1253.
11. Sabelli, R.; Mahin, S.; Chang, C. Seismic Demands on Steel Braced Frame Buildings with Buckling-restrained Braces. *Eng. Struc.* **2003,** *25* (5), 655–666.
12. Shete, P.; Madhekar, S.; Fayeq Ghowsi, A. Numerical Analysis of Steel Buckling-Restrained Braces with Varying Lengths, Gaps, and Stoppers. Pract. Period. Struct. Des. Constr. **2022,** *27* (1), 04021051.

CHAPTER 12

Non-Linear Dynamic Analysis of Multi-Story Reinforced Cement Concrete (RCC) Building Having Different Geometry

MAHESH PATHARE and RAMCHANDRA APTE

Department of Civil Engineering, Vishwakarma Institute of Information Technology, Pune, India

ABSTRACT

A major goal of earthquake engineers is to design and construct structures in such a way that damage to the structure and its structural components is minimized during an earthquake. The goal of this study is to perform a nonlinear dynamic analysis on a multistorey RCC building with variable geometry. The model of G + 15 RCC with a varying geometry plan is used for the analysis. ETABS, a finite element-based software, is used to do the analysis. Base shear, storey drift, storey displacement, and time period are some of the response parameters that can be determined. For dynamic analysis, the time-history technique is used. A time-history study is a step-by-step examination of a structure's dynamical response to a specific loading that may change over time. Nonlinear dynamic time-history analysis is one type of dynamic analysis. In this study, ETABS is used to do a nonlinear time-history analysis on a G + 15 RCC building frame utilizing the time history of the El Centro earthquake of 1940 and also check all three models with a Mivan structure having a wall thickness of 200 mm.

12.1 INTRODUCTION

The structural response to an earthquake is a dynamic event that depends on the dynamic characteristics of the structures and the intensity, duration, and frequency of the exciting ground motion.[1] Although seismic action is dynamic in nature, building codes recommend a single static load analysis for the design of earthquake-resistant buildings because of its simplicity. This is done by focusing on the dominant first-mode response and developing the equivalent seismic analysis is a subset of structural analysis that calculates the response of a building (or nonstructural) structure to an earthquake.[2,3,4] In earthquake-prone areas, it is part of the structural design, seismic engineering, or structural evaluation and retrofit process.

Because nonlinear dynamic analysis combines ground motion data with a thorough structural model, it can produce results with a low level of uncertainty.

Nonlinear properties of the structure are evaluated as part of a time domain analysis in nonlinear dynamic analysis.[5]

This is the most accurate way, and it is required by some building codes for unusual configurations or historically significant structures.

The following are the objectives of this study:

1. To study the seismic response of buildings by nonlinear dynamic analysis method.
2. To study the seismic response of buildings in terms of storey shear, storey drift, storey displacement, time period, base shear, base moments, storey displacement, etc.
3. To determine the optimal building shape that has the best seismic response.

12.2 METHODOLOGY OF WORK

12.2.1 PROBLEM STATEMENT OF RECTANGULAR GEOMETRY ANALYSIS

A G + 15-storey rectangular building with a 3.2 m floor-to-floor height has been analyzed by using Etabs software in zone III. It is an architectural plan, not the plan of any existing building.

12.2.2 MODEL DESCRIPTION OF RECTANGULAR GEOMETRY ANALYSIS

A. Preliminary data required for Rectangular Geometry Analysis

TABLE 12.1 Parameters to be Considered for Rectangular Geometry Analysis.

Sr. no.	Parameter	Values
1.	Number of storeys	G + 15
2.	Base to plinth	1.5 m
3.	Floor height	3.2 m
4.	Wall	200 mm thick
5.	Materials	Concrete M40 and Reinforcement Fe 500
6.	Frame size	18 m × 24 m building size
7.	Grid spacing	4.5 m grids in the X-direction and 6m grids in Y-direction.
8.	Size of column	600 mm × 600 mm
9.	Size of beam	300mm × 600 mm
10.	Depth of slab	150 mm
11.	Total height	49.5 m
13.	Mivan wall	200 mm

B. Load details

TABLE 12.2 Load Details.

a.	Dead load	In ETABS, the software itself calculates the dead loads by applying a self-weight multiplier factor, which is taken by the structure, and the rest of the load cases are kept zero. It is defined in the load section.
b.	Live load on the roof and floors	2 kN/m^2 (roof) and 4 kN/m^2 (floors) as per IS:875 (part-2)
c.	Floor finishes on the roof and floors	1.5 kN/m^2 as per IS:875 (part-2)
d.	Wall load on all levels	7.8 kN/m

C. Wind data required for analysis.

TABLE 12.3 Wind Data Required for Analysis.

Sr. no.	Parameter	Values as per IS 875:2015 (Part 3)	Reference
1.	Basic wind speed (Vb)	Pune = 39 m/sec,	Annex A
2.	Risk coefficient k1	1	Table 1, Clause 6.3.1
3.	Terrain Roughness Factor k2	1.22	Table 2, Clause 6.3.2.2
4.	Topography Factor k3	1	Table 3, Clause 6.4.2
5.	Importance Factor k4	1	Clause 6.3.4
6.	Windward coefficient cp	0.8	Clause 7.3.3
7.	Leeward coefficient cp	0.5	Clause 7.3.3

D. Seismic data required for analysis

TABLE 12.4 Seismic Data Required for Analysis.

Sr. no.	Parameter	Values as per IS 1893:2016 (Part 1)	Reference
1.	Type of structure	Special RC moment-resisting frame	Table 9, Clause 7.2.6
2.	Seismic zone	III	Table 3, Clause 6.4.2
3.	Location	Pune	Annex E
4.	Zone factor (Z)	0.16	Table 2, Clause 6.4.2
5.	Type of soil	Rock or Hard Soil	Clause 6.4.2.1
6.	Damping	5%	Clause 7.2.4
7.	Response spectra	As per IS 1893 (part 1):2016	Figure 2, Clause 6.4.6
8.	Load combinations	1) 1.5 (DL + IL)	Clause 6.3.1
		2) 1.2 (DL + IL+ EL)	
		3) 1.5 (DL + EL)	
9.	Response reduction	4	Table 9, Clause 7.2.6
10.	Importance factor (I)	1.2	Table 8, Clause 7.2.3

12.2.3 PROBLEM STATEMENT OF SQUARE GEOMETRY ANALYSIS

A G + 15-storey building in the shape of a square with a 3.2 m floor-to-floor height has been analyzed by using Etabs software in zone III. It is an architectural plan, not the plan of any existing building.

12.2.4 MODEL DESCRIPTION OF SQUARE GEOMETRY ANALYSIS

A. Preliminary data required for Square Geometry Analysis

TABLE 12.5 Parameters to be Considered for Square Geometry Analysis.

Sr. no.	Parameter	Values
1.	Number of storeys	G + 15
2.	Base to plinth	1.5 m
3.	Floor height	3.2 m
4.	Wall	200 mm thick
5.	Materials	Concrete M40 and Reinforcement Fe 500
6.	Frame size	18 m × 18 m building size
7.	Grid spacing	4.5 m grids in the X-direction and 4.5 m grids in the Y- direction.
8.	Size of column	600 mm × 600 mm
9.	Size of beam	300 mm × 600 mm
10.	Depth of slab	150 mm
11.	Total height	49.5 m
13.	Mivan wall	200 mm

B. Load details

TABLE 12.6 Load details.

a.	Dead load	In ETABS, the software itself calculates the dead loads by applying a self-weight multiplier factor of one, which is taken by the structure and the rest of the load cases are kept zero. It is defined in the load case section.
b.	Live load on the roof and floors	2 kN/m^2 and 4 kN/m^2 as per IS:875 (part-2)
c.	Floor finishes on the roof and floors	1.5 kN/m^2 as per IS:875 (part-2)
d.	Wall load on all levels	7.8 kN/m

C. Wind data required for analysis

TABLE 12.7 Wind Data Required for Analysis.

Sr. no.	Parameter	Values as per IS 875: 2015 (Part-3)	Reference
1.	Basic wind speed (Vb)	Pune=39m/sec	Annex A
2.	Risk coefficient k1	1	Table 1, Clause 6.3.1
3.	Terrain Roughness Factor k2	1.22	Table 2, Clause 6.3.2.2
4.	Topography Factor k3	1	Table 3, Clause 6.4.2
5.	Importance Factor k4	1	Clause 6.3.4
6.	Windward coefficient cp	0.8	Clause 7.3.3
7.	Leeward coefficient cp	0.5	Clause 7.3.3

D. Seismic data required for analysis.

TABLE 12.8 Seismic Data Required for Analysis.

Sr. no.	Parameter	Values as per IS 1893:2016 (Part 1)	Reference
1.	Type of structure	Special RC moment-resisting frame	Table 9, Clause 7.2.6
2.	Seismic zone	III	Table 3, Clause 6.4.2
3.	Location	Pune	Annex E
4.	Zone factor (Z)	0.16	Table 2, Clause 6.4.2
5.	Type of soil	Rock or Hard Soil	Clause 6.4.2.1
6.	Damping	5 %	Clause 7.2.4
7.	Response spectra	As per IS 1893 (part 1):2016	Figure 2, Clause 6.4.6
8.	Load combinations	1) 1.5 (DL + IL) 2) 1.2 (DL+ IL+ EL) 3) 1.5 (DL + EL) 4) 0.9 DL + 1.5 EL	Clause 6.3.1
9.	Response reduction factor (R)	4	Table 9, Clause 7.2.6
10.	Importance factor (I)	1.2	Table 8, Clause 7.2.3

12.3 PROBLEM STATEMENT OF CIRCULAR GEOMETRY ANALYSIS

A G + 15-storey circular building with 3.2 m floor-to-floor height has been analyzed by using Etabs software in zone III. It is an architectural plan, not the plan of any existing building.

12.3.1 MODEL DESCRIPTION OF CIRCULAR GEOMETRY ANALYSIS

A. Preliminary data required for circular geometry analysis

TABLE 12.9 Parameters to be Considered for Circular Geometry Analysis.

Sr. no.	Parameter	Values
1.	Number of storeys	G + 15
2.	Base to plinth	1.5 m
3.	Floor height	3.2 m
4.	Wall	200 mm thick

TABLE 12.9 *(Continued)*

Sr. no.	Parameter	Values
5.	Materials	Concrete M40 and Reinforcement Fe 500
6.	Frame size	25.45 m × 25.45 m building size
7.	Grid spacing	4.5 m grids in the X-direction with an end grid of 3.725 m and 4.5 m grids in the Y-direction with an end grid of 3.725 m.
8.	Size of column	600 mm 600 mm
9.	Size of beam	300 mm × 600 mm
10.	Depth of slab	150 mm
11.	Total height	49.5 m
13.	Mivan wall	200 mm

B. Load details

TABLE 12.10 Load Details.

a.	Dead load	In ETABS, the software itself calculates the dead loads by applying a self-weight multiplier factor of one, which is taken by the structure, and the rest of the load cases are kept zero. It is defined in the load case section.
b.	Live load on the roof and floors	2 kN/m^2 and 4 kN/m^2 as per IS:875 (part-2)
c.	Floor finishes on the roof and floors	1.5 kN/m^2 as per IS:875 (part-2)
d.	Wall load on all levels	7.8 kN/m

C. Wind data required for analysis.

TABLE 12.11 Wind Data Required for Analysis.

Sr. no.	Parameter	Values as per IS 875: 2015 (Part-3)	Reference
1.	Basic wind speed (Vb)	Pune = 39 m/s,	Annex A
2.	Risk coefficient k1	1	Table 1, Clause 6.3.1
3.	Terrain Roughness Factor k2	1.22	Table 2, Clause 6.3.2.2
4.	Topography Factor k3	1	Table 3, Clause 6.4.2
5.	Importance Factor k4	1	Clause 6.3.4
6.	Windward coefficient cp	0.8	Clause 7.3.3

D. Seismic data required for analysis.

TABLE 12.12 Seismic Data Required for Analysis.

Sr. no.	Parameter	Values as per IS 1893:2016 (Part 1)	Reference
1.	Type of structure	Special RC moment-resisting frame	Table 9, Clause 7.2.6
2.	Seismic zone	III	Table 3, Clause 6.4.2
3.	Location	Pune	Annex E
4.	Zone factor (Z)	0.16	Table 2, Clause 6.4.2
5.	Type of soil	Rock or Hard Soil	Clause 6.4.2.1
6.	Damping	5%	Clause 7.2.4
7.	Response spectra	As per IS 1893 (part 1):2016	Figure 2, Clause 6.4.6
8.	Load combinations	1) 1.5 (DL + IL)	Clause 6.3.1
		2) 1.2 (DL+ IL+ EL)	
		3) 1.5 (DL + EL)	
		4) 0.9 DL + 1.5 EL	
9.	Response reduction factor (R)	4	Table 9, Clause 7.2.6
10.	Importance factor (I)	1.2	Table 8, Clause 7.2.3

12.4 RESULTS AND DISCUSSION

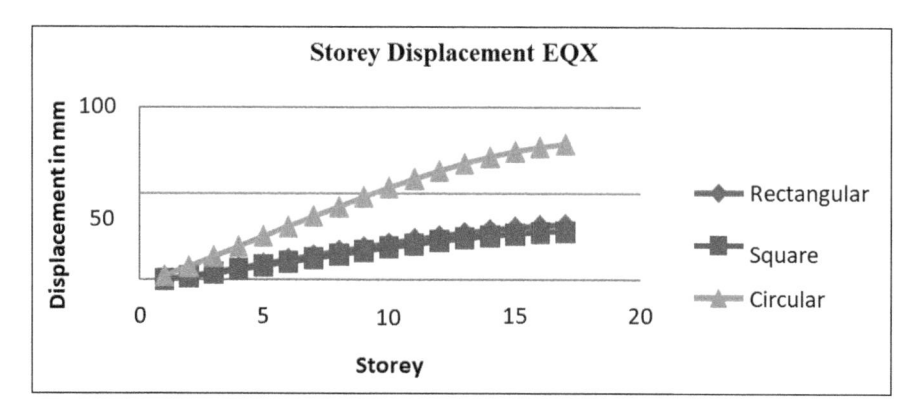

FIGURE 12.1 Storey displacement EQX.

From the above graph, we can observe that the percentage variation for Storey Displacement EQX in square geometry is less than that in rectangular and circular geometry. The variation is found to be 5–9%, for square to rectangular, and on further variation, for rectangular to circular, it is found to be 20–30%.

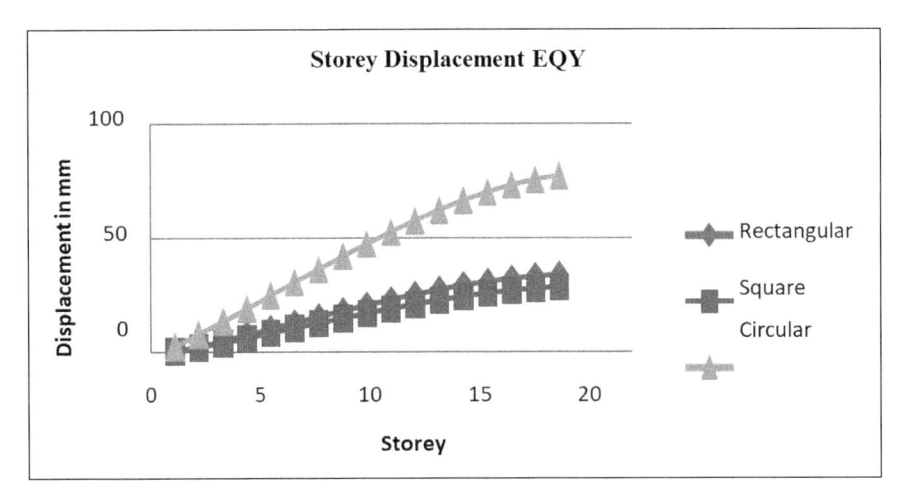

FIGURE 12.2 Storey displacement EQY.

The percentage variation for Storey Displacement EQY for square geometry is less than that for rectangular and circular geometry, as shown in the graph above. The variation is found to be 8–10% for square to rectangular and on further variation, for rectangular to circular, it is found to be 20–30%.

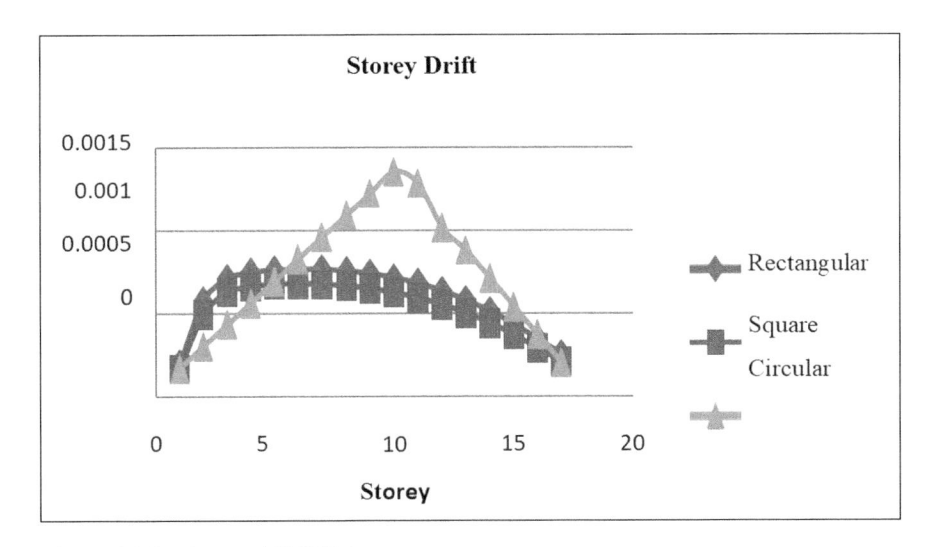

FIGURE 12.3 Storey drift EQX.

From the above graph, we can observe that the percentage variation for Storey Drift EQX for square geometry is less than that for rectangular and circular geometry. The difference between square and rectangular is determined to be 10–15%, and the difference between rectangular and circular is found to be 30–35%.

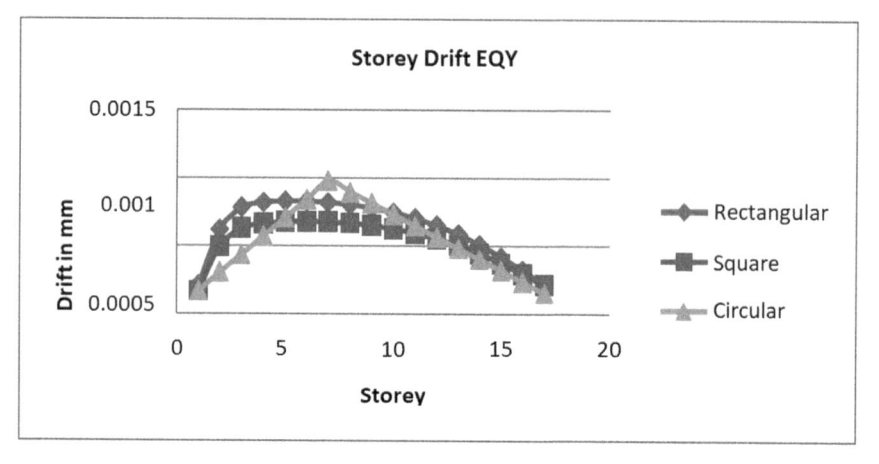

FIGURE 12.4 Storey drift EQY.

The percentage variation for Storey Drift EQY for square geometry is less than that for rectangular and circular geometry, as shown in the graph above. The variation is found to be 5–10%, for square to rectangular, and on further variation, for rectangular to circular, it is found to be 10–20%.

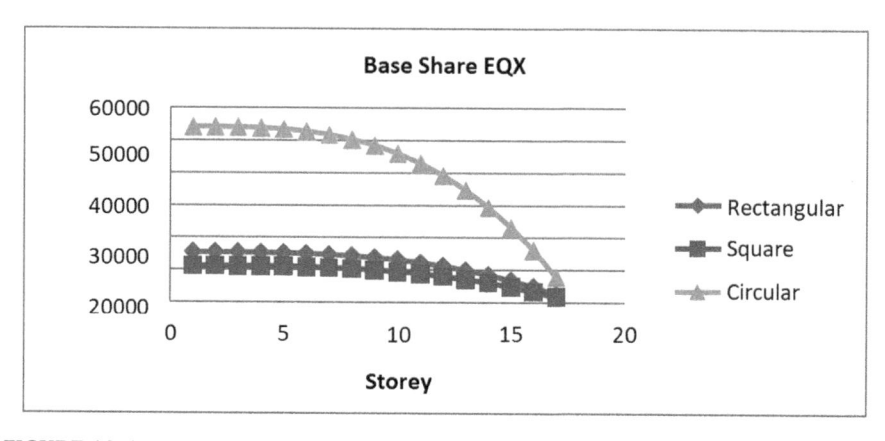

FIGURE 12.5 Base share EQX.

From the above graph, we can observe that the percentage variation for Base Share EQX for square geometry is less than that for rectangular and circular geometry. The variation is found to be 1–5%, for square to rectangular and on further variation, for rectangular to circular, it is found to be 50–55%.

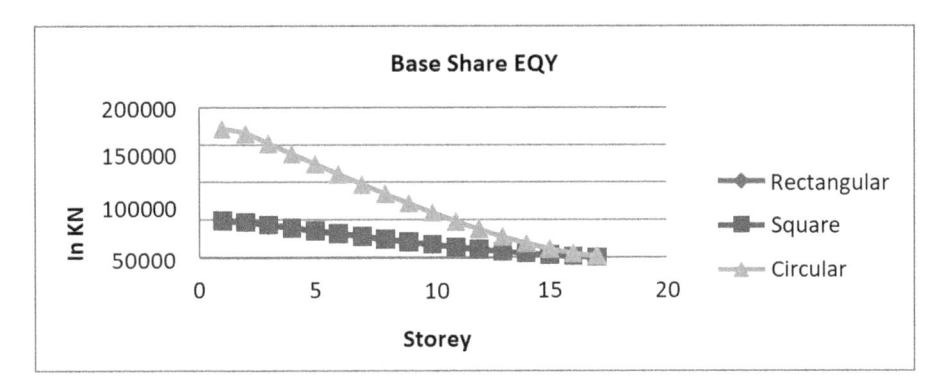

FIGURE 12.6 Base share EQY.

From the above graph, we can observe that the percentage variation for Base Share EQY for square geometry is less than that for rectangular and circular geometry. The variation is found to be 1–5%, for square to rectangular, and on further variation, for rectangular to circular, it is found to be 50–55%.

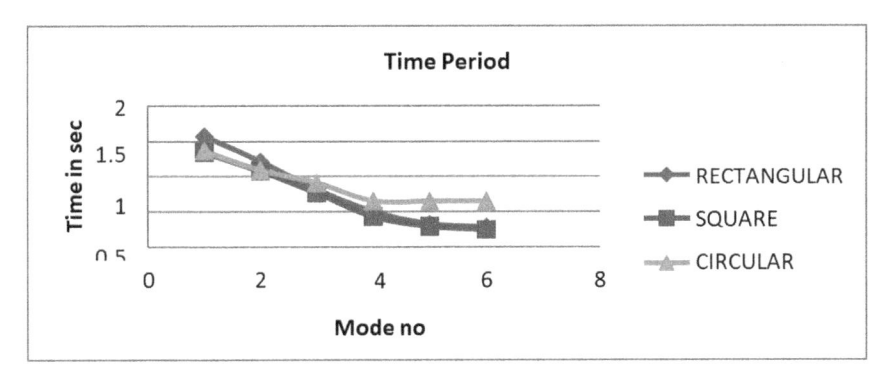

FIGURE 12.7 Time period.

From the above graph, we can observe that the percentage variation for Time Period for square geometry is less than that for rectangular and circular geometry. The variation is found to be 1–5%, for square to rectangular, and on further variation, for rectangular to circular, it is found to be 10–15%.

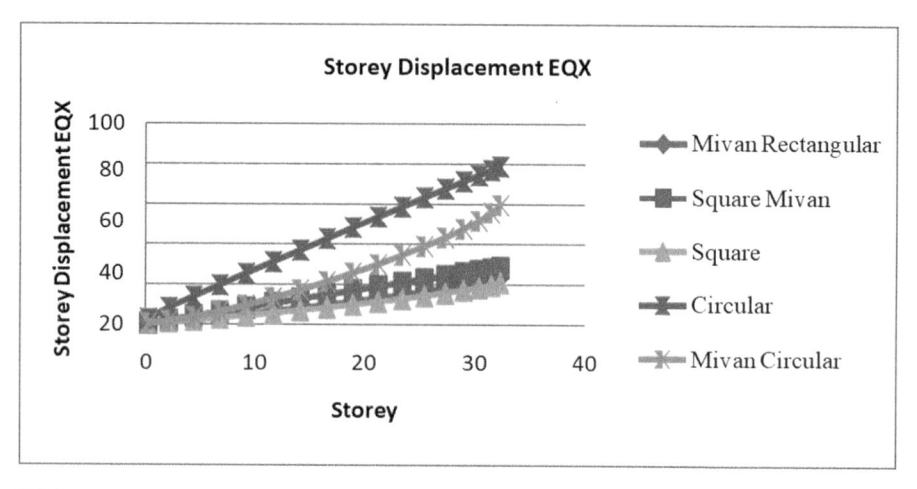

FIGURE 12.8 Storey displacement EQX.

From the above graph, we can observe that the percentage variation for Storey Displacement in the X direction after using Mivan for all models square geometry is less than that for rectangular and circular geometry with and without Mivan. The variation is found to be 5–9% for square to rectangular, and on further variation, for rectangular to circular, it is found to be 20–30%.

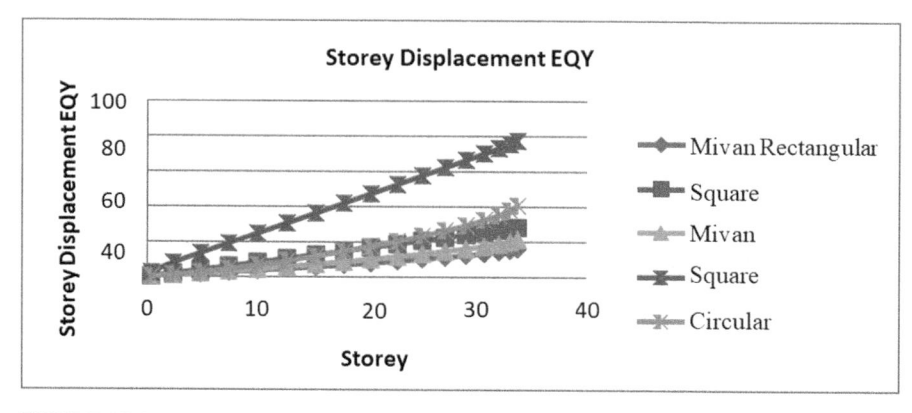

FIGURE 12.9 Storey displacement EQY.

From the above graph, we can observe that the percentage variation for Storey Displacement in the Y direction after using Mivan for all models square geometry is less than that for rectangular and circular geometry with and without Mivan. The variation is found to be 5–9% for square to rectangular, and on further variation, for rectangular to circular, it is found to be 25–30%.

12.5 CONCLUSIONS

From the following study, it is concluded that square RCC geometry is preferable for the dynamic analysis than rectangular and circular geometry. The percentage variance for Storey Displacement EQX for square geometry is less than that for rectangular and circular geometry, as can be shown. The variation is found to be 5–9% for square to rectangular, and on further variation, for rectangular to circular, it is found to be 20–30%. Square geometry has a lower percentage variation for Storey Displacement EQY than that for rectangular and circular geometry. The variation is found to be 8–10% for square to rectangular, and on further variation, for rectangular to circular, it is found to be 20–30%. Square geometry has a lower percentage variation for Storey Drift EQX than that for rectangular and circular geometry. The difference between square and rectangular is determined to be 10–15%, and the difference between rectangular and circular is found to be 30–35%. It can be observed that the percentage variation for Storey Drift EQY for square geometry is less than that for rectangular and circular geometry. The variation is found to be 5–10% for square to rectangular, and on further variation, for rectangular to circular, it is found to be 10–20%. It can be observed that the percentage variation for Base Share EQX for square geometry is less than that for rectangular and circular geometry. The variation is found to be 1–5% for square to rectangular, and on further variation, for rectangular to circular, it is found to be 50–55%. We can observe that the percentage variation for Base Share EQY for square geometry is less than that for rectangular and circular geometry. The variation is found to be 1–5% for square to rectangular, and on further variation, for rectangular to circular, it is found to be 50–55%. It can be observed that the percentage variation for Time Period for square geometry is less than that for rectangular and circular geometry. The variation is found to be 1–5% for square to rectangular, and on further variation, for rectangular to circular, it is found to be 10–15%. Maximum moment findings are less for square geometry. Maximum moment results are as follows:

- The rectangular building's maximum moment is roughly 19.31 KN-m.
- The square building's maximum moment is roughly 16.41 KN-m.
- The circular building's maximum moment is roughly 22.12 KN-m.

Percentage variation for Storey Displacement in the X direction after using Mivan for all models square geometry is less than that for rectangular and circular geometry with and without Mivan. The variation is found to be 5–9% for square to rectangular, and on further variation, for rectangular to circular, it is found to be 20–30%. Percentage variation for Storey Displacement for the Y direction after using Mivan for all models square geometry is less than that for rectangular and circular geometry with and without Mivan. The variation is found to be 5–9% for square to rectangular, and on further variation, for rectangular to circular, it is found to be 25–30%.

KEYWORDS

- nonlinear dynamic analysis
- ETABS
- varying geometry

REFERENCES

1. Balaji, U. A.; Selvarasan, M. E. B. Design and Analysis of Multi-storeyed Building Under Static and Dynamic Loading Conditions Using ETABS. *Int. J. Tech. Res. Appl.* **2016,** *4* (4) 1–5.
2. Lavanya, C. V.; Pailey, E. P.; Md. Sabreen, M. Analysis and Design of G+4 Residential Building Using Etabs. *Int. J. Civil Eng. Technol.* **2017,** *8* (4), 1845–1850.
3. Sallal, A. K. Design and Analysis Ten Storied Building Using ETABS Software-2016. *Int. J. Res. Adv. Eng. Technol.* **2018,** *4* (2); 21–27.
4. Guleria, A. Structural Analysis of a Multistoried Building Using ETABS for Different Plan Configuration. *IJERT Int. J. Res. Eng. Technol.* **2014,** *3* (5).
5. Anagha, M.; Anoop, P. P. Earthquake Response of Different Shapes of Tall Vertically Irregular Mivan Wall Building. *IJERT Int. J. Res. Eng. Technol.* **2016,** *5* (8).

CHAPTER 13

Application of an Average Response Spectrum for Analysis of Structures

PRAVEEN ASHOK PATIL, SHRADUL JOSHI, and RAHUL JOSHI

Department of Civil Engineering, Vishwakarama Institute of Information Technology, Pune, India

ABSTRACT

An earthquake is a natural calamity responsible for the shaking of the ground due to which both living and nonliving losses occur. It is the responsibility of the structural designer to analyze the structure for earthquake loads and design the structural configuration, which will ensure minimum damage during an event of earthquake. The response spectrum method is one of the most commonly used seismic analysis methods. This method involves the calculation of peak responses such as displacement, velocity, acceleration, and member forces in each mode of vibration using smooth design spectra, which is the average of several earthquake time histories. In this paper, an attempt is made to obtain average response spectra for the northeast India region from the recent time history data of seven sites in the Assam region. The response spectra are generated using Seismo-signal software. A ten-storey RCC building is modeled in SAP 2000, and a comparative study is made for this structure considering average response spectra and response spectra defined in IS 1893 2016.

13.1 INTRODUCTION

Generally, structures are designed as per the response spectra given in IS 1893 2016.[2] Many earthquakes have occurred in the last five decades in the

---

Smart Innovations and Technological Advancements in Civil and Mechanical Engineering.
Satish Chinchanikar, Ashok Mache, Shardul Joshi, & Preeti Kulkarni (Eds.)
© 2025 Apple Academic Press, Inc. Co-publis hed with CRC Press (Taylor & Francis)

northeast India region. Northeast India is one of the most highly seismically active regions in the world with more than seven earthquakes on an average per year of magnitude 5.0 and above. Due to the frequent occurrence of earthquakes, it is necessary to study site-specific response spectra for the northeast India region. Response spectrum is a linear dynamic seismic analysis method. In this method, the structure is idealized as a series of infinite single-degree freedom structures. The natural period and mass participation factor of each single degree of freedom of the structure are evaluated from the modal analysis, and from the natural period, we can evaluate the spectral acceleration of lumped masses. The response of each single degree of freedom of the structure is added by modal combination methods such as the square root of sum of squares, the complete quadratic combination method, the absolute sum method, etc. In this paper, the response of structure is evaluated for the average response spectra and the response spectra given in the IS code method. Comparison is made by considering storey drift, displacements, and base reactions.

For the average response spectrum, a significant amount of work has been carried out. Boominathan et al.[7] have performed a site-specific seismic study considering the local site effects for the proposed power plant site near Samalkot town located in the Godavari Rift basin in Peninsular India. The deterministic seismic hazard analysis performed using three attenuation relationships identified the Vasishta—Godavari fault located at 50 km from the site with a magnitude Mw of 5.0 as the controlling earthquake source. Iyengar and Raghukanth[6] have studied the attenuation of strong ground motion in Peninsular India. Sil and Sitharaman[9] have developed a site-specific design response spectrum for the capital city of Agartala, Tripura. Sitharam and Sil[8] have a comprehensive seismic hazard assessment of Tripura and Mizoram states.

13.1.1 COSMOS EARTHQUAKE DATA

The COSMOS Strong-Motion Virtual Data Center, or VDC, is affiliated with COSMOS, the Consortium for Strong-Motion Observation Systems, a consortium of government agencies, private organizations, universities, and private individuals who have a common interest in promoting earthquake safety and education. The data are available for download as text files containing raw (digitized) acceleration recordings, processed acceleration, velocity, displacement, and Fourier and response spectra.

Application of an Average Response Spectrum for Analysis 211

The following figure shows the main page of the Cosmos earthquake data. Time histories for the following locations were obtained from the Cosmos Earthquake Data Centre. The earthquake data from the North Indian region is considered for the analysis. While performing the seismic analysis, the building is assumed to be in zone V.

TABLE 13.1 Earthquake Time History Data.

Site 1	Baithlongo site Assam
Site 2	Bomungao site Assam
Site 3	Haflong site Assam
Site 4	Hojai site Assam
Site 5	Kathakal site Assam
Site 6	Umrangso site Assam
Site 7	Shillong site Assam

13.1.2 SEISMO-SIGNAL SOFTWARE

From the Cosmos data bank, the time history data obtained is converted into the response spectrum using the software named Seismo-signal software. Seismo signal constitutes an easy and efficient way to process strong motion data, featuring a user-friendly interface and being capable of deriving a number of strong motion parameters often required by engineers, seismologists, and earthquake engineers.

Seismo signal calculates elastic and constant-ductility inelastic response spectra, Fourier and power spectra, Arias (Ia) and characteristic (Ic) intensities, Cumulative Absolute Velocity (CAV) and Specific Energy Density (SED), Root-mean-square (RMS) of acceleration, velocity, and displacement, number of effective cycles, and average spectral acceleration (Sa, avg).

Figure 13.2 indicates the screenshot of the Seismo-signal software for a particular earthquake data.

13.1.3 FORMATION OF RESPONSE SPECTRUM FOR AN EARTHQUAKE DATA

From the time history data of the recent earthquake, the response spectrum graph is generated as shown in the figure and by combining the graphs, the average of the seven sites is calculated as shown in Figure 13.3.

FIGURE 13.1 COSMOS data bank.

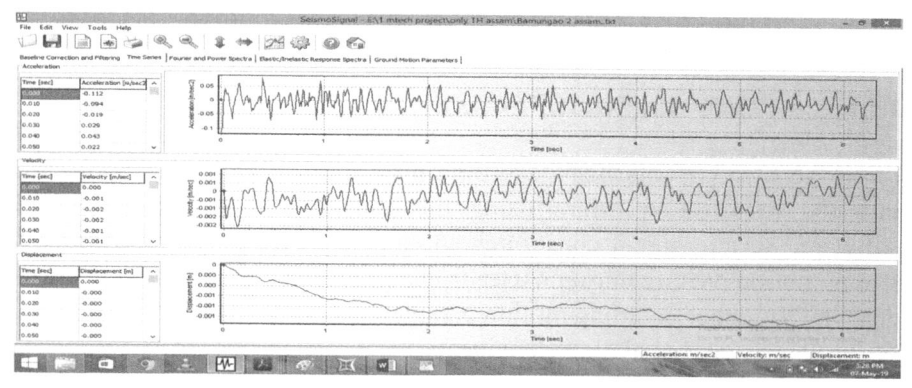

FIGURE 13.2 Seismo-signal software data.

The average response spectrum is obtained from the seven different time-history earthquake datasets. The Seismo-signal software has been used to obtain the response spectrum for each earthquake time history, and the average response spectra are then arrived at by taking the geometric mean.

13.2 NUMERICAL MODEL USING SAP 2000

13.2.1 GENERAL

The SAP software is used for modeling and analyzing the structure. The elevation and 3 D view are shown in Figure 13.5.

Application of an Average Response Spectrum for Analysis 213

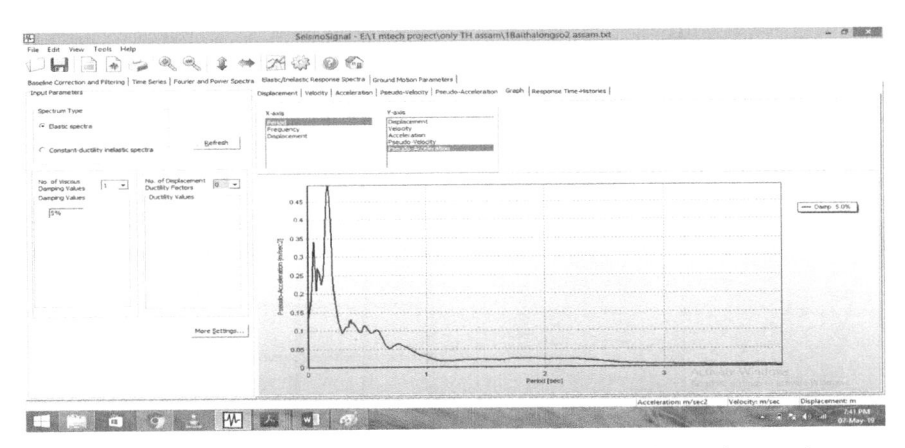

FIGURE 13.3 Response spectrum graph obtained from acceleration vs time graph.

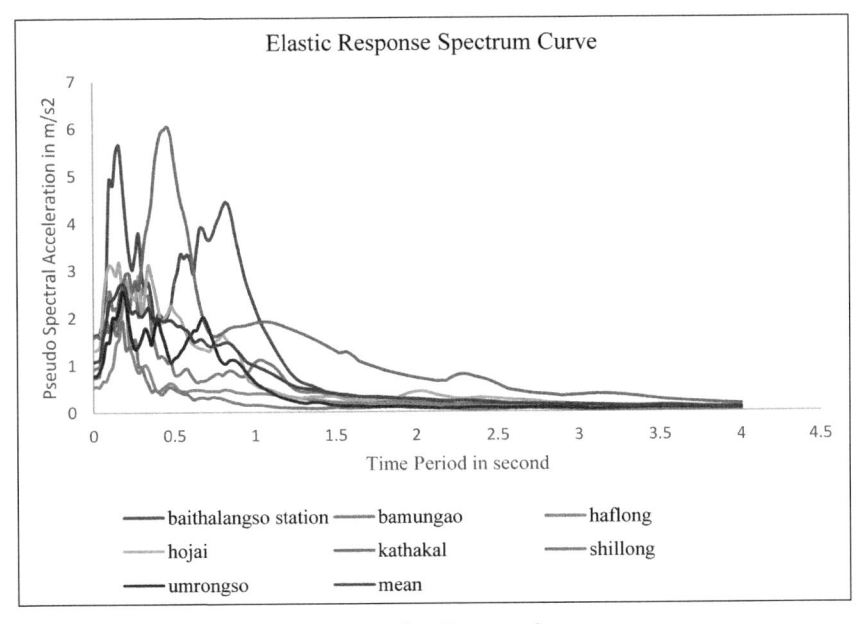

FIGURE 13.4 Average response spectra for all seven sites.

13.2.2 PARAMETERS USED IN SAP 2000 SOFTWARE

A ten-storey R.C. building is modeled in SAP 2000 software. The details of the building are presented in Table 13.2 below. The building is 30 m in height, symmetric in plan, and hence is free from torsion mode of vibration. The building is assumed to be located in seismic zone V.

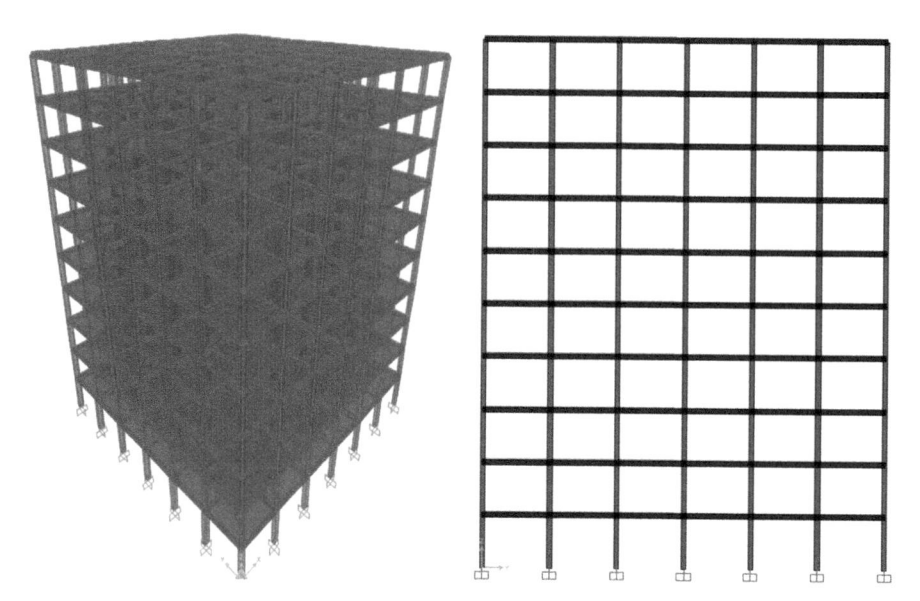

FIGURE 13.5 SAP model of a ten-storey building.

TABLE 13.2 Details of the Building.

Floor-to-floor height = 3.0 m.	The shape of the building = regular
Base-to-ground floor height = 5 m	S.B.C. of soil considered =180 kN/sq. m
Size of building = 24.2 × 21.2 m	Zone factor, Z = 0.36
Overall height of the building =30 m	Maximum displacement of the building: 0.04 × 33 = 1.32 m

13.3 RESULTS AND DISCUSSION

The analysis has been carried out using the response spectrum recommended by IS 1893 (Part1): 2016[2] and the average response spectrum obtained from the above-mentioned exercise.

The type of soil at these sites is not accounted for while averaging out the response spectra.

13.3.1 COMPARISON OF THE BASE SHEAR ALONG THE X AND Y AXIS

The comparison between the base shear forces obtained from the analysis is presented in Table 13.3. The base shear force computed from the average

Application of an Average Response Spectrum for Analysis 215

response spectrum is found to be greater than that obtained from the IS code-recommended response spectrum.

TABLE 13.3 Comparison of Base Shear.

	Base Shear along the X direction (KN)	Base Shear along the Y direction (KN)
IS CODE – response spectrum	1342	1184
AVERAGE – response spectrum	1628	1417

13.3.2 DISPLACEMENT ALONG X DIRECTION

Figure 13.6 depicts the displacement of floors in the building subjected to earthquake excitation along the X direction. The floor displacements are computed from the IS-recommended response spectrum as well as the average response spectrum obtained above. The displacements are computed on the higher side for the analysis performed using the IS-recommended response spectrum.

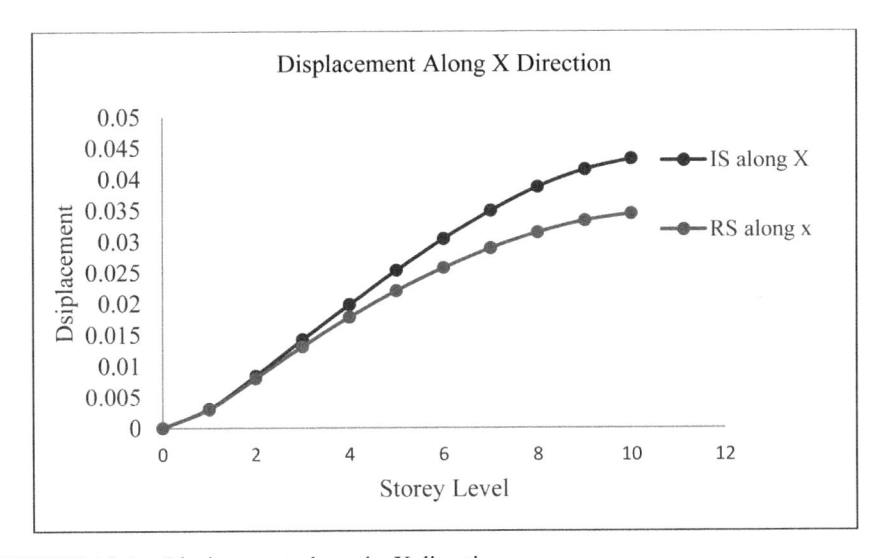

FIGURE 13.6 Displacement along the X direction.

13.3.3 DISPLACEMENT ALONG Y DIRECTION

Figure 13.7 depicts the displacement of floors in the building subjected to earthquake excitation along the Y direction. The floor displacements are computed from the IS-recommended response spectrum as well as the average response spectrum obtained as above. The displacements are computed on the higher side for the analysis performed using the IS-recommended response spectrum.

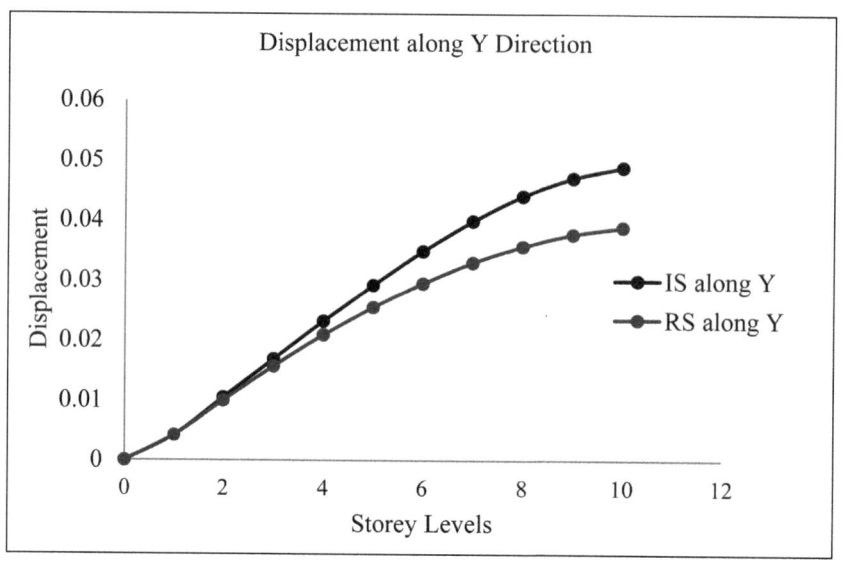

FIGURE 13.7 Displacement along the y direction.

13.3.4 DRIFT ALONG X DIRECTION IN ALL 3 ANALYSES

The drift of the building is also calculated from the average response spectrum and the IS-recommended response spectrum. The results are compared and shown in Figure 13.8.

13.4 RESULTS

Table 13.4 compares the results obtained through the two different analyses. It is observed that the base shear computed from the average response spectra is around 20% higher than that obtained from the standard response spectra. But in the case of floor displacements, the average response spectra resulted in fewer displacements as compared to the one obtained from the standard response spectra.

Drift in X Direction

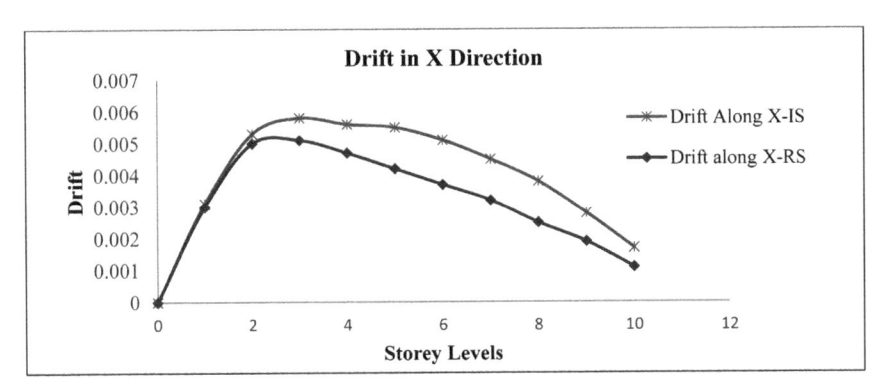

FIGURE 13.8 Drift along the X direction.

TABLE 13.4 Comparison between the Building Responses.

Sr. no.	Method	Base Shear (KN) Along X Dir	along Y	Maximum Displacement along X	along Y
1	IS CODE – response spectra	1342	1184	43.2 mm	48.9 mm
2	AVERAGE – response spectra	1628	1417	34.4 mm	38.9 mm
4	Difference in IS code and average response spectra	21.31%	19.42%	25.58%	25.70%

13.5 CONCLUSIONS

The present study aims at obtaining the building response from the average response spectra arrived at by taking the geometric mean of the seven different earthquake time history data. These building responses are compared with those obtained from the standard response spectra recommended by IS 1893 (part 1): 2016.[2] Based on the results obtained, the following conclusions are drawn:

1. Greater base shear is obtained for average response spectra in comparison with that obtained from IS code response spectra.
2. Greater floor displacements are obtained for IS code response spectra in comparison with those obtained from average response spectra.
3. To arrive at realistic conclusions, it is essential to obtain the average response spectra using the maximum number of earthquake time history data points.

KEYWORDS

- time history
- response spectrum
- push-over analysis
- IS 1893 2016
- SAP 2000
- Seismo-signal software

REFERENCES

1. Paz, M. *Structural Dynamics*, 2nd ed.; CBS Publishers and Distributors.
2. IS1893-2016, *Indian Standard Criteria for Earthquake Resistant Design of Structure*; Bureau of Indian Standards, 5th revision, 1979.
3. IS 456 Indian Standard for Plain and Reinforced Concrete. In *Code of Practice*; Bureau of Indian Standards: New Delhi; 2000.
4. Chopra, A. K. Dynamic of Structures. In *Theory And Application to Earthquake Engineering*, 2nd ed.
5. The Cosmos Virtual Data Center for thr Time History Data [Online]. https//strongmotion-center.org/COSMOSConverter.htm
6. Iyengar, R. N.; Raghukanth, S. T. G. Attenuation of Strong Ground Motion in Peninsular India. *Seismol. Res. Lett.* **2004,** *75* (1), 530–540.
7. Boominathan, A.; Vikshalakshie, M. G.; Subramanian, R. M. In *Site Specific Seismic Study for a Power Plant Site at Samalkot, Godavari Rift Basin in Peninsular India*, 12th World Conference on earthquake Engineering, 2012.
8. Sitharaman, T. G.; Sil, A. Comprehensive Seismic Hazard Assessment of Tripura and Mizoram states. *J. Earth Syst. Sci.* **2014,** *123,* 837–857.
9. Sil, A.; Sitharaman, T. G. Site Specific Design Response Spectrum Proposed for the Capital City of Agartala, Tripura. *Geomat. Nat. Hazards Risk* **2016,** *7* (5), 1610–1630.
10. Wayan, S, I.; Det, K. In Site-Specific Response Analysis (SSRA) and Pairs Of Ground-Motions Time-History Generation of a Site in Jakarta, *4th International Conference on Earthquake Engineering & Disaster Mitigation (ICEEDM 2019)*, paper no. 3009, 2019.

CHAPTER 14

Fluid Dynamics Analysis of Liquid Sloshing in a Rectangular Container under Lateral Excitation

SAURABH PATIL[1] and SHARDUL JOSHI[2]

[1]Student of Mechanical Engineering, Vishwakarma Institute of Information Technology, Pune, India

[2]Department of Civil Engineering, Vishwakarma Institute of Information Technology, Pune, India

ABSTRACT

Sloshing is a common phenomenon observed in liquid containers with a free surface under external excitation. The efforts to describe sloshing behavior using the analytical method are limited due to its highly nonlinear nature. In past studies, numerical methods have proven to provide accurate solutions to sloshing problems. This study provides fundamental information for Computational Fluid Dynamics analysis of sloshing phenomena using ready-to-use CFD codes. In this paper, a two-phase 2D rectangular sloshing model under harmonic excitation is studied using ANSYS FLUENT. The fluid motion is solved using the FVM method, and the free surface is tracked using the VOF method. The fluid model was excited at its fundamental frequency to observe sloshing behavior at resonant conditions. The free-surface elevation and hydrodynamic pressure values are then compared with those of the linear analytical solution.

14.1 INTRODUCTION

Sloshing is a common phenomenon observed in liquid containers with a free surface under external excitation. These free-surface oscillations are generally of low frequency and high amplitude. However, due to the larger mass of water and the low stiffness of container walls, small disturbances can result in large wave impacts on the walls. Excitation at a near-natural frequency of the fluid motion can lead to violent surface waves that can cause potential damage to the tank walls. On production conveyors, high sloshing heights can cause loss of fluid by spilling over the container. In the case of thin plastic walls, the high-energy fluid motion can force to topple the container. Thus, understanding the sloshing behavior is critical for determining the mechanical design and operational parameters of fluid containers under motion.

Sloshing is a highly nonlinear phenomenon. Hence, the early analytical formulations treated fluids as inviscid, irrotational, and incompressible. Housner used a lumped parameter approach to model slosh behavior, assuming rigid walls and a smaller wave height.[1,2] The theory divides the hydrodynamic effects into impulsive (effects due to motion) and convective (effects due to liquid oscillation). The theory can determine sloshing pressure forces due to each component (i.e., impulsive and convective), but it fails to accurately predict the sloshing wave height. Faltinsen derived a linear analytical solution for liquid sloshing in a horizontally excited 2-D rectangular tank,which has been widely used in the validation of numerical models.[3] It describes the fluid motion using the Laplace Equation based on the potential flow theory. Later, many presented analytical solutions based upon a similar theory with additional correcting functions to approximate analytical solutions for sloshing modes with higher h/a ratios and viscous damping.[4]

In the past couple of decades, the development of numerical solution algorithms has made it possible to solve the highly nonlinear sloshing models with higher accuracy. Since the analytical methods are restricted to describing sloshing for a few defined motions, the numerical method can analyze sloshing behavior for varied velocity or acceleration profiles. This has helped in understanding sloshing problems and solving real physical problems. In numerical methods, fluid motion inside the container has been represented with either Laplace, Euler, wave, or Navier-Stokes equations, which have been solved by employing the boundary element method (BEM), finite difference method (FDM), or finite-element method (FEM).[5] In this paper, a two-phase 2D rectangular sloshing model under harmonic excitation is studied using ANSYS FLUENT. The general finite volume method with

Fluid Dynamics Analysis of Liquid Sloshing in a Rectangular Container   221

cell-centered formulation is used to solve the sloshing model, and the free-surface motion is defined using the VOF method.

14.2 ANALYTICAL FORMULATION

The analytical formulation used to compare numerical studies is based on Faltinsen's[6] linear analytical model. It describes the sloshing for a 2D rectangular rigid tank system with a free-surface height of 'h' and a tank length of '2a' subjected to forced harmonic excitation in the horizontal direction. A Cartesian coordinate system (x, y) with the origin at the center of the free surface is established. Hence, the harmonic displacement function of the tank can be written as,

$$x = b \sin(\omega t) \tag{14.1}$$

where 'b' is the displacement amplitude and ω is the angular frequency of the excitation, assuming fluid to be inviscid and irrotational and neglecting the effect of surface tension. From the Laplace equation, the fluid inside the tank can be defined as $\nabla^2 \phi = 0$ where ϕ is the velocity potential function of the fluid. Solving the Laplace equation under the velocity and free-surface boundary conditions, the velocity potential of fluid flow inside the tank is given by the following expression:[4]

$$\phi(x,y,t) = \sum_{n=0}^{\infty} \sin\left\{\frac{(2n+1)\pi}{2a} \cdot x\right\} \cosh\left\{\frac{(2n+1)\pi}{2a} \cdot (y+h)\right\} \tag{14.2}$$
$$\left(A_n \cos(\omega_n t) + C_n \cos(\omega t)\right) - \omega A x \cos(\omega t)$$

where

$$A_n = -C_n - \frac{K_n}{\omega}, \quad C_n = \frac{\omega K_n}{\omega_n^2 - \omega^2},$$

$$K_n = \frac{\omega A}{\cosh\left\{\frac{(2n+1)\pi}{2a} h\right\}} \cdot \frac{2}{a}\left[\frac{2a}{(2n+1)\pi}\right]^2 (-1)^n \tag{14.3}$$

and ω_n is the natural frequency with sloshing mode 'n' given by the expression,

$$\omega_n = g \frac{(2n+1)\pi}{2a} \tanh\left\{\frac{(2n+1)\pi}{2a} h\right\}, \tag{14.4}$$

Hence, the first three natural frequencies for the model under study can be calculated to $\omega_0 = 6.0578$, $\omega_1 = 12.348$, and $\omega_2 = 16.439$ $rad\ s^{-1}$.

The free-surface displacement, η, can be further derived from the velocity potential function ϕ as follows:

$$\eta(x,t) = \frac{1}{g}\frac{d\phi}{dt} \qquad (14.5)$$

14.3 COMPUTATIONAL FLUID DYNAMICS ANALYSIS OF THE SLOSHING

For the numerical analysis model similar to the Liu and Lin[6] study, a 2D rectangular tank with a length of 0.57 m, height of 0.30 m, and water filled up to a height of 0.15 m (refer to Figure 14.1) is considered. In this study, sloshing analysis was performed using ANSYS FLUENT.[7] The analysis domain is discretized using linear quad elements with a mean mesh size of 5 mm. The domain is split into two equal parts by the line $y = 0$ (free surface). The upper and lower discretized domains are assigned air and liquid (water) properties. The fluid properties of the two phases are mentioned in Table 14.1.

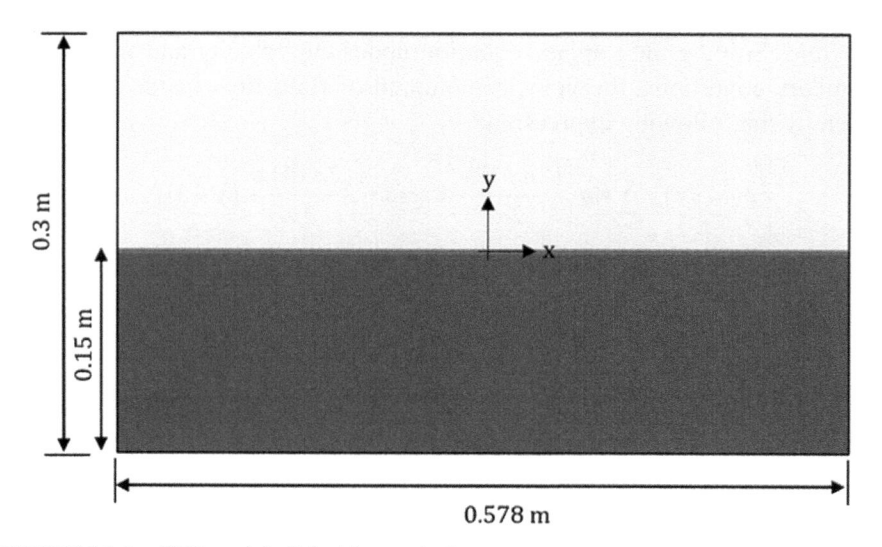

FIGURE 14.1 CFD model of sloshing analysis.

TABLE 14.1 Fluid Properties.

Property	Water	Air
μ (kg/ms)	100.3×10^{-5}	1.7894×10^{-5}
ρ (kg/m³)	998.2	1.225

The VOF method is used to track the free surface with the surface tension coefficient of the water–air interaction set to 0.0725 N/m. The standard $k \cdot \varepsilon$ model is used for predicting the sloshing motion. The tank walls are set to no-slip conditions, and the top edge is set equal to a gauge pressure of 0 Pa.

The Pressure-Implicit with Splitting of Operators (PISO) pressure–velocity coupling scheme is used as it provides faster convergence and is recommended for transient calculations.[8] The tank was set to total rest at $t = 0$ and excited to the first fundamental resonant frequency of $\omega_o = 6.0578\ rad\ s^{-1}$. This excitation is imparted by assigning the harmonic acceleration given by the expression $a(t) = 0.185 \sin(6.0578\ t)$ m/s² as shown in Figure 14.2.

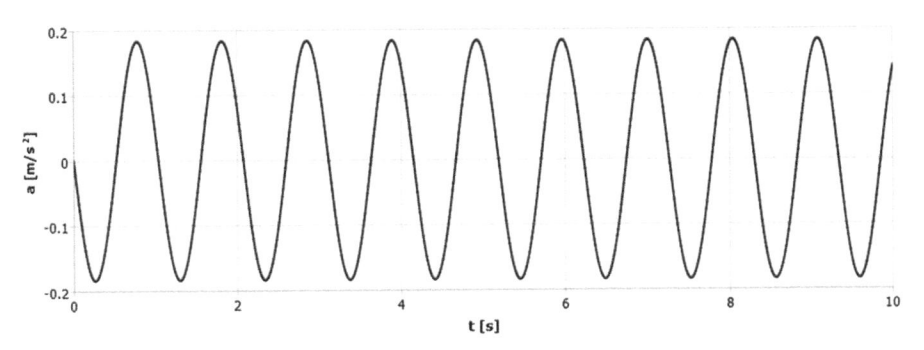

FIGURE 14.2 Acceleration input of the sloshing model.

14.4 RESULTS

The analysis is run for 10 s of physical time with free-surface elevation recorded at $x = -0.265$ and $x = 0$. The free-surface plot at both locations is shown in Figure 14.1. Since the model is under resonant frequency loading, the wave amplitude increases with time. High amplitude waves are observed on the left and right walls of the tank. A maximum elevation of 0.14 m is observed on the right wall at the end of the run. The fluid behavior at different time steps is shown in Figure 14.2. A similar increase is observed in the total pressure, reaching its maximum at the bottom of the tank. Though hydrostatic pressure values are higher in comparison to hydrodynamic pressure, a significant increase is observed in total pressure by the resonant waves.

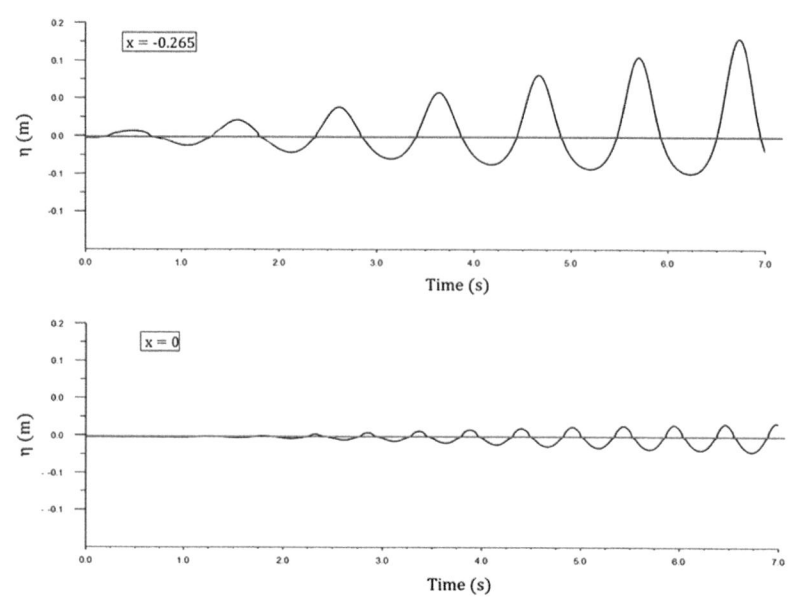

FIGURE 14.3 Sloshing wave elevation time history in the first fundamental mode $\omega_o = 6.0578$ $rad\ s^{-1}$.

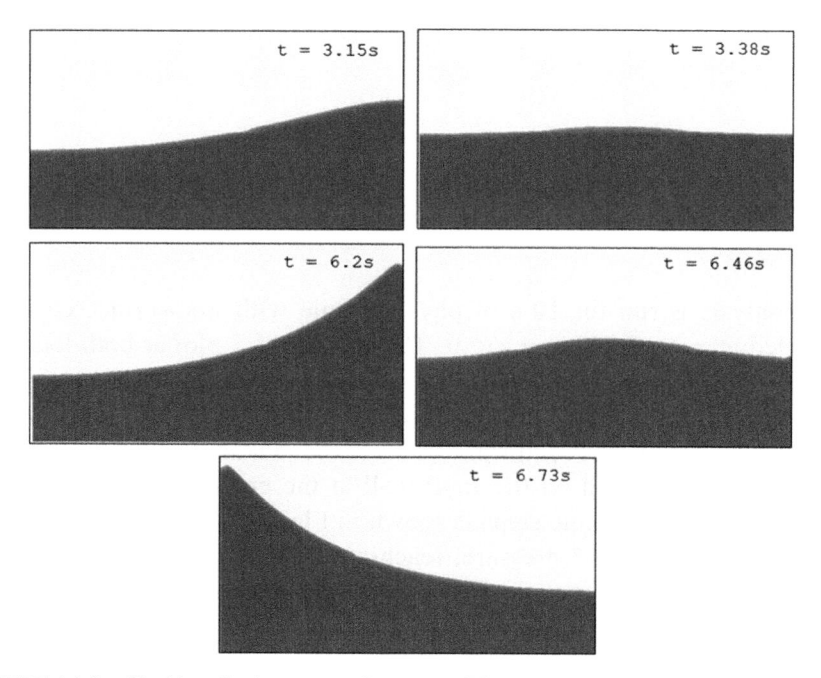

FIGURE 14.4 Sloshing displacement of water at different analysis times.

14.5 CONCLUSIONS

The fluid motion was solved using the FVM method, and the free surface was tracked using the VOF method. The fluid model was excited at its fundamental frequency to observe sloshing behavior at resonant conditions. From the numerical study of a tank model, it can be inferred that sloshing at resonant loading results in high-amplitude waves that increase in magnitude over time. The sloshing motion also creates high hydrodynamic pressure waves. Such high-energy waves can cause damage to the wall and top surfaces. Though this study does not extend more to the slosh dynamics and suppression methods, it provides fundamental information for Computational Fluid Dynamics analysis of sloshing phenomena.

KEYWORDS

- **sloshing**
- **computational fluid dynamics**
- **volume of fluid**
- **numerical analysis**
- **potential flow theory**
- **free surface**
- **wave height**

REFERENCES

1. Housner, G. W. Dynamic Pressures on Accelerated Fluid Containers. *Bull. Seismological Soc. Amer.* **1957,** *47* (1), 15–35.
2. Housner, G. W. The Dynamic Behavior of Water Tanks. *Bull. Seismological Soc. Amer.* **1963,** *53* (2), 381–387.
3. Starnes, J. Effect of a Circular Hole on the Buckling of Cylindrical Shells Loaded by Axial Compression. *AIAA J.* **1972,** *10*, 1466–1472.
4. Faltinsen, O. M. A Numerical Nonlinear Method of Sloshing in Tanks with Two-Dimensional Flow, *J. Ship Res.* **1978,** *22*, 193–202.
5. Ozdemir, Z.; Souli, M.; Fahjan, Y. M.; FSI Methods for Seismic Analysis of Sloshing Tank Problems, *Mécanique Indus.* **2010,** *11* (2), 133–147.
6. ANSYS Fluent, 2009. *User's Guide Release 12.0/12.1*; Ansys Inc.: USA
7. Liu, D.; Lin, P.; A Numerical Study of Three-Dimensional Liquid Sloshing in Tanks, *J. Comput. Phys.* **2008,** *227*, 3921–3939.

CHAPTER 15

Study of Mechanical and Micro Structural Properties of Fly Ash and GGBS-Based Geopolymer Concrete

SHIVDATTA B. BHOSALE[1], S. G. JOSHI[1], R. A. JOSHI[1], and J. P. WATVE[2]

[1]Department of Civil Engineering, Vishwakarma Institute of Information Technology, Pune, India

[2]Department of Mechanical Engineering, Vishwakarma Institute of Information Technology, Pune, India

ABSTRACT

This study investigates the mechanical and microstructural properties of alkali-activated fly ash/GGBS concrete with various proportions of fly ash to GGBS (0:100, 50:50, and 0:100) and the effect of alkaline activators on the compressive strength. Sodium hydroxide and sodium silicate with a modulus ratio of 1 were used as alkaline activators to alkali-activate various fly ash/GGBS ratios with an alkaline activator-to-binder ratio of 0.35. A compressive strength test with CTM and scanning electron microscopy (SEM) analysis were conducted. Test results reveal that both the fly ash/GGBS ratio and the composition of alkaline activators are significant factors influencing the mechanical and microstructural characterization of geopolymers.

15.1 INTRODUCTION

Cement is the most generally utilized manmade material on earth. When it is mixed with water and aggregates, it produces concrete that is used in the construction of everything from roads, buildings, dams, and bridges to all kinds of other infrastructures. But while cement has a major importance in the construction industry, it is also a major source of carbon dioxide emissions into the atmosphere. It is responsible for approximately 7% of the world's greenhouse gas emissions according to the International Energy Agency.[1] For every ton of cement production, one ton of CO_2 gets released into the environment. The manufacturing process of cement involves very high temperatures (1400–1500°C), the destruction of quarries to remove crude materials, and the emission of greenhouse gases like CO_2 and NOx.[2]

Easily-available industrial byproducts like fly ash and ground-granulated blast furnace slag have been adopted to satisfy these needs. To this end, it was assessed that the quantity of fly ash produced by the year 2010 was about 780 million tons,[3] providing a way to meet growing demand. The reused utilization of this powder material in development will ease the expense of transfer somewhere else and decrease the expense of solid assembling by and large.

The expression "geopolymer" is conventionally used to depict the amorphous to crystalline reaction products from the combination of alkali aluminosilicates with alkali hydroxide/alkali silicate solution. Geopolymeric composites are also commonly referred to as alkali-activated cement, geocement, low-temperature aluminosilicate glass, alkali-bonded ceramic, inorganic polymer concrete, and hydroceramic.[5] A geopolymer paste of fly ash/GGBS is used to bind the loose aggregates and other nonreacted materials together to form geopolymer concrete.[6] Geopolymer binders can provide comparable performance to conventional cementitious binders in various applications with the added advantage of significantly decreasing greenhouse gas emissions.[4]

15.2 LITERATURE REVIEW

Subhash V. Patankar,[7] in his paper, has proposed the guidelines for the structure of fly ash powder, which put together geopolymer concrete with respect to the premise of amount and fineness of fly slag, amount of water, and evaluation of fine total by keeping up water to geopolymer cover proportion of 0.35, answer for fly fiery debris proportion of 0.35, and sodium silicate to sodium hydroxide proportion of 1 with convergence of sodium hydroxide as 13 M. Warmth-relieving was done at 60°C for a span of 24 h and tried following 7 days after stove warming. Trial consequences of M20, M25, M30, M35, and M40 evaluations of geopolymer cement blends utilizing proposed techniques for blend configuration show promising aftereffects of functionality and compressive quality. Thus, these rules help in the plan of fly-fiery debris-based geopolymer cement of conventional and norms reviews as referenced in IS456: 2000. Vijaya Rangan[4] studied fly ash powder-based geopolymer solid plan procedures and the components that impact its crisp and solidified repairman properties. Testing was conducted on 100 × 200 mm chambers that had been both dry and steam-relieved in 60°C temperatures from 4 to 96 h, and the first pressure was tried at 21 days. It was expressed that more drawnout relieving times brought about more prominent compressive qualities. Also, the initial 24 h created a quick quality increase, while restoring past 24 h resulted in a slower quality addition for every unit time. It was additionally noticed that dry relieving brought about 15% more noteworthy compressive quality than steam restoring for a similar time frame. The total blend was composed of coarse SSD totals (7–20 mm) and fine sands. Low-calcium, class F fly fiery remains were added to the totals and blended for three minutes. A superplasticizer was added to the sodium hydroxide/sodium silicate activator to improve its functionality. The last testing reasoned that the crisp cement could be worked for as long as 120 min without influencing the by-and-large compressive quality of the example. Besides, the antacid/fly slag proportion for the fastener configuration was prescribed to be 0.30–0.45, and the sodium silicate/sodium hydroxide proportion (by mass) was proposed to be 2.5 for the best outcomes. Little shrinkage was seen in the geopolymer examples following a one year timeframe. The deliberate incentive for geopolymer was 100 miniaturized scale strains, when contrasted with 500–800 smaller scale strains for Portland bond concrete. One-year creep coefficients were determined at 0.6–0.7 for examples having compressive quality qualities between 40 MPa and 57 MPa. It was noticed that these qualities are just 50% of the prescribed coefficients for Portland bond concrete (per Australian Standard AS3600).

Khale et al.[2] have discussed the fundamentals of the geopolymerization process and the experimental evidence that helps in the development of geopolymerized concrete. It also finds source materials used to induce polymerization, chemical reactions within the matrix, concentration of alkaline solution, curing temperatures and period of curing, pH balance, and water/solid ratios. Curing temperatures and relieving periods are both useful for increasing unconfined compressive quality. Longer relieving occasions bring about expanded compressive qualities; in any case, periods past 48 h produce immaterial quality increments. The temperature guideline in the

2–5 h fix period is basic to potential quality improvement. It was expressed that restoring temperatures in the range of 30–90°C is required for a sufficient concoction response. By and large, the quality of the geopolymer can likewise be influenced by fixing the proportions and pH levels of the initiating arrangement. Tests infer that an expanded water-to-strong proportion diminishes quality, while high proportions of Si:Al or Na:Al increment the geopolymer mechanics. Experimental information has demonstrated that a pH level of the initiating glue more prominent than 13 (5–10 M alkaline solution) advances expanded responses and expanded quality.

Katz, A[8] has observed that reactivity of fly ash within the alkaline condition is used to develop polymerization methods. The activator used for testing was a strong sodium hydroxide (NaOH) solution of various concentrations and various water/fly ash ratios. The highest temperature of curing was 90°C, and the highest concentration level of utilized NaOH was 4 M.

Testing was done in three stages, which tended to the accompanying impacts: NaOH focus, water/fly fiery debris proportion, and restoring temperatures. Test examples comprised 25 mm³ molds submerged in 1–4 M NaOH solution showers. Compressive testing for all fixation examples was conducted on day seven and yielded results demonstrating an immediate connection between alkalinity and compressive quality: the 4 M NaOH solution gave pressure estimations of about 6 MPa (at 7 days), while the 1 M solution tests yielded 0.2 MPa.

Analysis using SEM was additionally looked into for all focus levels. No fly ash remains, responses were apparent in the 1 M solution; however, the higher concentrations (3 M NaOH) brought about surface drawing and needle-like precious stones. The 4 M arrangement was seen to have experienced microstructural change, bearing cubic crystalline developments unlike those found in the 3 M arrangement. It was presumed that the given lattice had turned out to be less permeable and that densification had happened at the higher fixations.

15.3 OBJECTIVES

The objective of this paper is to evaluate geopolymer production methodologies and assess their potential as viable construction materials for industrial use.
- Identification of factors influencing mechanical properties.
- Investigation of pozzolanic materials and their reactivity to various alkalines.

- Mineralogical and microstructural analysis of products.

15.4 METHODOLOGY

1. Conceptual understanding (literature review) of the chemical composition of Ordinary Portland Cement, Fly Ash, and Ground-Granulated Blast Furnace Slag.
2. Arrangement of required concrete materials, geopolymer materials, and alkaline activators.
3. Based on the outcomes and findings of the literature survey, a test matrix was developed aimed at ascertaining the performance of the geopolymer concrete mix.
4. Casting of conventional and geopolymer concrete mixtures.
5. Laboratory testing of the fresh properties (workability) and hardened properties (compressive strength) of produced conventional and geopolymer concrete.
6. Microstructural characterization of the alkali-activated binders using a scanning electron microscope.

15.5 EXPERIMENTAL WORK

15.5.1 ALKALINE ACTIVATORS

In the experiment, sodium-based alkaline activators are used. A single activator, either sodium hydroxide or sodium silicate, alone is not much effective as clearly seen from past investigations.[7] So, a combination of sodium hydroxide and sodium silicate solutions is used for the activation of fly ash-based geopolymer concrete. The concentration of sodium hydroxide was maintained at 7.5 M and 13 M, while the concentration of sodium silicate solution contains Na_2O of 16.37%, SiO_2 of 34.35%, and H_2O of 49.72% as alkaline solutions. The ratio of sodium silicate to sodium hydroxide by mass was maintained, which set cubes within 24 h after casting and gave good compressive strength results.[7]

As the alkaline activator solution-to-fly ash ratio increases, strength also increases. But the rate of gain of strength is not much significant beyond the solution-to-fly ash ratio of 0.35. Similarly, the mix was more viscous with the higher ratios and unit cost increased. So, in the present mix design method, the solution-to-fly ash ratio was maintained at 0.35.[8] Figure 15.1 shows the solutions prepared for performing the experiment.

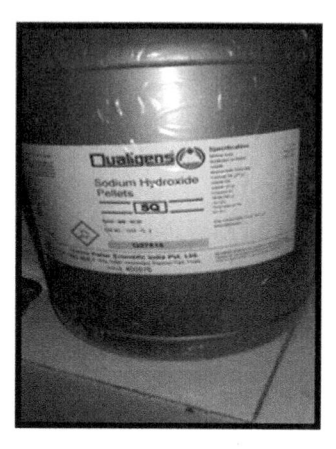

FIGURE 15.1 Solution of alkaline activators.

15.5.2 FLY ASH AND GGBS

Two materials are used as a substitute of cement in geopolymer concrete with variable proportions. Fly ash is obtained from thermal power plants (Mahagenco), Parli (Maharashtra), and GGBS is obtained from JSW Steel Ltd. (JSW Cement), Khar, Raigad (Maharashtra).

15.5.3 MIX DESIGN

To satisfy the mix design for M30 grade as per the IS10262:2009 code,[10] the materials were selected as coarse aggregates, fine aggregates confirming to IS383:1970 code,[11] Ordinary Portland Cement 53 grade confirming to IS12269, and cementitious material as stated in Section 15.3.2. Tables 15.1 and 15.2 indicate the proportion of ingredients and cementitious materials, respectively.

15.6 RESULTS AND DISCUSSION

15.6.1 MECHANICAL PROPERTIES

The average compressive strength is measured for each sample after 7 days. The concrete cubes are tested for compressive strength in accordance with

IS516:1959.[8] The compressive strength data is summarized in Tables 15.3–15.5. The testing measured the strength of three cubes for each sample. The concrete specimens are tested for compressive strength in a compression testing machine as shown in Figure 15.2.

TABLE 15.1 Material Proportion.

Sr. no.	Material	Unit	Quantity/m³
1	Cement/ Fly Ash/ GGBS	Kg	475
2	Fine Aggregate (crushed sand/ natural sand)	Kg	640.01
3	Coarse Aggregate	Kg	1188.59
4	Alkaline Activator	Lit	166.25
5	Water	Lit	12.86

TABLE 15.2 Cementitious Material Proportion with Alkaline Activators and Curing Temperature.

Sr. no.	Sample	Cement %	Fly ash %	GGBS %	Alkaline activators	Curing
1	1	100	0	0	-	Normal
2	2	0	100	0	7.5 M	24 hr at 60°C
3	3	0	50	50	7.5 M	24 hr at 60°C
4	4	0	0	100	13 M	24 hr at 60°C

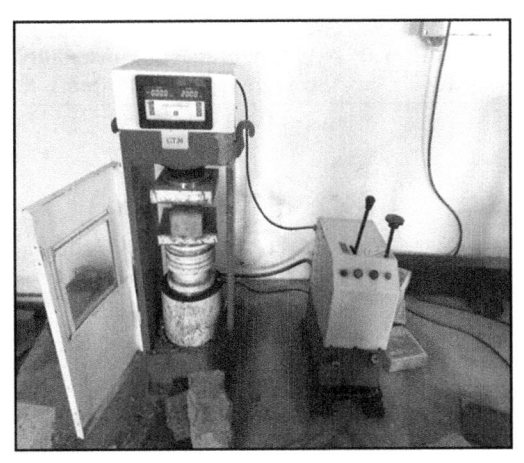

FIGURE 15.2 Compressive strength test.

TABLE 15.3 Results of Conventional Concrete.

Sr. no.	Sample 1	Compressive strength (MPa) at 7 days	Avg. compressive strength (MPa)	Workability (mm)
1	Cement + Crushed Sand	22.22		
		24.22	22.75	45
		21.82		
2	Cement + Natural Sand	17.50	18.7	55
		20.50		
		18.10		

TABLE 15.4 Results of Geopolymer Concrete with 100 % Fly Ash.

Sr. no.	Sample 2	Compressive strength (MPa) at 7 days after curing at 60°C	Avg. Compressive Strength (MPa)	Workability (mm)
1	Fly Ash + Crushed Sand	3.91	3.08	20
		2.67		
		2.67		
2	Fly Ash + Natural Sand	2.87	3.20	15
		3.22		
		3.51		

TABLE 15.5 Results of Geopolymer Concrete with 50% Fly Ash + 50% GGBS.

Sr. no.	Sample 3	Compressive strength (MPa) at 7 days after curing at 60°C	Avg. Compressive Strength (MPa)	Workability (mm)
1	Fly Ash + GGBS + Crushed Sand	21.82	19.94	25
		18.62		
		19.38		
2	Fly Ash + GGBS + Natural Sand	18.84	17.79	15
		15.87		
		18.67		

Figure 15.3 indicates the comparison of the compressive strength of the concrete specimen for different mix proportions of fly ash/GGBS with the conventional concrete specimen.

TABLE 15.6 Results of Geopolymer Concrete with 100% GGBS.

Sr. no.	Sample 4	Compressive strength (MPa) at 7 days after curing at 60°C	Avg. Compressive Strength (MPa)	Workability (mm)
1	GGBS + Crushed Sand	18.13	17.45	25
		17.08		
		17.13		
2	GGBS + Natural Sand	14.68	15.43	20
		16.24		
		15.38		

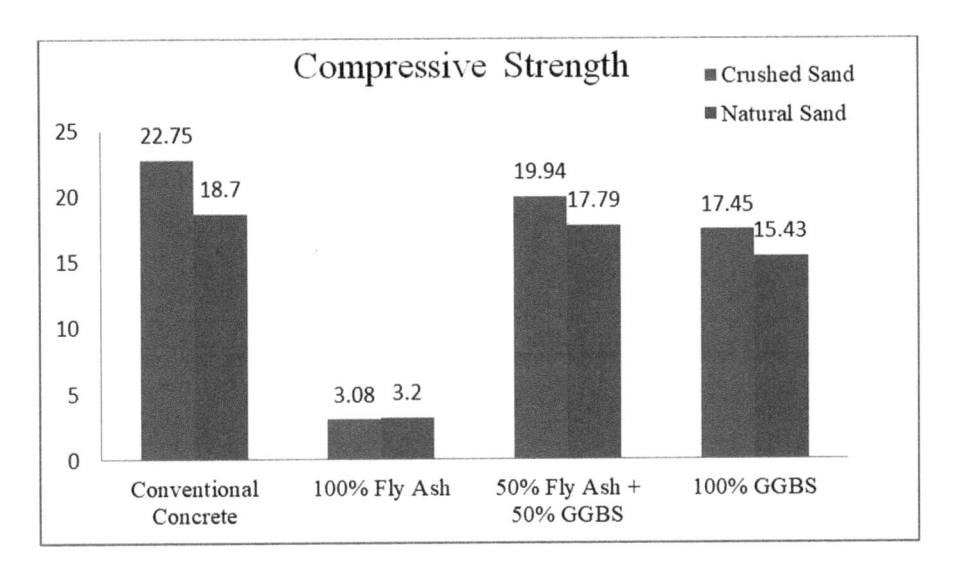

FIGURE 15.3 Compressive strength of conventional concrete and geopolymer concrete.

15.6.2 MICROSTRUCTURAL PROPERTIES

This section of the chapter discusses the microstructural characteristics of the hardened geopolymer cement concrete and identifies influencing factors contributing to improved mechanical characterization. Detailed analysis using Scanning Electron Microscopy (SEM) is used to quantify minerals and measure their inherent properties at a microlevel. It is a qualitative analysis at the microstructural level.

15.6.3 FLY ASH + CRUSHED SAND, FLY ASH + NATURAL SAND

As we did not get the promising results of compressive strengths from these two mix designs, we did not go for their microstructural analysis. From the compressive strength test, we found that the specimens of these mixes were too brittle and did not have proper bonding between cement paste and aggregate. Figures 15.4–15.7 show the images obtained under a scanning electronic microscope.

15.6.4 FLY ASH + GGBS + CRUSHED SAND

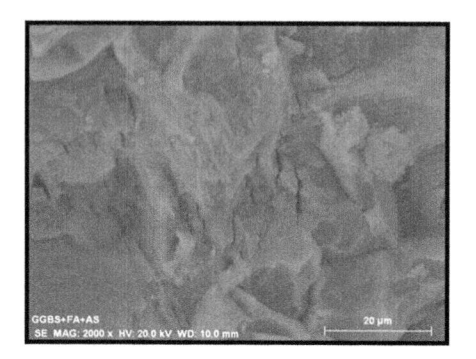

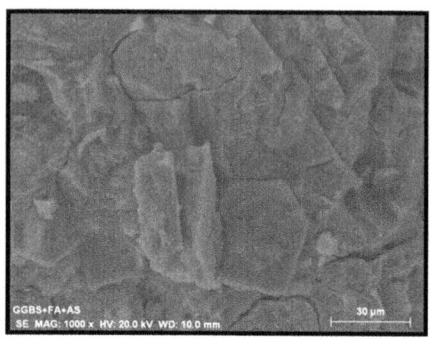

FIGURE 15.4 SEM analysis of fly ash + GGBS + crushed sand.

There was a proper bonding between cement paste and aggregates as compared to other specimen mixes. There were very small cracks of smaller than 20 micron even after compression testing. The structure was very impervious. Overall, the compressive strength was nearly equal to the target strength.

15.6.5 FLY ASH + GGBS + NATURAL SAND

There was no proper bonding between cement paste and natural aggregates as compared to the specimen mix of Fly Ash + GGBS + Crushed Sand. Therefore, the compressive strength was less than that of the mix with crushed sand. The cracks were not developed, and hence, there was not much decrease in the compressive strength. The structure was impervious.

FIGURE 15.5 SEM analysis of fly ash + GGBS + natural sand.

15.6.6 GGBS + CRUSHED SAND

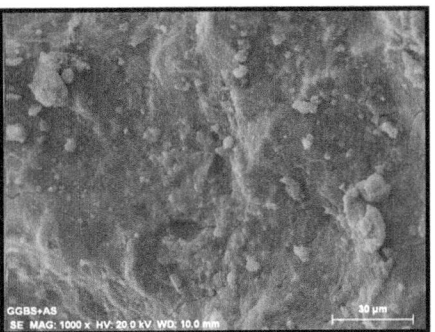

FIGURE 15.6 SEM analysis of GGBS + crushed sand.

There was a proper bonding between cement paste and aggregates, but minor cracks of 20–30 microns were observed even after increasing the concentration of the alkaline solution. But due to proper bonding, a higher concentration of alkaline solution, and fewer cracks, the compressive strength was not much less than the target strength.

15.6.7 GGBS + NATURAL SAND

There was no proper bonding between cement paste and natural aggregates. Also, the cracks were more prominent in the specimen. Hence, the compressive strength was lower than that of other specimen mixes.

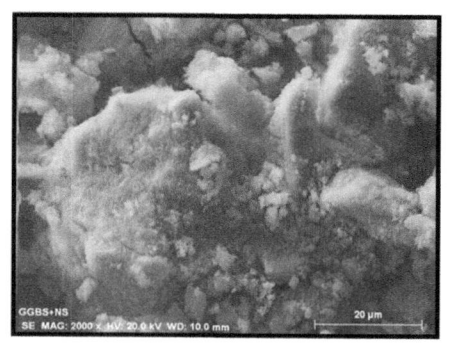

FIGURE 15.7 SEM analysis of GGBS + natural sand.

15.7 CONCLUSIONS

Based on the experimental results, the following conclusions are drawn:

1. Different proportions of FA and GGBS have been tested by maintaining the constant ratios of solution-to-fly ash and sodium silicate-to-sodium hydroxide. The conclusions on the compressive strength and workability of geopolymer concrete with FA and GGBS are as follows:

    a) Replacement of cement with 100% FA gives very little compressive strength and workability as compared to the expected results.
    b) Replacement of cement with 50% FA and 50% GGBS gives promising results in compressive strength but not in workability.
    c) There is no significant difference in the compressive strengths of geopolymer concrete with 50% FA and 50% GGBS and conventional concrete.
    d) Replacement of cement with 100% GGBS gives promising results in compressive strength but not in case of workability.

2. Geopolymer concrete promotes environmentally friendly concrete production by replacing carbon-producing cement by using byproducts from two different industries, thus reducing the question of their dumping.

KEYWORDS

- alkaline activator
- compressive strength
- fly ash
- geopolymer
- GGBS (ground-granulated blast furnace slag)
- SEM analysis

REFERENCES

1. Carboncure Homepage [Online]. https://www.carboncure.com/; Fernandez-Jimenez, A. M.; Palomo, A.; Lapez-Hombrados, C. Engineering Properties of Alkali—Activated Fly Ash Concrete. *ACI Mater. J.* http://findarticles.com/p/articles/mi_qa5360/is_200603/ai_n21395768 (accessed Mar/Apr 2006).
2. Khale, D.; Chaudhary, R.; Mechanism of Geopolymerization and Factors Influencing Its Development: A Review. *J. Mater. Sci.* **2007,** *42,* 729–746.
3. Duxson, P.; Fernandez-Jimenez, A.; Provis, J. L.; Lukey, G. C.; Palomo, A.; van Deventer, J. S. J. Geopolymer Technology: The Current State of the Art. *J. Mater. Sci.* DOI: 10.1007/s10853-006-0637-z
4. Vijaya Rangan, B. Fly Ash-Based Geopolymer Concrete [Online]. http://www.your-building.org/display/yb/Fly+Ash-Based+Geopolymer+Concrete
5. Skvara, F.; Dolezal, J.; Svoboda, P.; Kopecky, L.; Pawlasova, S.; Lucuk, M.; Dvoracek, K.; Beksa, M.; Myskova, L.; Sulc, R. Concrete Based on Fly Ash Geopolymers. Research Project CEZ:MSM 6046137302: Preparation and Research of Functional Materials and Material Technologies using Microand Nanoscopic Methods and Czech Science Foundation Grant 103/05/2314 Mechanical and Engineering Properties of Geopolymer Materials Based on Alkali-Activated Ashes
6. Patankar, S. V.; Jamkar, S. S.; Ghugal, Y. M. Effect of Water-to-Geopolymer Binder Ratio on the Production of Fly Ash Based Geopolymer Concrete. *Int. J. Adv. Technol. Civ. Eng.* **2013,** *2* (1):79–83.
7. Patankar, S. V.; Jamkar, S. S.; Ghugal, Y. M. Effect of Grading of Fine Aggregate on Flow and Compressive Strength of Geopolymer Concrete. In *UKEIRI Concrete Congress-Innovations in Concrete*, 2014; pp 1163–1172.
8. Katz, A. Microscopic Study of Alkali-Activated Fly Ash. *Cement Concrete Res.* **1998,** *28* (2), 197–208.
9. IS 456-2000. *Plain and Reinforces Concrete-Code of Practice*, 4th revision; Bureau of Indian Standards: New Delhi.

10. IS 10262-1970. *Recommended Guidelines for Concrete Mix Design on for Coarse and Fine Aggregates from Natural Sources for Concrete*; Bureau of Indian Standards: New Delhi.
11. IS 383-1970. *Specification for Coarse and Fine Aggregates from Natural Sources for Concrete*; Bureau of Indian Standards: New Delhi.

PART IV
Environmental and Water Resources Engineering

CHAPTER 16

Application of Solar Dryer for Drying of Agricultural Products

TANVI SHAH[1] and KRISHNAKEDAR GUMASTE[2]

[1]Department of Civil-Environmental Engineering, Walchand College of Engineering, Sangli, India

[2]Faculty of Civil-Environmental Engineering, Walchand College of Engineering, Sangli, India

ABSTRACT

Due to higher prices and shortages of fossil fuels and to reduce the fuel consumption used in the drying process, more importance is given to solar energy sources as it is freely available. For these purposes, several attempts were made in developing an indirect type solar dryer to dry agricultural products. In this study, an indirect solar dryer was constructed with natural circulation, using waste plastic bottles as solar air collector. The experiments were conducted on turmeric, onions, and grapes. Moisture content of grapes was reduced to 85.35% in indirect solar drying (IDSD), 81.95% in direct solar dryer (DSD) and 79.10% in open sun drying (OSD). While moisture content of turmeric was reduced to 92.10% in IDSD, 90.60% in DSD and 85.40% in OSD. Moreover, moisture content of onions was reduced to 92.90% in IDSD, 84.30% in DSD and 77.60% in OSD. The efficiency of solar collector was 23% and that of DSD was 27%.

16.1 INTRODUCTION

According to Mathias Aarre Maehlum, Energy Informative, the type of energy that reaches the surface of the earth, twice the amount of all nonrenewable resources, is solar energy. This immense solar energy can be used as an alternative for fossil fuels that are depleting day by day.

One of the problems that developing countries are facing nowadays is food security. The developing countries are finding it difficult to provide food to all their citizens due to the imbalance between the increasing population and the amount of food that is in eatable condition by the time it reaches the consumer. Every green and grown crop has its own shelf life. They start deteriorating after a certain time. The spoilage of food might also occur during its transportation or due to the interference of microorganisms or chemical changes that start occurring within the crop after a certain period of time due to natural processes such as enzyme actions. The quality of food is also deteriorating because of poor processing techniques and a shortage in the storage facilities in developing countries. For example, according to the government of Maharashtra, 70% of the fresh fruits and vegetables reach the consumers, while 30% gets wasted. To maintain the right balance between food supply and population growth, preserving food is necessary. By preserving food, an increase in the shelf-life of the crops is ensured, and nutritional value is maintained.

Drying has proved to be a very promising type of food preservation that is often practiced by the farmers post harvesting stage. Drying vegetables and fruits using heat energy ensures longer storage times and easier transportation. The term drying basically refers to the minimization of moisture content in a substance. It is one of the oldest and most commonly used energy-consuming unit operations in the process industries. It is popularly used in various industries, especially chemical industries. The fundamental nature of all drying processes is the removal of moisture from mixtures to yield a solid product. It is generally achieved by convection from the current of the air.

During drying, it is observed that two processes occur simultaneously, i.e., the transfer of energy from the vicinity of the dryer (heat transfer) and the transfer of moisture from within the substance to be dried (mass transfer). The hot air causes the water content present in the food to evaporate and carries it into the environment. Moisture is not fully eliminated during drying as it is desirable to retain a little moisture in the solid. Even though there are many drying techniques and equipments present in the market, most products (i.e., fruits and vegetables) are still dried by the method of hot air drying as this is regarded as the simplest and most economical drying method.

Traditionally, the open sun drying method has been used to dry perishable crops as it is the most economical and environment-friendly method. But there are many disadvantages of this method like the fact that agricultural products may get contaminated due to interference of insects, birds, and animals, color loss, and loss of nutrients due to uncontrolled exposure to sunlight. So, to overcome these disadvantages, a variety of closed-cabinet dryers were invented. Depending on the quantity of moisture present initially and the final moisture content to be retained in the food, the type of dryer is chosen.

Amer et al.[1] examined and worked an integrated solar system for drying chamomile during the summer season 2013 in Germany. Banout et al.[2] compared performances of a new designed Double-pass solar drier (DPSD) with those of a typical cabinet drier (CD) and a traditional open-air sun drying for drying of red chilli in central Vietnam. Bena and Fuller[3] combined a direct-type natural convection solar dryer and a simple biomass burner to demonstrate a drying technology suitable for small-scale processors of dried fruits and vegetables in non-electrified areas of developing countries.

Bennamoun and Belhamri[4] studied a simple efficient and inexpensive solar batch dryer for agriculture products. It is reported that the quality of final banana dried in the biomass-fuelled dryer and that of banana dried in natural sun are similar although the product dried in the dryer is browner. The quality of chili from different modes of dryer operation is almost the same as that of chili openly dried in the sun.[5] Chandrasekar et al.[6] eliminated the use of electricity in the indirect solar dryers by utilizing split A/C condenser unit that is placed outdoors.

Janjai and Tung[7] developed a solar dryer for drying herbs and spices using hot air from roof-integrated solar collectors. The dryer was a bin type with a rectangular perforated floor. Goud et al.[8] developed an indirect type solar dryer (ITSD) and its air flow was encouraged by inlet fans that operated by solar photovoltaic (PV) panels. Drying experiments were performed with green chili (Capsicum Annum) and okra (Abelmoschus Esculentus). The drying kinetics and the performance parameters of ITSD were estimated. Kishk et al.[9] designed and fabricated an efficient and cheap solar air collector from recyclable aluminum cans. Two dryers of different configurations (closed and open modes) were constructed and examined for drying tomatoes under different operating conditions.

Due to higher prices and shortages of fossil fuels and to reduce the fuel consumption used in the drying process, more importance is given to solar energy sources as it is freely available. For these purposes, an indirect type solar dryer was designed and developed to dry agricultural products.[10-13] Efforts were made in designing and fabricating a hybrid dryer consisting of a solar flat plate collector, a biomass heater, and a drying chamber for to dry agricultural products.[14]

Shanmugam and Natarajan[15] built and tested an indirect forced convection with desiccant integrated solar dryer. Their study observed approximately in all the drying experiments 60% of moisture was removed by air heated using solar energy and the remainder by the desiccant. The inclusion of reflective mirror on the desiccant bed made faster regeneration of the desiccant material. Wang et al.[16] presented an indirect forced convection solar dryer (IFCSD) with auxiliary heating device for drying mango.

This study found few studies on design and fabrication of an indirect solar dryer with natural circulation, using waste plastic bottles as solar air collectors. With this view, this study constructed an indirect solar dryer with natural circulation, using waste plastic bottles as solar air collectors, and further evaluated the moisture content of turmeric, onions, and grapes using indirect solar drying (IDSD), direct solar dryer (DSD), and open sun drying (OSD).

16.2 EXPERIMENTAL METHODS AND MATERIALS

16.2.1 SOLAR COLLECTOR

The collector was made up of an Aluminum Composite Panel (ACP) and was covered with a plastic sheet. The Solar Air Collector (SAC) was constructed using waste-corrugated plastic water bottles. Eight bottles were glued to each other to form a pipe of 132 cm, with an internal diameter of 7.5 cm. A total of six pipes were used in this collector. They were painted black to absorb maximum solar energy. The collector was then covered with a plastic sheet in order to reduce heat losses. The dimensions of the collector are $1.32 \times 0.6 \times 0.22$ m.

The dried material was sieved at 125 μm. The sieved adsorbent was stored in an airtight container for further experimentation.

16.2.2 DRYING CHAMBER

The drying chamber frame was made up of steel pipes of 1.25" and 0.75, which were then covered with ACP sheets. The gross dimension of the chamber was $0.6 \times 0.4 \times 0.7$ m. The ACP panels were chosen as they are of high strength, light in weight, heat-resistant, and waterproof.

Various research papers were referred prior to finalizing the dimensions of the dryer.

16.2.3 DIRECT SOLAR DRYER

The direct solar dryer was constructed with the same material as the indirect dryer. This dryer was used as a control. The area of the dryer was similar to that of the indirect type of solar dryer. Holes of 13 mm in diameter were provided along both the lengths to provide ventilation within the dryer.[1-7]

At first, the bottles taken for making SAC were plain in structure and were painted completely black in color. But, they were later replaced with plain bottles that were painted partially or half black. But, as they could not withstand high temperatures, they were replaced with corrugated bottles that were completely painted black.

Then no-load test was carried out using three different solar collector angles, namely, 10°, 17°, and 25°. This was carried out to finalize the optimum tilt angle. The temperature within the drying chamber and the ambient temperature were recorded. The test was carried out from 9:00 am to 5:00 pm.

The tests were also carried out with different orientations of the solar dryer, i.e., in the south direction and in the southwest direction.

Drying experiments were conducted on grapes, raw turmeric, and onion in the month of May. The experiments were carried out from 9:00 am to 5:00 pm. A Handheld Temp/RH/CO_2 meter was used to measure relative humidity and temperature, hourly, within the dryers along with the ambient temperature. A Texla digital weighing scale with a capacity of 1 gm to 10 kg was used to measure the weight of the product after every couple of hours.

16.2.4 GRAPES

The grape samples were dried in Open Sun Drying (OSD), Indirect Solar Drying (IDSD), and Direct Solar Drying (DSD). Six kg of grapes were brought from the local market, thoroughly washed, and laid on a cotton cloth to remove any excess water left after washing. Two kg of grapes were kept in direct, indirect, as well as open sun drying. In IDSD, the grapes were uniformly spread on all the four trays, whereas in DSD and OSD, the grapes were spread uniformly over the surface.

16.2.5 ONIONS

For conducting experiments, 2 kg of onions were bought from a local market and washed thoroughly. The onions were then sliced and kept for drying.

One thousand grams of onion slices were spread evenly on each tray in the indirect solar dryer, whereas 500 gm and 1000 gm of onion slices were kept for drying in the OSD and direct dryer, respectively.

16.2.6 TURMERIC

Raw turmeric was bought from a local farmer of Sangli, India. The turmeric was then boiled and sliced into thin pieces. Each tray, in the indirect solar dryer, had 250 gm of turmeric spread uniformly over the surface. One and a half kg of turmeric slices were kept for drying in DSD and OSD, respectively.

16.3 RESULTS AND DISCUSSION

16.3.1 SOLAR COLLECTOR ORIENTATION TEST

The solar collector was first oriented to face in the southwest direction, and the hourly temperature inside the drying chamber was recorded along with the ambient temperature. The tests were carried out by placing the solar collector at different angles (10°, 17°, and 25°) with respect to the ground. The results are shown in Figures 16.1–16.5. It was observed that the solar dryer achieved a maximum temperature of 57.9°C, at 4:00 PM, when the collector was placed at an angle of 17° w.r.t. the ground.

The solar collector was then angled in the north direction. The results showed that the solar collector, when placed facing the southwest direction, proved to be much more efficient as compared to the solar collector when placed facing the south direction.

16.3.2 NO-LOAD TEST

No-load test was carried out for three consecutive days. A maximum temperature of 64.40°C was attained by the dryer at 2:00 PM at the top tray (i.e., tray 4). The collector reached its peak value when the ambient temperature was 34.1°C. This indicates that the maximum rise in temperature of the dryer was about 65.9% (21.75°C) more as compared to the ambient temperature at that time.

The temperature within the indirect kind of sun dryer on days 1, 2, and 3 is illustrated in Tables 16.1, 16.2, and 16.3, respectively.

Application of Solar Dryer for Drying of Agricultural Products 249

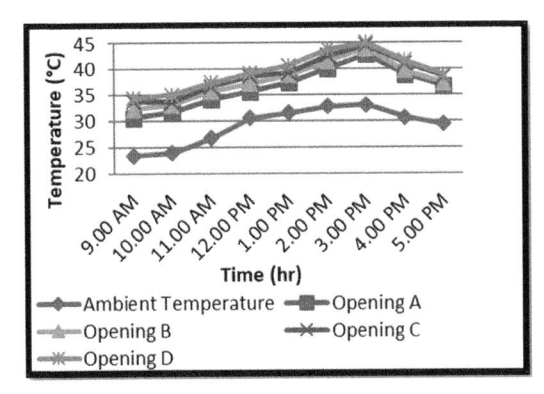

FIGURE 16.1 Time vs temperature graph when the collector was placed at angle of 10° in southwest direction.

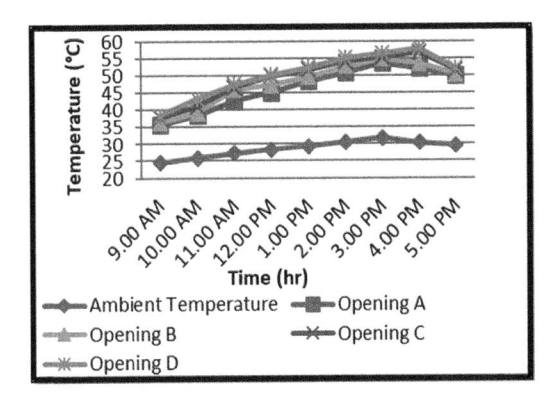

FIGURE 16.2 Time vs temperature graph when the collector was placed at angle of 17° in southwest direction.

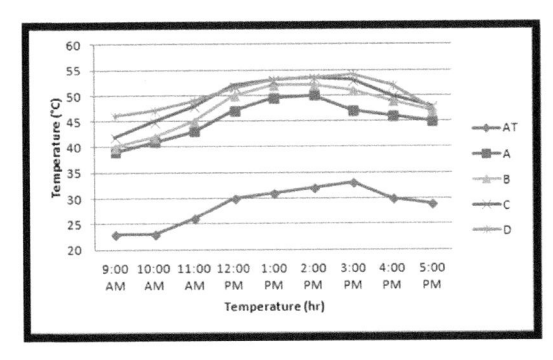

FIGURE 16.3 Time vs temperature graph when the collector was placed at angle of 10° in north direction.

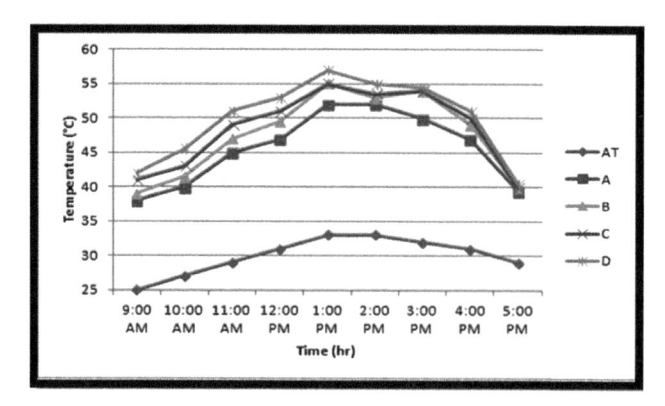

FIGURE 16.4 Time vs temperature graph when the collector was placed at angle of 17° in north.

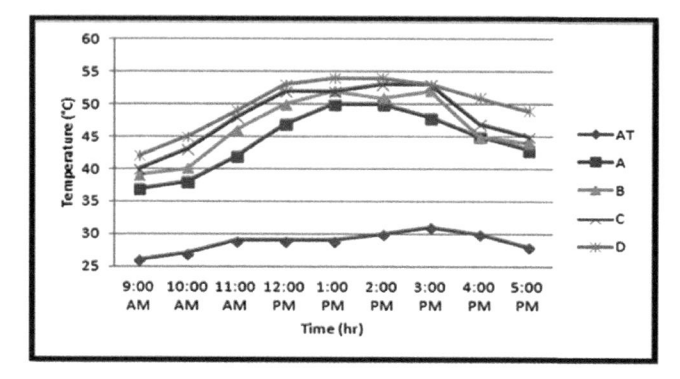

FIGURE 16.5 Time vs temperature graph when the collector was placed at angle of 25° in north.

TABLE 16.1 Temperature within the Indirect Type of Solar Dryer on Day 1.

Time	Ambient temperature (°C)	Opening A (°C)	Opening B (°C)	Opening C (°C)	Opening D (°C)
9:00 AM	25.6	38.3	39.7	41.2	43.2
10:00 AM	26.8	41.1	42.5	43.2	44.7
11:00 AM	28.4	45.8	46.5	47.2	48.6
12:00 PM	29.7	49.2	50.7	51.8	52.6
1:00 PM	31.2	53.4	54.6	55.3	56.9
2:00 PM	32.8	55.8	56.5	57.9	58.3
3:00 PM	33.5	56.7	57.1	57.5	58.4
4:00 PM	32.1	54.3	55.1	55.9	56.5
5:00 PM	30.6	49.9	48.3	47.5	46.1

Application of Solar Dryer for Drying of Agricultural Products 251

When drying agricultural products, relative humidity (RH) was measured on an hourly basis using a handheld/RH/Co meter. The RH was high when the air temperature was low and viceversa. In IDSD, RH was high until the air reached tray 1, after which it started decreasing. RH in dryers is less than that of ambient air as the air is heated when it enters the dryers. Figure 16.6 shows that while drying grapes, the indirect solar dryer (IDSD) was able to reduce relative humidity (RH) from 32.4% to 21.2% around 2 pm, while the direct solar dryer (DSD) was able to reduce the relative humidity to 25.4% at the same time. On the other hand, Figure 16.7 shows that while drying onions, IDSD was able to reduce the RH to 18.7% and DSD was able to reduce the RH to 25.8% when the RH of ambient air was 34.2%. It is observed that IDSD was able to achieve 24.4% RH, while DSD was able to achieve 29.6% RH when the RH of ambient air was 37.6%. The temperature within and around the dryers was measured on an hourly basis using a

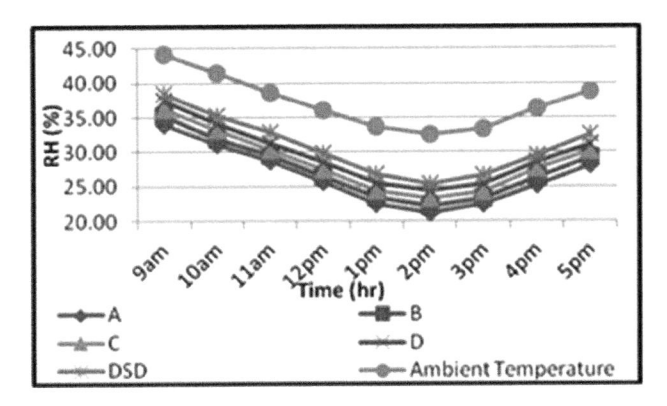

FIGURE 16.6 Time vs RH for grapes.

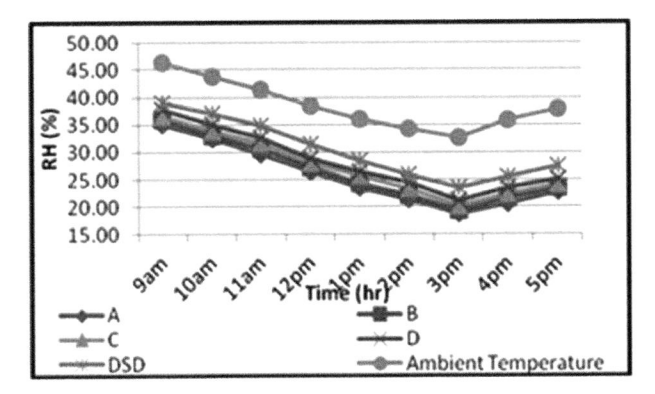

FIGURE 16.7 Time vs RH for onions.

Handheld Temp/RH/CO meter. It was observed that in IDSD, temperature decreased from tray A, which was at the bottom of the drying chamber, to tray D, which was placed at the top of the drying chamber. The reason is that the hot air entering the drying chamber initially comes in contact with tray A, where it loses some of its thermal energy for drying the food products kept on that tray. A similar mechanism is observed for tray B, C, and D, and hence, the temperature of the air in tray A is higher than that in tray D. As shown in Figure 16.9, while drying grapes, DSD was able to achieve 43.1°C, and the IDSD was able to achieve a temperature of 42.2°C when the ambient temperature was 32.8°C. When drying onions, the DSD was able to raise the temperature to 44.9°C, and the IDSD reached 44.5°C, while the ambient temperature was 32.9°C, as shown in Figure 16.11. When turmeric was subjected to drying (Figure 16.8), the DSD was able to raise the temperature to 41.9°C, and 41.85°C was observed in the IDSD when the ambient temperature was 30.2°C. Refer to Figure 16.11.

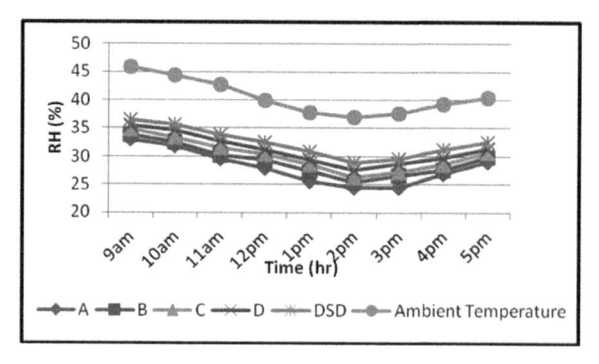

FIGURE 16.8 Time vs RH for turmeric.

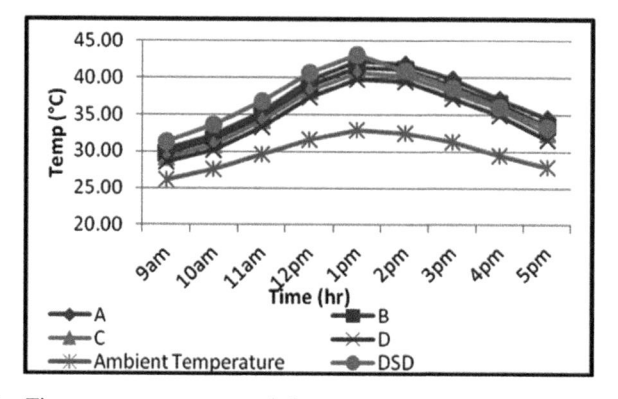

FIGURE 16.9 Time vs temperature graph for grapes.

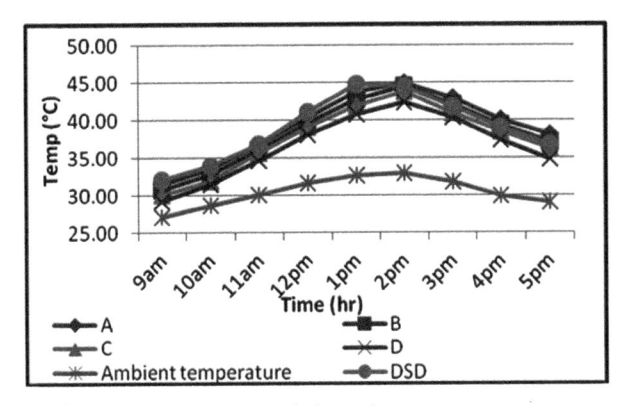

Application of Solar Dryer for Drying of Agricultural Products 253

FIGURE 16.10 Time vs temperature graph for onions.

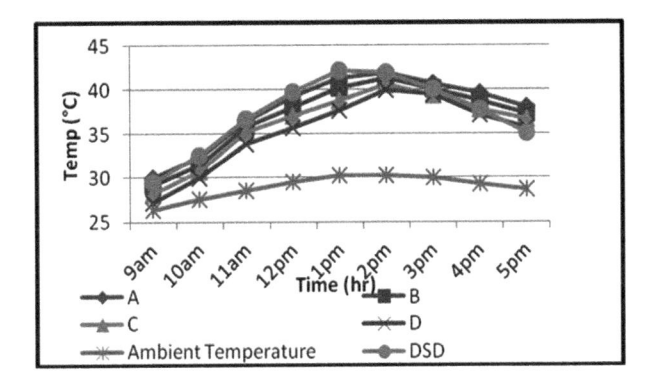

FIGURE 16.11 Time vs temperature graph for turmeric.

Moisture content (MC) was calculated by weighing the product after every couple of hours. Refer to the data shown in Tables 16.4–16.6. It was observed that the water content evaporated easily in the initial phase, but after a certain point, the rate of drying decreased.

16.4 CONCLUSIONS

1. The conduit channels in the solar collector prepared using waste-corrugated plastic water bottles displayed better stability compared to that of uncorrugated plastic bottles.
2. The maximum temperature was observed when the solar collector was placed at an angle of 17° with respect to the ground (equal to the latitude of the region).

3. The drying chamber exhibited a maximum average temperature of 60°C against the ambient temperature of 33°C.
4. In IDSD, grapes took 11 days to reach the desired moisture level due to their high-water content and were subjected to drying in their whole form, while turmeric and onion took only 2 and 3 days, respectively, as they were introduced to drying in sliced form, which shows better drying kinetics. However, in DSD, grapes took 12 days, turmeric took 3 days, and onions took 4 days to reach safe storage levels. Moreover, in OSD, grapes took 15 days, turmeric took 4 days, and onions took 5 days to reach a safe storage level.

TABLE 16.2 Temperature within the Indirect Type of Solar Dryer on Day 2.

Time	Ambient temperature (°C)	Opening A (°C)	Opening B (°C)	Opening C (°C)	Opening D (°C)
9:00 AM	26.1	40.1	41.5	42.9	43.7
10:00 AM	27.4	43.4	44.8	46.2	47.7
11:00 AM	28.6	45.8	46.3	47.8	48.2
12:00 PM	30.2	49.3	50.2	51.7	52.8
1:00 PM	33.3	57.4	58.7	59.8	60.3
2:00 PM	34.1	60.7	61.4	62.8	64.4
3:00 PM	32.3	57.8	58.1	58.6	59.2
4:00 PM	31.5	54.1	55.3	56.8	57.9
5:00 PM	29.6	53.2	54.3	55.4	56.6

TABLE 16.3 Temperature within the Indirect Type of Solar Dryer on Day 3.

Time	Ambient temperature (°C)	Opening A (°C)	Opening B (°C)	Opening C (°C)	Opening D (°C)
9:00 AM	25.4	37.6	38.5	39.1	40.3
10:00 AM	27.7	39.4	40.7	41.3	42.8
11:00 AM	29.6	41.4	41.9	42.6	43.8
12:00 PM	31.2	44.1	45.4	46.3	47.7
1:00 PM	33.6	48.2	49.1	50.3	51.6
2:00 PM	33.9	53.5	54.8	55.6	56.7
3:00 PM	32.5	50.7	51.4	52.8	53.1
4:00 PM	31.1	48.3	49.6	50.2	51.7
5:00 PM	29.7	46.3	47.2	48.6	49.1

TABLE 16.4 MC Removed from Grapes.

Day	Osd	idsd	dsd
1	9.25%	10.45%	9.95%
2	18.65%	24.30%	21.35%
3	25.70%	37.05%	35.00%
4	35.40%	50.05%	48.20%
5	42.55%	62.90%	60.85%
6	50.80%	71.05%	70.25%
7	57.85%	76.35%	75.70%
8	70.70%	79.95%	77.45%
9	72.85%	82.80%	79.40%
10	74.35%	85.45%	80.30%
11	76.45%	85.35%	81.05%
12	76.65%		81.95%
13	77.30%		
14	78.55%		
15	79.10%		

TABLE 16.5 MC Removed from Turmeric.

Day	Osd	idsd	dsd
1	16.60%	58.40%	53.90%
2	56.40%	92.10%	81.20%
3	81.80%		90.60%
4	85.40%		

TABLE 16.6 MC Removed from Onions.

Day	Osd	idsd	dsd
1	20.80%	34.00%	38.80%
2	48.60%	74.70%	76.80%
3	65.40%	83.00%	92.90%
4	76.40%	84.30%	
5	77.60%		

KEYWORDS

- **natural convection**
- **indirect-type solar dryer**
- **plastic solar air**

REFERENCES

1. Amer, B.; Gottschalk, K.; Hossain, M. Integrated Hybrid Solar Drying System and its Drying Kinetics of Chamomile. *Renew. Energy* **2018**, 1–23.
2. Banout, J.; Ehl, P.; Havlik, J.; Lojka, B.; Polesyny, Z.; Verner, V. Design and Performance Evaluation of a Double-Pass Solar Drier for Drying of Red Chilli (Capsicum annum L.). *Sol. Energy* **2011**, *85*, 506–515.
3. Bena, B.; Fuller; R. Natural Convection Solar Dryer with Biomass Back-up Heater. *Sol. Energy* **2002**, *72*, 75–83.
4. Bennamoun, L.; Belhamri, A. Design and Simulation of a Solar Dryer for Agriculture Products. *J. Food Eng.* **2003**, *59*, 259–266.
5. Bhattacharya, S.; Ruangrungchaikul, T.; Pham, H. Design and Performance of a Hybrid Solar/Biomass Energy Powered Dryer for Fruits and Vegetables. *World Renewable Energy Congress VI*, 2000, 1161–1164.
6. Chandrasekar, M.; Senthilkumar, T.; Kumaragurubaran, B.; Fernandes, J. Experimental Investigation on a Solar Dryer Integrated with Condenser Unit of Split Air Conditioner (A/C) for Enhancing Drying Rate. *Renew. Energy* **2018**, 1–36.
7. Janjai, S.; Tung, P. Performance of a Solar Dryer using Hot Air from Roof-Integrated Solar Collectors for Drying Herbs and Spices. *Renew. Energy* **2005**, *30*, 2085–2095.
8. Goud, M.; Reddy, M.; Chandramohan, V.; Suresh, S. A Novel Indirect Solar Dryer with Inlet Fans Powered by Solar PV Panels: Drying kinetics of Capsicum Annum and Abelmoschus esculentus with Dryer Performance. *Sol. Energy* **2019**, *194*, 871–885.
9. Kishk, S.; ElGamal, R.; ElMasry, G. Effectiveness of Recyclable Aluminum Cans in Fabricating an Efficient Solar Collector for Drying Agricultural Products. *Renew. Energy* **2019**, *133*, 307–316.
10. Lingayat, A.; Chandramohan, V.; Raju, V. R. K. Design, Development and Performance of Indirect Type Solar Dryer for Banana Drying. *Energy Procedia* **2016**, *109*, 409–416.
11. Pangavhane, D.; Sawhney, R.; Sarsavadia, P. Design, Development and Performance Testing of a New Natural Convection Solar Dryer. *Energy*, **2002**, *27*, 579–590.
12. Prasad, J.; Vijay, V. Experimental Studies on Drying of Zingiber Officinale, Curcuma Longa l. and Tinospora Cordifolia in Solar-Biomass Hybrid Drier. *Renew. Energy* **2005**, *30*, 2097–2109.
13. Sallam, Y.; Aly, M.; Nassar, A.; Mohamed, E. Solar Drying of Whole Mint Plant under Natural and Forced Convection. *J. Adv. Res.* **2013**, 1–8.
14. Saravanan, D.; Wilson, V.; Kumarasamy, S. Design And Thermal Performance Of The Solar Biomass Hybrid Dryer For Cashew Drying. *Mech. Eng.* **2014**, *12*, 277–288.

15. Shanmugam, V.; Natarajan, E. Experimental study of Regenerative Desiccant Integrated Solar Dryer with and without Reflective Mirror. *Appl. Therm. Eng.* **2007,** *27,* 1543–1551.
16. Wang, W.; Li, M.; Hassanien, R.; Wang, Y. Thermal Performance of Indirect Forced Convection Solar Dryer and Kinetics Analysis of Mango. *Appl. Therm. Eng.* **2018,** 1–20.
17. Yaha, M. Design and Performance Evaluation of a Solar Assisted Heat Pump Dryer Integrated with Biomass Furnace for Red Chilli. *Int. J. Photoenergy* **2016,** *2016,* 14.
18. Zomorodian and Zamanian. *Designing and Evaluating an Innovative Solar Air Collector with Transpired Absorber and Cover*; Elsevier, 2012.

CHAPTER 17

Flow Estimation and Flood Forecasting over Narmada River Using Three Data-Driven Techniques

ROHIT GAIKWAD[1], RAMCHANDRA KAVANEKAR[1], PRADNYA DIXIT[2], and PREETI KULKARNI[2]

[1]*Student of MTECH-Water Resources and Environmental Engineering, Department of Civil Engineering, Vishwakarma Institute of Information Technology, Pune, India*

[2]*Department of Civil Engineering, Vishwakarma Institute of Information Technology, Pune, India*

ABSTRACT

Designing a system that can be useful for early detection of floods is important for controlling and minimizing the flood-related losses. In the present study, daily, one-day, and two-day advance flow values at Mandaleshwar station on the Narmada River are predicted. For the prediction of discharge, three data-driven techniques, namely, Model Tree (MT), Artificial Neural Networks (ANNs), and Support Vector Regression (SVR), are employed for estimation. The daily discharge (Q) and water level (WL) values from the year 1999 to 2007 for 9 years at the three stations, Mandleshwar, Kogaon, and Mortakks, are considered for analysis. The daily discharge and water stage data is collected from the Water Resources Information System (WRIS) platform. Data-driven models are developed to estimate the daily discharge data at Mandleshwar station using the measured water level (WL) and discharge (Q) values of the remaining two stations. The outcomes of the MT, SVR, and ANN models are appraised over the observed value of discharge based on performance indicators like Correlation Coefficient (r), Root Mean Square Error (RMSE), and scatter plots. The results of the evaluation showed that the SVR model outperformed the ANN and MT models. In addition, it is seen that by merely increasing input values, this data-driven technique will not give you improved results. This proves that data-driven techniques understand the data based on a physics-based concept.

17.1 INTRODUCTION

The study of the water level and discharge of rivers is a crucial factor in flood detection and flow forecasting. Generally, the records of the water stage and discharge of a river are maintained by time-to-time measurements occurring at a particular place. The measure of elevation of the water surface above the datum line is called the stage, and discharge is the total volume of water passing through the gauging station at a particular time. Measurement of water stage and discharge is done with the help of self-recording gauges or nonrecording gauges, which are located at specific stations. A flood is usually a high discharge and stage in a river at which the river overflows in its surrounding area.[1] The damage caused by floods is well-known to all of us. To prevent and minimize flood damages, it is important to focus on flood forecasting. Flood forecasting is necessary to save the lives of people. Therefore, one and two days in advance flood forecasting is useful for the society. In the present paper, the authors have tried to forecast the flood at the Narmada River station (Mandleshwar station) for one to two days in advance using three data-driven techniques.

A little information about earlier attempts for forecasting the flood using the soft computing techniques is explained below. Attempts were made to calculate daily river-flow forecasting with the help of data-driven techniques, which showed high accuracy of these data-driven techniques.[2,3] The evapotranspiration was found by using the M5 Model Tree and Artificial Neural Network (ANN), which proved the high accuracy of both the models.[4] This successful application of machine learning in water management inspired the exploration of the applicability of these approaches in modeling, and it is concluded that the ANN and M5 model tree-based models are superior in accuracy than the traditional models.[5] Hybrid techniques like neurowavelet with multilevel decomposition are used for forecasting significant waves using the previous wave heights.[6] The prediction of extreme events for a univariate wave model can be made accurately using a hybrid Neurowavelet Technique.[7]

The use of ANN in the prediction and forecasting of a number of water resource variables, such as rainfall, water level, flow, and various water quality parameters, has increased in recent years.[8] Across all disciplines of ocean engineering, this soft computing technique is also being extensively used.[9] The end results obtained from MT are found to be more understandable than the results obtained from other tools, and it allows users to build a number of models of varying complexity and accuracy.[10] In hydrological modeling for forecasting the river flow, data-driven techniques based on machine learning are often the first choice.[11] Hydraulic structures like dams, rivers, and the sufficient supply of water to different agencies are some examples where river-flow forecasting in advance is important.[12]

Many researchers worked with the data-driven techniques such as M5 Model Trees, Support Vector Regression (SVR), and Artificial Neural Network (ANN) in the hydrology and water resources fields. Taking into consideration these many applications in Hydrology and Water Resources, the authors have presented a study on correlating discharge and water level measurement to estimate and forecast the flood at Mandleshwar station using three data-driven techniques.

17.2 STUDY AREA AND DATA

Narmada is one of the largest west-flowing rivers in India. It flows across a large area of Madhya Pradesh, Maharashtra, and Gujarat. The Narmada River basin is situated between the Vindhya and Satpura mountain ranges. After flowing through three states of India, this river finally disappears into the Gulf of Camby in the Arabian Sea. The Narmada basin is situated between longitudes 72°38′–81°43′E and latitudes 21°27′–23°37′N. The total catchment area covered is 98,796 km². The source of the Narmada River is the Maikala Range near Amarkantak, Madhya Pradesh. This range has an elevation of about 1057 m.

The three stations, namely, Mandleshwar, Kogaon, and Mortakka, are considered in the present study. The observations of daily average streamflow values related to Mandaleshwar, Mortakka, and Kogaon stations for the years 1999–2007 were obtained from river-monitoring stations from the India Water Resources Information System web portal. The Kogaon station on the Kundi River (Tributary of the Narmada River) is located at 6 km upstream from Mandleshwar. The Mortakka station on the Narmada River is located around 42 km from Mandleshwar and on the upstream side. Daily water level and discharge data for a continuous period of 9 years are used in the present study. Figure 17.1 represents the location map of the study area. Table 17.1 below represents the statistical parameters of daily water level and discharge at three stations.

TABLE 17.1 Data Statistics of Three Stations.

Stations	Mortakka		Kogaon		Mandleshwar	
	Discharge Q (in cumecs)	Water Level (m)	Discharge Q (in cumecs)	Water Level (m)	Discharge Q (in cumecs)	Water Level (m)
Min	2.00	154.48	0.003	152.08	3.46	138.82
Max	19300.00	167.85	5900.00	161.85	30150.00	151.66
Average	1560.14	157.38	69.11	153.08	1752.27	141.22
Median	740.00	157.02	8.92	152.90	730.33	140.70
Standard Deviation	2551.17	1.62	302.32	0.70	3068.03	1.73
Kurtosis	18.03	9.57	175.24	35.84	30.76	6.83
Skewness	3.84	2.48	11.78	4.52	4.75	2.18

*Daily water level and discharge data from 1999 to 2007 were used for the present study.

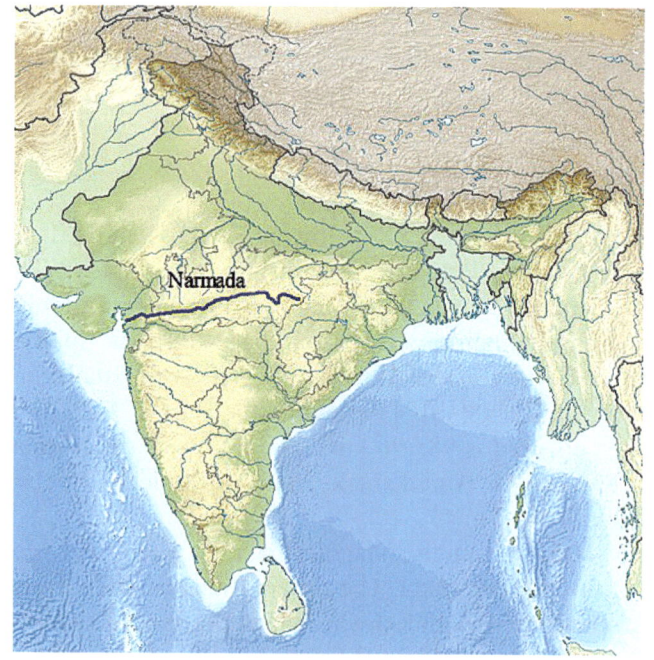

FIGURE 17.1 Location map of Narmada River (Source: Google Map).

17.3 TOOLS USED

For building and testing the model with ANN, MT, and SVR, the WEKA software was used. Weka is used for data mining purposes. In simple language, it is a group of machine learning algorithms. We can apply these algorithms directly on data sets. We can use Weka for different tasks like data processing, regression, visualization, classification, and clustering of data. It can also be used for processing new machine learning tools.

17.3.1 THE M5 MODEL TREE (MT)

The M5 model tree algorithm was originally developed by Quinlan in 1992.[13] The idea of the M5 model tree is based on a decision tree, a data-driven method. This method divides the input parameters into subspaces and builds a linear regression model for each of them. The splitting of data follows the concept of a decision tree, but here, at leaves instead of class labels, it uses a linear regression function that can produce continuous numeric attributes. The benefit of model trees is that they can tackle tasks with very high efficiency and also learn effectively. The trees generated in MT are much smaller as compared to regression trees and also show clear decision strength. The regression function normally does not involve any variables.

17.3.2 ARTIFICIAL NEURAL NETWORKS (ANNS)

ANN is designed by programing computers to behave like interconnected brain cells; it has the capacity of storing experimental knowledge and making use of it. It used an input layer and an output layer for supervised training; these input and output layers are connected by weights and biases by one or more hidden layers. The initial trail weights and bias values are then used for the further testing process, in which input values are multiplied with trail weights and summed with a bias value. Either linear or sigmoid transfer functions are used to transfer this sum to get the output, and then this sum is transferred either by linear or sigmoid transfer functions to get an output. This outcome is then used again in the hidden layers to continue the process until the final output layer is obtained. The error function transforms the value obtained from the difference between the output and the target, and then it propagates the resulting error back to correct the values of weights and biases to lower the error by using an optimization technique. The process continues until it gets outputs with the desired accuracy. Then the trained model can be used for testing unseen data.

17.3.3 SUPPORT VECTOR REGRESSION (SVR)

SVR is a classification and regression method (nonlinear regression method), which has been derived from statistical learning theory by Vapnik in 1995. The primary idea is to map the data into a higher-dimensional space. The linear regression problems are solved by using nonlinear mapping. SVR includes the Structural Risk Minimization Principle. The Kernel functions are used to achieve nonlinear mapping.

In this study, these three techniques are used with the help of the Weka 3.9 platform, which is available freely through the given link, to develop data-driven models for correlating water level and discharge at the mentioned three stations(https://waikato.github.io/weka-wiki/downloading_weka/).

17.4 METHODOLOGY AND MODEL DEVELOPMENT

In this study, estimation as well as forecasting is done using ANN, MT, and SVR. The initial data processing was carried out to get continuous records of the data. For all the three stations in the current study period, the records selected are from 1999 to 2007. To find out the discharge at Mandleshwar station (output), previously measured values of the water stage and discharge at Kogaon and Mortakka were used as input to the model. Table 17.2 represents the input–output combinations of each model to find out the discharge (Q) values at the Mandleshwar station for different time intervals. A total of three models were prepared, giving the output of discharge (Q) at different time intervals from 24 h to 48 h. A total of 1189 records were available for the said event; out of these 1189 records, 70% of data (833 records) were used for training the model, and 30% of data (356 records) were used for the testing purpose. Model Tree (M5 algorithm), Support Vector Regression (SMO-reg), and Artificial Neural Network (Multilayer Perceptron) implemented in WEKA 3.9 software are used for calibrating and testing the models. To check the working accuracy of the model, the dataset was divided for training and testing purposes.

Three models were developed, considering different combinations of discharge and water levels at three stations. The Model-1 architecture consists of 4:1 (Input: Output), Model-2 consists of 8:1, and Model-3 consists of 10:1, as shown in Table 17.2, which shows the model setup of the three models.

TABLE 17.2 Model Setup.

	Input	Output
Model 1	MO_Q(t), MO_WL(t), KO_Q(t), KO_WL(t)	MD_Q(t)
Model 2	MO_Q(t-1), MO_Q(t), MO_WL(t-1), MO_WL(t), KO_Q(t-1), KO_Q(t), KO_WL(t-1), KO_WL(t)	MD_Q(t+1)
Model 3	MO_Q(t-1), MO_Q(t), MO_WL(t-1), MO_WL(t), KO_Q(t-1), KO_Q(t), KO_WL(t-1), KO_WL(t), MD_Q(t), MD_WL(t)	MD_Q(t+2)

*Names starting with a prefix indicate station names, and the suffix represents discharge and water level data for previous and current time records and future forecasts, like

'MO' is used for Mortakka station,

'KO' for Kogaon station, and

'MD' for Mandaleshwar station.

'Q(t-1) and WL(t-1)' represents previous day discharge and water level records,

'Q(t) and WL(t)' for current discharge and water level records, and

'Q(t+1) and Q(t+2)' indicates one-day and two-day forecasts of discharge in advance.

17.5 RESULTS AND DISCUSSION

In this section, the results of the models are discussed and compared with the MT, ANN, and SVR models.

All the models were tested, and their performance was checked by comparing the correlation coefficient (r) and root mean square error (RMSE). Hydrographs and scatterplots were also plotted to observe the behavior of the forecasting model.

TABLE 17.3 MT Model Results.

Model	r ANN	RMSE ANN	r MT	RMSE MT	r SVR	RMSE SVR
Model-1	0.95	789.21	0.92	1143.39	0.96	756.29
Model-2	0.81	1657.5	0.73	2137.86	0.81	1580.12
Model-3	0.58	2481.77	0.49	2586.3	0.51	2370.46

*r: correlation coefficient; RMSE: root mean square error.

Model-1 shows good outputs in testing. The 'r' value obtained in Model-1 for observed and forecasted discharges is 0.96. Model-2 shows a reasonable performance as compared to others, with an 'r' value of 0.81, and Model-3 showed an 'r' value of 0.58, which clearly indicates less accuracy.

The results confirm the better performance of all the models in testing except Model 3. The most effective tool is found out on the basis of the performance indicators using the Correlation Coefficient and Root Mean Square Error of the models. The estimated results show that SVR gives better performance for models 1 and 2. For Model 3, ANN is showing good results.

Comparing the overall results, it seems that SVR is superior compared with the other two data-driven techniques, MT and ANN. It indicates that the results are influenced by the variability in the data and the data interval.

17.5.1 HISTOGRAMS FOR R AND RMSE

The histogram for correlation coefficient and root mean square error shows the comparison of MT, SVR, and ANN for Model-1, Model-2, and Model-3.

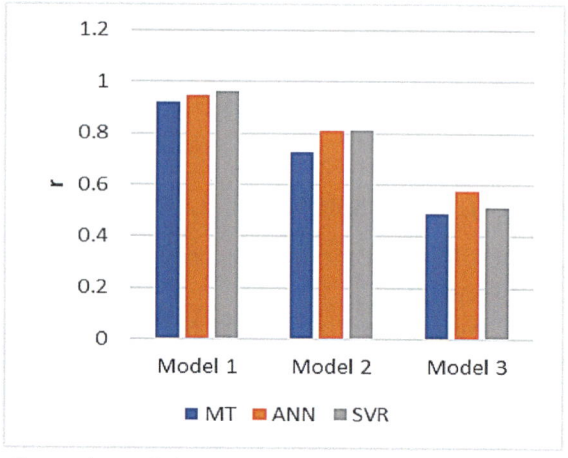

FIGURE 17.2a Comparison of r by MT, SVR, ANN.

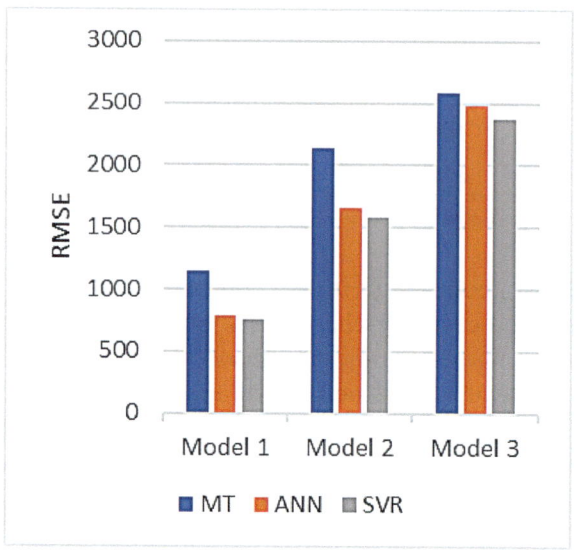

FIGURE 17.2b Comparison of RMSE by MT, SVR, ANN.

Flow Estimation and Flood Forecasting over Narmada River 267

The correlation coefficient is a normalized measurement of the covariance and is used to measure the relationship between two variables. The result always has a value between −1 and 1. It is obvious that the value of the coefficient of correlation decreases because of losses such as evaporation, seepage, friction, etc. The values of the correlation coefficient of the actual and predicted quantities are determined, which represent the correlation between the observed and predicted stages of each model. Figure 17.2B shows the comparison of the relative mean square error for all three models, which indicates that SVR is the best tool fitted for the data.

17.5.2 SCATTER PLOTS AND HYDROGRAPHS

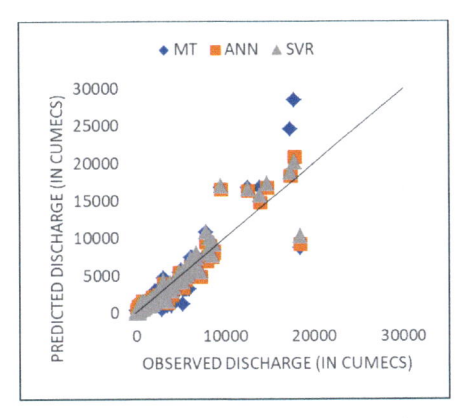

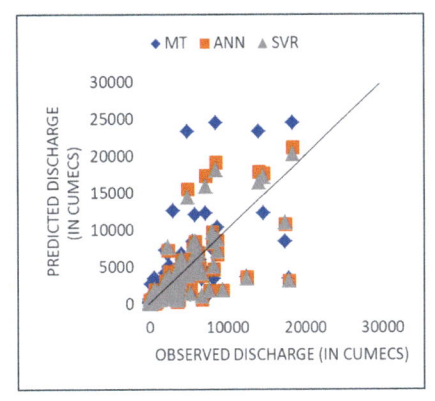

FIGURE 17.3a Scatter plots for model 1.   **FIGURE 17.3b** Scatter plots for model 2.

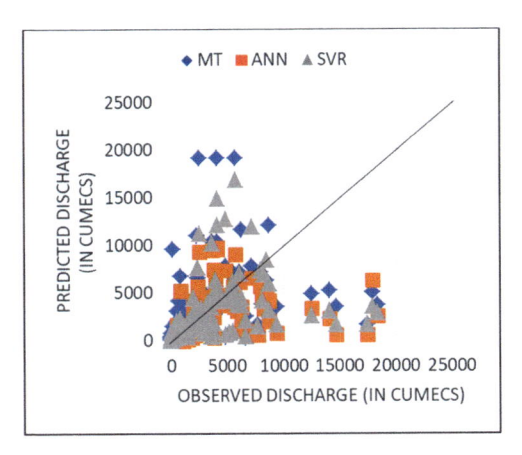

FIGURE 17.3c Scatter plots for model 3.

Figure 17.3A–C shows scatter plots for models 1, 2, and 3, respectively. Data variation is being observed through these graphs. The point lying on the 45° line ($y = x$) represents the exact prediction of the data (Exact Prediction). The point lying below the 45 line ($y = x$) represents a predicted value less than the measured value (Underpredicted), and the point lying above the 45° line ($y = x$) represents a predicted value greater than the measured value (Overpredicted).

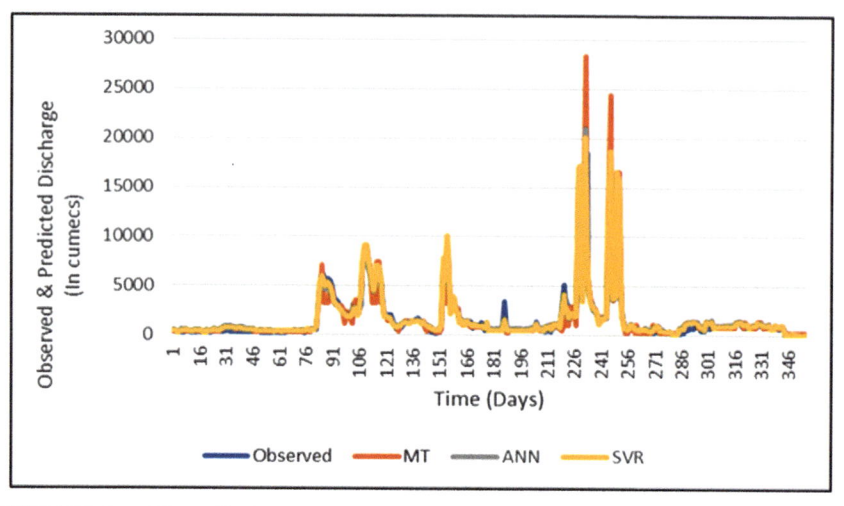

FIGURE 17.4a Water discharge graph for model 1.

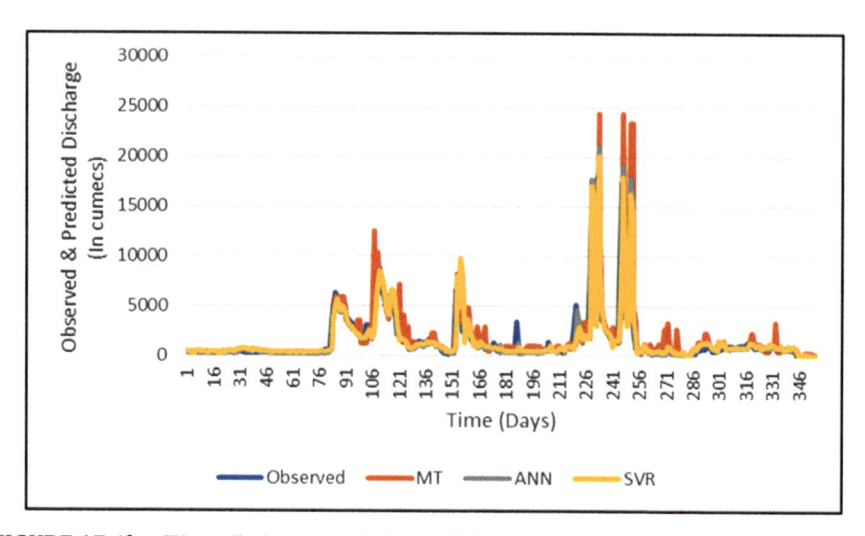

FIGURE 17.4b Water discharge graph for model 2.

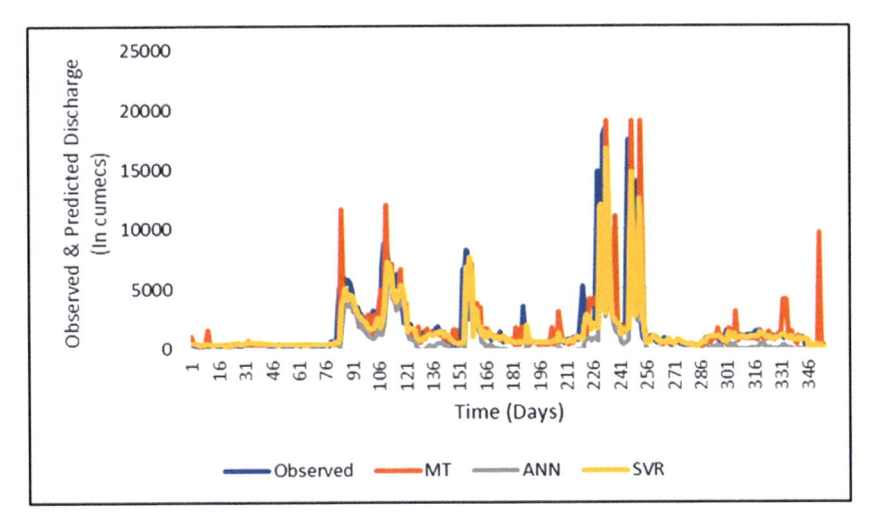

FIGURE 17.4c Water discharge graph for model 3.

Figure 17.4A–17.4C shown above represents the water discharge graphs for models 1, 2, and 3, respectively. The above water stage graphs show the variation of measured discharge and predicted discharge with data-driven techniques. In figure 17.4C, the peak value shows the highest rainfall between the 226th and 256th days. The profile after the peak shows less rainfall. From this, we can see the spatial and temporal variation of water flow.

17.5.3 MODEL TREE

The following figure shows model trees of testing data, in which the model splits on the basis of values of the standard deviation. It has linear regression features at the leaves, which can predict nonstop numeric attributes.

17.6 CONCLUSIONS

The present paper aims in estimating and forecasting the discharge of the Narmada River at Mandleshwar station. The tools used are MT, ANN, and SVR. To train the model, water stage and discharge data from all the three stations are used in different combinations. A total of three models were prepared. Out of these three models, Model 1 performed well with SVR, having r = 0.96. It suggests that SVR can effectively be used to predict the daily river discharge for the dataset used in this study. A comparison of correlation coefficients and RMSE values indicates that Model 2 also performs well with SVM with the dataset used in this study. But in Model 3, although we have increased the number of inputs, the output is showing less accuracy. From the results, it is clear that increased techniques will not improve the prediction of discharge. This activity is evident for techniques that are actually showing you performance; their performance is like they are understanding the physics involved in that particular phenomenon. Therefore, by merely increasing inputs, this data-driven technique will not give you improved results. This proves that the data-driven techniques understand the data based on a physics-based concept.

ACKNOWLEDGMENT

The water level data is collected from the India Water Resource Information System (WRIS Platform). The authors are thankful to the Central Warehousing Corporation (CWC), India, for providing access to download the data free of cost.

KEYWORDS

- **floods**
- **model tree (MT)**
- **artificial neural networks (ANNs)**
- **support vector regression (SVR)**
- **Narmada River**

REFERENCES

1. Subramanya, K. *Engineering Hydrology*, 3rd ed.; Tata McGraw-Hill Publishing: New Delhi, 2008.
2. Sattari, M.; Pal, M.; Apaydin, H.; Ozturk F. M5 Model Tree application in Daily River Flow Forecasting in Sohu Stream, Turkey, *Water Resources*; Springer, 2013, vol 40; pp 233–242.

3. Solomatine, D. P.; Yunpeng, X. M5 Model Trees and Neural Networks: Application to Flood Forecasting in the Upper Reach of the Huai River in China. *J. Hydrol. Eng.* @ ASCE/Nov/Dec 2004.
4. Armin Alipour, A.; Yarahmadi, J.; Maryam Mahdavi, M. Comparative Study of M5 Model Tree and Artificial Neural Network in Estimating Reference Evapotranspiration using MODIS Products. *Hindawi Publishing Corporation, J. Climatol.* **2014**, *11*, A-ID 839205.
5. Bhattacharya, B.; Solomatine, D. P. In *Neural Networks and M5 Model Trees in Modeling Water Level-Discharge Relationship for an Indian River*, Proceedings of the 11th European Symposium on Artificial Neural Network, Bruges, Belgium, d-side, Evere Belgium; Verleysen, M., Eds.; 2005. pp 407–412.
6. Dixit, P.; Londhe, S.; Dandawate, Y. Removing Prediction Lag in Wave Height Forecasting using Neuro- Wavelet Modelling Technique used Artificial Neural Networks. *Ocean Eng.* **2015**, *93*, 74–83.
7. Dixit, P.; Londhe, S. Prediction of Extreme Wave Heights using Neuro Wavelet Technique. *Appl. Ocean Res.* **2016**, *58*, 241–252.
8. Maier and Dandy. Neural Networks for Prediction and Forecasting of Water Resources Variable. *Environ. Model. Softw.* **2000**, *15*, 101–124.
9. Jain, P.; Deo, M. C. Neural Networks in Ocean Engineering.
10. Solomatine, D. P.; Dulal, K. Model Tree as an Alternative to Neural Network in Rainfall-Runoff Modeling. *Hydrol. Sci. J.* **2003**, *48* (3), 399–411.
11. Solomatine, D. P.; Maskey, M.; Durga Lal Shrestha, D. L. Instance-based Learning Compared to other Data-Driven Methods in Hydrological Forecasting. *Hydrol. Process* **2008**, *22*, 275–287, Wiley Inter-Science.
12. Londhe, S.; Charhate, S. Comparison of Data-Driven Modelling Techniques for River Flow Forecasting. *Hydrol. Sci. J.* **2010**, *55* (7), 1163–1174.
13. Quinlan, J. R. In *Learning with Continuous Classes*, Proc. AI'92 (*Fifth Australian Joint Conf. on Artificial Intelligence*), Adams, A., Sterling, L., Eds.; World Scientific: Singapore, 1992; pp 343–348.
14. Solomatine, D. P.; Maskey, M.; Durga Lal Shrestha, D. L. Instance-based Learning Compared to other Data-Driven Methods in Hydrological Forecasting. *Hydrol. Process* **2008**, *22*, 275–287.

CHAPTER 18

Correlating Stage Measurement Stations Using Three Data-Driven Techniques: A Comparative Assessment

AALISHA LANJEWAR[1], VISHAKHA KONDHALKAR[1], PRADNYA DIXIT[2], and PREETI KULKARNI[2]

[1]*Student of MTECH-Water Resources and Environmental Engineering, Department of Civil Engineering, Vishwakarma Institute of Information Technology, Pune, India*

[2]*Department of Civil Engineering, Vishwakarma Institute of Information Technology, Pune, India*

ABSTRACT

Modeling the stage in river flow is more important in the prevention of floods, sustainable growth planning, management of water assets and development of trade and industry, etc. In the current work, three data-driven techniques, namely, model tree (MT), artificial neural networks (ANNs), and support vector regression (SVR), are employed for daily water level correlation assessment. The data of diurnal water stage values from January 2000 to December 2020 for 21 years at the three stations, namely, Somanpalli, Mancherial, and Bhatpalli, situated in the Godavari basin, Telangana, are considered for analysis. The daily stage data is collected from the Water Resources Information System (WRIS) platform. Data-driven models are developed to forecast the one-day ahead stage values at Bhatpalli station using the measured stage values at the remaining two stations. The outcomes of the MT, SVR, and ANN models are evaluated over the actual water level based on performance indicators, viz., Correlation Coefficient, Mean

Absolute Error, Relative Absolute Error and Root Mean Square Error, and scatter plots. The outcome of the estimation indicated that the MT model outperformed the ANN and SVR models with the combination of four inputs. In addition, the MT model was observed to be very effective and resilient compared with the remaining models and showed a significant contribution for the management of hydrologic sources at the analyzing site.

18.1 INTRODUCTION

The study of the stage of water is an important integrant for flood hazards, surface water discharge, variability of water content, etc. Recently, awareness of the spatial and temporal variation of water stage (water level) has increased because of the increased flood events all over the world, which are attributed to the effects of climate change. For applications of most of the hydrological design and modeling, it is necessary to have records of the stage (water level) for a long period of time, especially for flood monitoring, discharge measurement, and many more. Generally, water stage records of a river are maintained by time-to-time measurements of water level occurring at a particular place. The water stage is the rise of the water level above the datum. Water stage (WL) measurement is done with the use of gauges, which are installed at specific stations. The gauges may be self-registering gauges or nonself-registering gauges. The collection, launching, and operational procedures of the gauging station are described in IS: 18365–2013. The unavailability of the water stage data at a particular station has several reasons. Some of those are the malfunctioning of gauges and related equipments, variations in parameters such as rainfall pattern, humidity, temperature, wind speed, the effects of natural disasters (hurricanes, floods, and landslides), human-related issues, etc. These parameters are the errors which affect the hydrological models which use water level data as an input parameter. Hard computing depends on deterministic formulas and requires a lot of computational work and time, whereas data-driven techniques work on available data, and no specific formula is required. Data-driven techniques have been widely used since the last two decades for hydrological modeling.

Many researchers have worked with the data-driven techniques such as M5 Model Trees, Support Vector Regression (SVR), and Artificial Neural Network (ANN) in the study of water distribution and its management; a few of those are enlisted here. Maier and Dandy[11] used neural networks for the estimation and foretelling of variables of water resources. Solomite and Xue[19] have done a study on the application of flood forecasting in China over the Upper Reach of the Huai River by using MT Model Trees and Neural Networks and discussed the applicability and performance of so-called M5 Model Tree machine learning, which is then compared to Artificial Neural Networks. Bhattacharya and Solomatine[4] used Model Trees and ANN, which showed the relation between the water level and discharge in rivers. Dawson and Wilby[6] considered the application of ANN to rainfall-runoff modeling and forecasting. The paper on the prediction of river stage on the basis of distributed support vector regression by Wu et al.[23] proposed a novel distributed SVR model on the water stage series data collected to perform the forecast. Londhe and Dixit[8] employed the Neuro Wavelet technique for the prediction of extreme wave heights, which is a combination of the Discrete Wavelet Transform and ANN. In 2015, Dixit, Londhe, and Dandawate[7] removed prediction lag in wave height forecasting using Artificial Neural Networks. Londhe and Panchang (2018) used ANN techniques for the survey of coastal applications. Jain and Deo used neural networks in ocean engineering.

Taking into consideration these many studies in water science and water conservation, the authors have presented a study on correlating stage measurement stations using three data-driven techniques by predicting the water stage at three different stations of the Godavari River located in the Telangana catchment area. As mentioned above, MT, SVR, and ANN, which are the data-driven techniques, are used in the present work for comparative assessment by correlating stage measurements of three stations of the Godavari River. The three stations selected are, namely, Somanpalli, Mancherial, and Bhatpalli. The results obtained by these soft computing tools are judged by the standard error measures to determine the capability of MT, SVR, and ANN in predicting the water stage at all the three stations. Also, it can be noticed that 20–25% of the data from three selected stations in the Godavari basins have inconsistent water stage records. Of the selected stations, Somanpalli is a manually operated station, while the other two are telemetry stations. Though there is ample work and research available in the same area of context, no paper can give a generic solution for the hydrological problem as hydrometeorological parameters vary a lot from region-to-region concepts. The three stations chosen for the present study are situated in Telangana state in the Pranhita sub-basin. The present work is an attempt to give a solution toward a practical field (real-time) problem in this area. The organization of the paper is as follows: The next part contains information on the research field and statistical data, as well as information regarding the data-driven methodologies used in previous research. Next to that, methodology and model development are explained, and finally, the results are presented along with the histograms, hydrographs, and scatter plots.

18.2 STUDY AREA AND DETAILS

Godavari, the Dakshin Vahini Ganga, is the third leading river in India. The river has a drainage area of 3,12,812 sq.km. and is spread in the states of Andhra Pradesh (23.4%), Maharashtra (48.6%), Chhattisgarh (10.9%), Madhya Pradesh (10%), Orissa (5.7%), and Karnataka (1.4%). The basin is located in the Deccan plateau and is between latitude 16° 16′ 00 north and 22° 36′ 00″ north and longitude 73° 26′ 00″ east and 83° 07′ 00″ east. The elevation of the Godavari River near Brahmagiri Hills in the Western Ghats in the Nasik district in the state of Maharashtra is 1067 m.

The three stations, namely, Somanpalli, Mancherial, and Bhatpalli, are considered in the present study for correlating stage measurements. The Somanpalli and Bhatpalli River points are located at 47 kms and 133.34 kms from the Mancherial River point, respectively. Daily water level data for the period of 21 years from 2000 to 2020 is used in the existing study. Figure 18.1 represents the location map of the study area in detail. Table 18.1 below represents the statistical parameters of daily water level data at the three stations.

TABLE 18.1 Data Statistics for All Three Stations.

Stations	Minimum value	Maximum value	Standard deviation	Kurtosis
Mancherial	0.59 m	133.46 m	22.42 m	24.90
Somanpalli	0.26 m	182.5 m	25.74 m	16.16
Bhatpalli	2.43 m	194.15 m	35.88 m	13.01

18.3 TECHNIQUES USED

The present age is an age of data-driven techniques, and applications of soft computing techniques are spread all over the world in each and every sector. Consequently, hydrological research is not out of the box for these applications, and therefore, the present paper aims in exploring the capabilities of three data-driven techniques for correlating the stage measurement stations in Telangana, India. A short explanation of these techniques is given in this section.

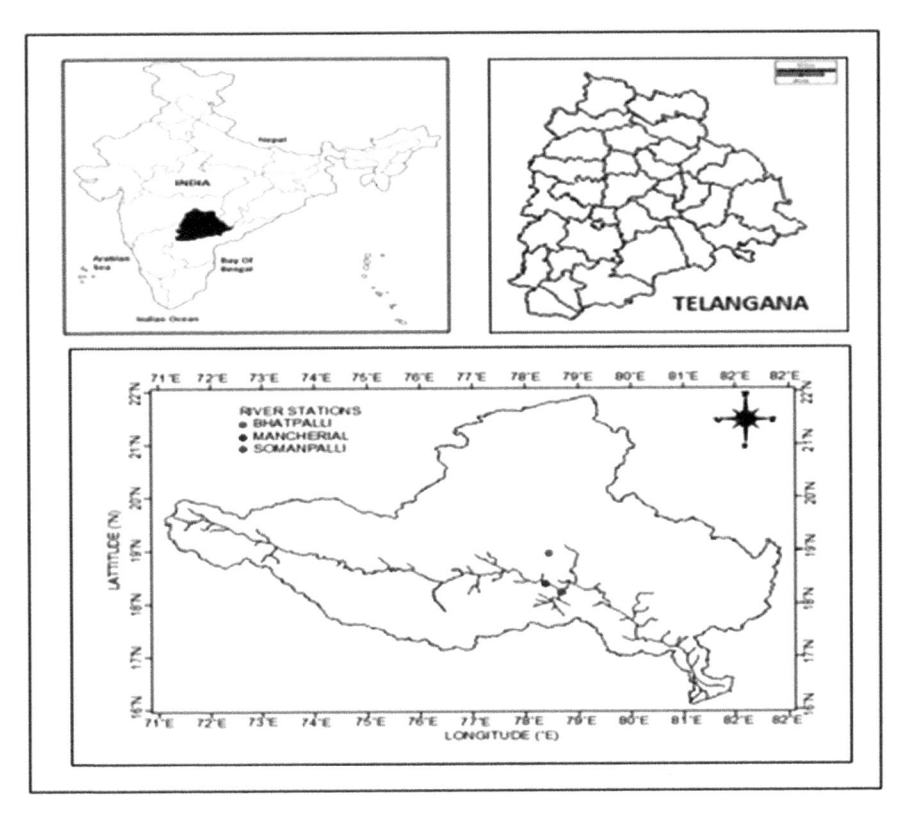

FIGURE 18.1 Location map of the minimized river catchment
Source: Google Map.

18.3.1 M5 MODEL TREES (MTS)

In 1992, Quinlan developed M5 Model Trees. It is a method accredited to the area of machine learning. At the leaves, a conventional decision tree is combined with the risk of generating linear regression functions by model trees. This representation is comparably visible since the decision structure is vivid and the reversion functions generally do not contain many variables. A piecewise linear model is developed by the M5 tree, and its structure resembles that of decision trees. The standard deviation of the class values is used in the M5 model tree algorithm such that it reaches a node, measuring the error at the node, and the expected error reduction is calculated as a result of which each attribute is tested at that node. In the beginning, the building of the initial tree is done, and then the overfitting problem is overcome by pruning the initial tree. Lastly, at the leaves of the pruned tree, sharp discontinuities are compensated by employing a smoothing process between adjacent linear models.

18.3.1.1 MODEL TREE PRUNING AND SMOOTHING

18.3.1.1.1 Pruning

If a developed tree has numerous leaves, it may become more precise and hence overfit and be generalized poorly. Simplifying a tree can make it possible to make it robust. To put it another way, by pruning or merging some of the lower subtrees into a single node.

18.3.1.2 Smoothing

It is done to compensate the sharp discontinuities which may occur unavoidably between neighboring linear models at the leaves of the trees, which are pruned. This particular problem seems to be observed in models which are made up of a limited number of training examples. By producing linear models, smoothing can be accomplished for each internal node along with the leaves at the moment of building the tree. It was observed from the experiments that the accuracy of predictions increased substantially with smoothing. For a detailed discussion of the M5 Model Tree, readers are heading for Quinlan[13] and Solomatine and Xue.[19]

18.3.2 SUPPORT VECTOR REGRESSION (SVR)

The support vector machine, introduced by Vapnik (1999), is a technique usually applied for sorting and reversion purposes. And in recent years, it has become one of the most eye-catching forecasting tools. The Principle of Structural Risk Minimization is represented in their formulation, which has been made known to be superior to the principle of traditional Empirical Risk Minimization, employed by ANN. In contrast to the ERM, which reduces error on the training data, the SRM minimizes an upper bound on the estimated risk. The SVM has a better capability to simplify, which is the goal of arithmetic learning. The margin between the two input classes included in the two data sets is maximized by the hyperplane. Simply, the larger the margin, the better the simplification of the inaccuracy of the classifier. In the situation of reversion, the kernel used in the SVM fitting of the curve is attempted on the data points in such a way that as many as possible points should be there between the two margins of hyperplanes so that the regression error is minimized. In the reversion of SVM, firstly, the input x is recorded onto an n-dimensional feature space with the help of a Kernel function using some fixed mapping, and previously, the construction of a linear model was done in this feature space. Different categories of kernels used are Radial Basis Function, Linear, Multilayer Perceptron, Polynomial, Splines, Fourier Series, etc. For further details on SVR, readers are directed to Wu et al.[23] and Londhe and Gavraskar (2018).

18.3.3 ARTIFICIAL NEURAL NETWORKS (ANNs)

ANN is a commanding data-driven technique which is widely used for numerical prediction and grouping. The ANN tool is also used to operate models of nonlinear systems, and in comparison to customary mathematical approaches, as inputs, it requires a small amount of data. Biological neural systems inspired artificial neural networks, which work on the basis of processing a set of data. Data implementation is done through neurons and their connections by an interconnected group of neurons present in neural networks, which are placed in layers adjacent to each other. ANN consists of one input and output layer and one or more hidden layers where weighted connections join the neurons of each layer to all neurons of the previous layer. The input vectors are received by neurons in the input layer, and the values can be carried across connections to the next layer of processing elements. In each hidden layer, the user determined the number of neurons. Up to the output layer, this process is being continued. Two-stage modeling of ANN is carried out with training and testing. After getting input data, the inputs are converted into desired outputs during the training stage, and connecting weights are also determined. In the testing stage, by using different datasets, the weights are observed. For further details on ANN, readers are directed to ASCE (2000), Bose and Linang (1990), Mair and Dandy (2000), Dawson and Wilby,[6] and Londhe (2008).

In the present study, these three techniques are used with the help of the Weka 3.9 platform, which is available freely through the given link, to develop data-driven models for the correlating stage at the mentioned three stations (https://waikato.github.io/weka-wiki/downloading_weka/).

18.4 METHODOLOGY AND MODEL DEVELOPMENT

As discussed earlier, the present study aims to correlate the water stage (water levels) measurement at 3 stations. For this, daily water stage data from 2000 to 2020 at three stations of the Godavari River, namely, Somanpalli, Mancherial, and Bhatpalli catchments, is used to develop the data-driven models. To find out the water stage at Bhatpalli station (output), previously measured values of the water stage at Somanpalli and Mancherial were used as input to the model. Table 18.2 represents the input–output combinations of each model to find out the stage (WLs) values at the Bhatpalli station for different time intervals. Consequently, Model 1 is developed to estimate water levels at Bhatpalli (B_t) at time "t" using water level values at the same time "t" at Somanpalli (S_t) and Mancherial (M_t) stations as inputs. Similarly, in Model 2, the previous day water level of the "t–1" time step at Somanpalli (S_{t-1}) and Mancherial (M_{t-1}) and the water level of the present day at time "t" at Somanpalli (S_t) and Mancherial (M_t) are used as inputs to give the output as today's stage values at Bhatpalli (B_t). Further, Model 3 is developed to forecast stage values one day in advance ("t+1") at Bhatpalli station using water level values at "t" at Somanpalli (S_t) and Mancherial (M_t) stations. Thus, Model 4 forecasts one-day ahead rainfall at Bhatpalli (B_{t+1}) using the inputs of previous day water level "t–1" at Somanpalli (S_{t-1}) and Mancherial (M_{t-1}) and water levels of present day at time "t" at Somanpalli (S_t) and Mancherial (M_t) stations.

TABLE 18.2 Model Development for the Bhatpalli Station.

MODELS	INPUT	OUTPUT
MODEL 1	S_t, M_t	B_t
MODEL 2	$S_t, M_t, S_{t-1}, M_{t-1}$	B_t
MODEL 3	S_t, M_t, B_t	B_{t+1}
MODEL 4	$S_{t-1}, M_{t-1}, S_t, M_t, B_t$	B_{t+1}

The first two models (Models 1 and 2) are developed to estimate the stage values at Bhatpalli station using the previously measured values of the stages at the nearby two stations. The next two models (nos. 3 and 4) are developed to forecast the stage value 24 h ahead in time (i.e., one day ahead forecast) at the Bhatpalli station using the same concept. Thus, in the present paper, a network of three stations is prepared to find out the stage value at one particular station using the stage values of the remaining two stations in that group of three stations.

TABLE 18.3 Combined Results of MT, SVR, and ANN.

Tools used	Model no.	Input	Output	r	MAE (in m)	RMSE (in m)	RAE (in %)
MT	Model 1	S_t, M_t	B_t	0.9771	1.0269	8.0879	6.0723
	Model 2	$S_t, M_t, S_{t-1}, M_{t-1}$	B_t	0.9765	1.0473	8.1672	6.1927
	Model 3	S_t, M_t, B_t	B_{t+1}	0.9723	1.006	8.784	5.9514
	Model 4	$S_{t-1}, M_{t-1}, S_t, M_t, B_t$	B_{t+1}	0.9675	1.0089	9.6045	6.0253
SVR	Model 1	S_t, M_t	B_t	0.9098	2.1507	15.8135	12.7124
	Model 2	$S_t, M_t, S_{t-1}, M_{t-1}$	B_t	0.9098	2.1488	15.8111	12.7062
	Model 3	S_t, M_t, B_t	B_{t+1}	0.9868	0.4162	6.094	2.4621
	Model 4	$S_{t-1}, M_{t-1}, S_t, M_t, B_t$	B_{t+1}	0.9672	0.7795	9.6511	4.6552
ANN	Model 1	S_t, M_t	B_t	0.9116	2.0242	15.6428	11.9694
	Model 2	$S_t, M_t, S_{t-1}, M_{t-1}$	B_t	0.9145	2.0995	15.3998	12.4149
	Model 3	S_t, M_t, B_t	B_{t+1}	0.9822	1.1522	7.0859	6.8159
	Model 4	$S_{t-1}, M_{t-1}, S_t, M_t, B_t$	B_{t+1}	0.9739	0.9766	8.5212	5.8321

The Model Tree (M5) algorithm, Support Vector Regression (SMO-reg), and Artificial Neural Network (Multilayer Perceptron) implemented in WEKA 3.9 software are used for calibrating and testing the models. The dataset was partitioned for training and testing purposes to assess the model's working accuracy and resilience. From the available data, 70% is used for training purposes, and the remaining 30% is used for testing purposes. An autosplit option (percentage split) provided in WEKA software is used for testing of models of all three data-driven techniques.

18.5 RESULTS AND DISCUSSION

As mentioned earlier, in the present study, a total of 12 data-driven models are developed to correlate the stage values of Bhatpalli station using the previously measured stage values of two nearby stations in the Telangana state in the Godavari River basin. All the developed models were tested with conspicuous data, and their performance is analyzed by traditional error measures like RMSE (Root Mean-Squared Error), MAE (Mean Absolute Error), and r (Correlation Coefficient) along with the histograms and scatter plots. The results of all these models are presented in this section of the paper along with the results of three data-driven techniques, viz., Model Trees, Support Vector Regression, and Artificial Neural Networks, for the four models presented in the above table. The aim of the study is to compare these three data-driven techniques for correlating stage measurements at three stations. All the models performed reasonably well in testing. The most effective tool is found out on the basis of the performance indicators, viz., Correlation Coefficient, Mean Absolute Error, Relative Absolute Error, and Root Mean Square Error. The estimated results show that the M5 model tree gives better performance for models 1 and 2. Similarly, SVR gives an accurate performance for Model 3 compared with MT and ANN. For Model 4, ANN is showing good performance. Comparing the overall results, it seems that MT is superior to the other two data-driven techniques, SVR and ANN. There seems to be an influence of variability in the data and data interval on the results of the developed models.

18.5.1 HISTOGRAMS FOR R AND RMSE

The histogram for correlation coefficient and root mean square error is showing a comparison of MT, SVR, and ANN for Model No. 1, Model No. 2, Model No. 3, and Model No. 4.

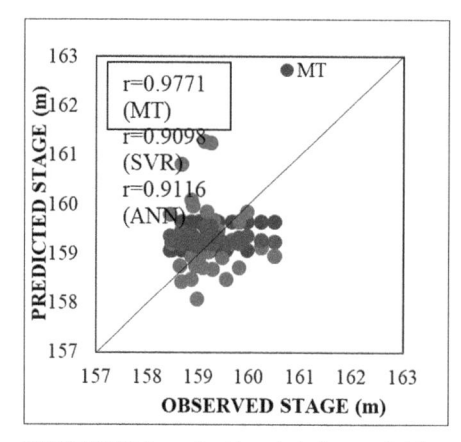

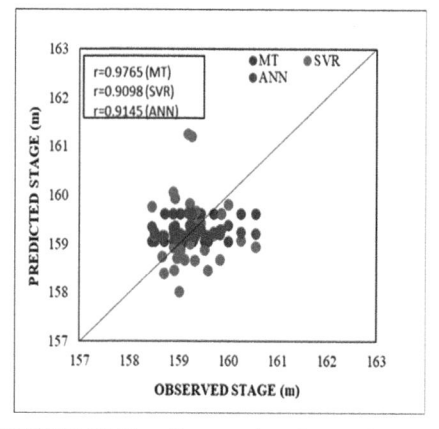

FIGURE 18.2a Scatter plots for model 1. **FIGURE 18.2b** Scatter plots for model 2.

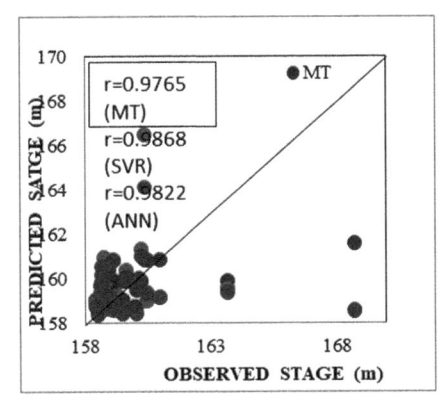

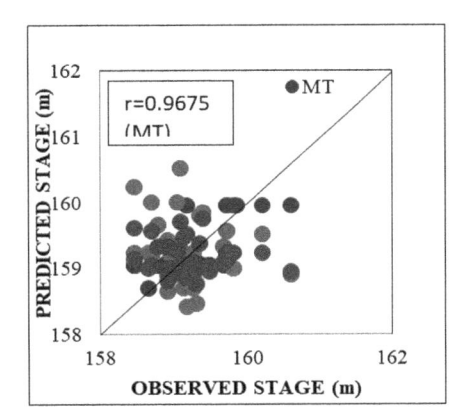

FIGURE 18.2c Scatter plots for model 3. **FIGURE 18.2d** Scatter plots for model 4.

The correlation coefficient is a normalized measurement of the coefficient of variation, and its result value always lies between −1 and 1. It is evident that the correlation coefficient decreases because of losses such as evaporation, seepage, friction, etc. The values of the correlation coefficient of the actual and predicted quantities are determined, which represent the relation between the practical and predicted stages of each model. Figure 18.2 shows a comparison of the relative mean square error for all four models, which indicates that the Model Tree is the best tool fitted for the data.

18.5.2 SCATTER PLOTS AND HYDROGRAPHS

Figure 18.3a–d shows scatter plots for Model 1, Model 2, Model 3, and Model 4, respectively. Data variation is being observed through these graphs. The point lying on the 45° line ($y = x$) represents the exact prediction of the data (Exact Prediction). The point lying below the 45° line ($y = x$) represents a predicted value less than the measured value (Underpredicted), and the point lying above the 45° line ($y = x$) represents a predicted value greater than the measured value (Overpredicted).

Figure 18.4a–d shown above represents the water stage graph for models 1, 2, 3, and 4, respectively.

The above water stage graphs show the variation of the measured and predicted water stages using data-driven techniques. The peak value shows the highest rainfall between the 20th and 24th days. The profile before and after the peak shows very little rainfall. From this, we can see the dimension-based and time-based variation of the water stage.

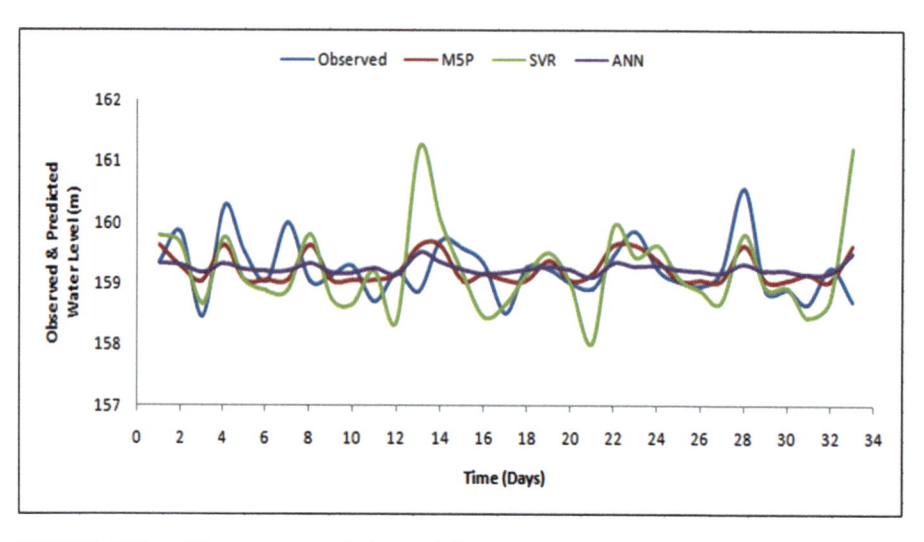

FIGURE 18.3a Water stage graph for model 1.

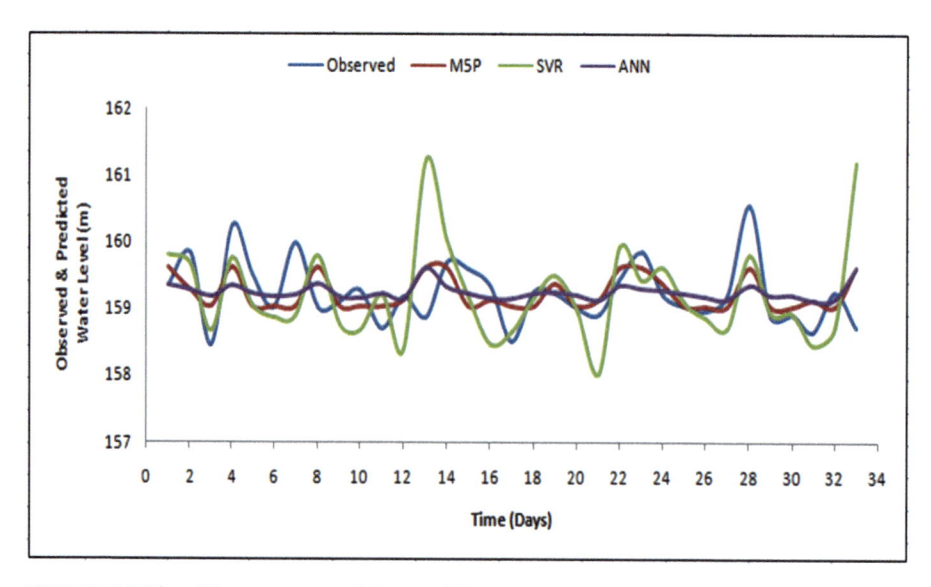

FIGURE 18.3b Water stage graph for model 2.

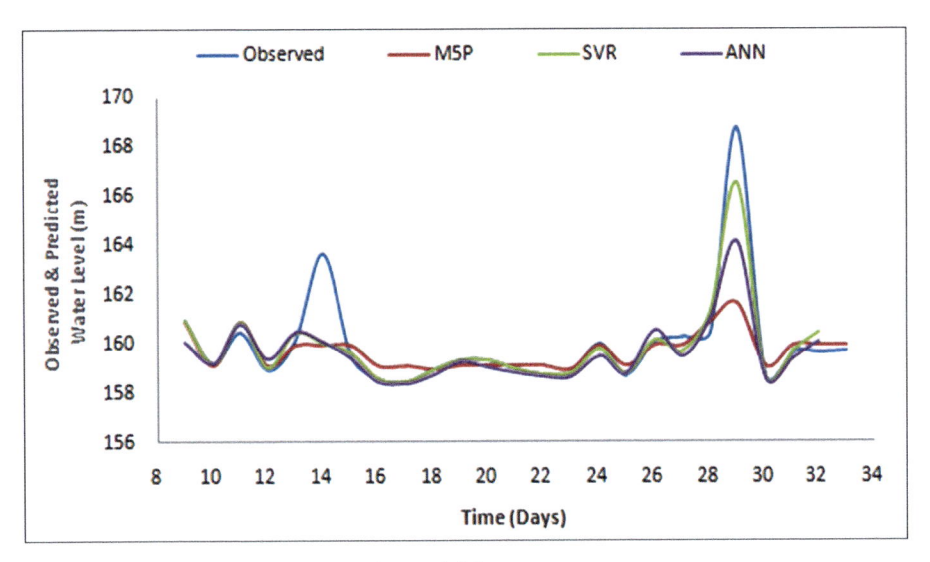

FIGURE 18.3c Water stage graph for model 3.

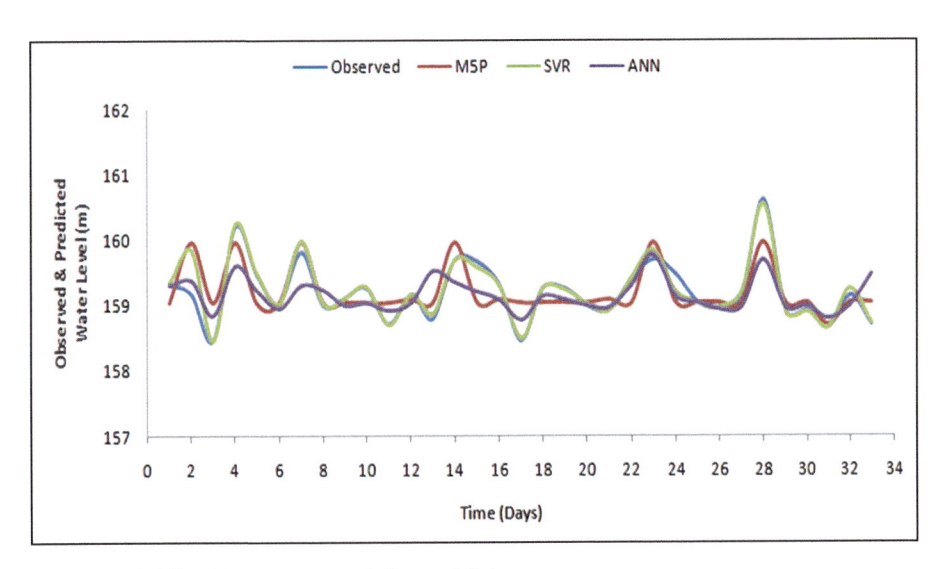

FIGURE 18.3d Water stage graph for model 4.

Note: For all models, a model tree is generated, but due to space limitations, it is not presented in this paper.

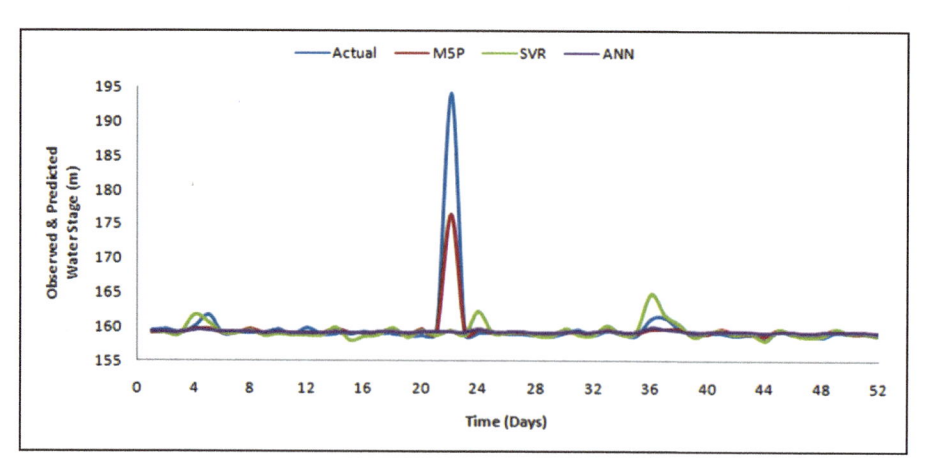

FIGURE 18.3e Water stage graph for model 1.

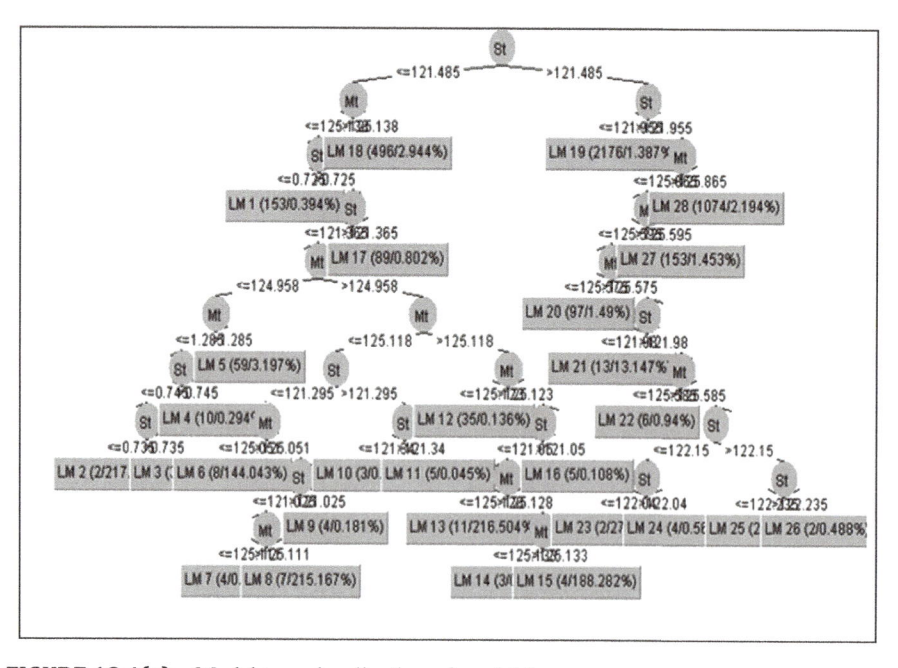

FIGURE 18.4(a) Model tree visualization of model 1.

ACKNOWLEDGMENT

The water level data is collected from the India Water Resource Information System (WRIS Platform). The authors are thankful to the Central

Warehousing Corporation (CWC), India, for providing access to download the data free of cost.

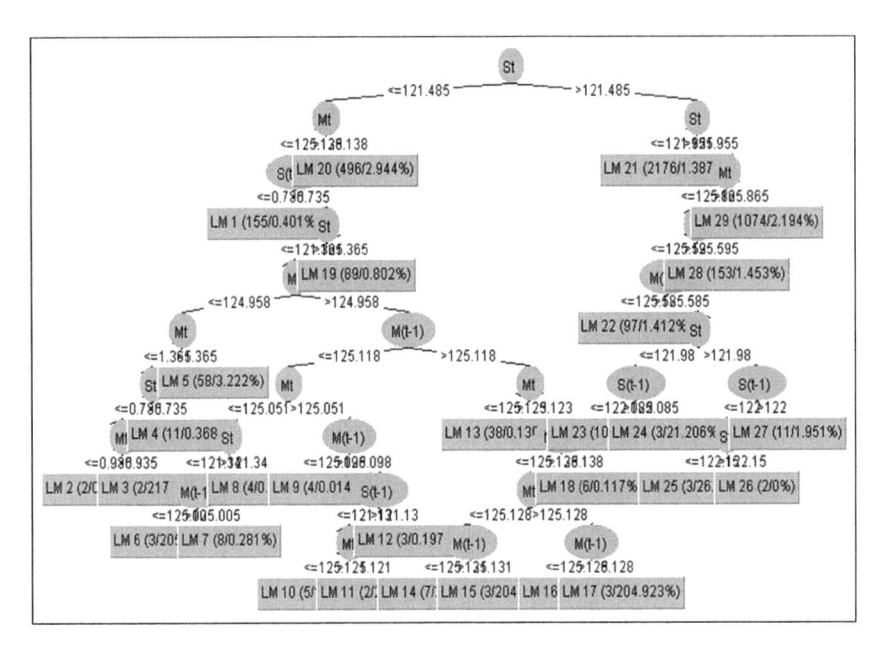

FIGURE 18.4(b) Model tree visualization of model 2.

KEYWORDS

- **floods**
- **model tree**
- **support vector regression**
- **artificial neural networks**
- **Water Resources Information System (WRIS)**

REFERENCES

1. Alipour, A., Yarahmadi, J.; Maryam Mahdavi, M. Comparative Study of M5 Model Tree and Artificial Neural Network in Estimating Reference Evapotranspiration using MODIS Products. *J. Climatol.* **2014,** *11,* A-ID 839205.

2. ASCE. *Hydrology Handbook*, 2nd ed.; American Society of Civil Engineers: New York, 1996.
3. Baldonado, M.; Chang, C.-C. K.; Gravano, L.; Paepcke, A. The Stanford Digital Library Meta data Architecture. *Int. J. Digit. Libr.* **1997**, *1*, 108–121.
4. Bhattacharya, B.; Solomatine, D. P. In *Neural Networks and M5 Model Trees in Modeling Water Level-Discharge Relationship for an Indian River*, Proceedings of the 11th European Symposium on Artificial Neural Network, Bruges, Belgium, d-side, Evere Belgium, Verleysen, M., Eds.; 2005, pp 407–412.
5. Bhattacharya, B.; Price, R. K. Solomatine, D. P. Machine Learning Approach to Modeling Sediment Transport. *J. Hydraul. Eng.* **2007**, *133* (4).
6. Dawson, C. W.; Wilby, R. L. Hydrological Modelling using Artificial Neural Networks. *Prog. Phys. Geogr.* **2001**, *25* (1), 80–108.
7. Dixit, P.; Londhe, S.; Dandawate, Y. Removing Prediction Lag in Wave Height Forecasting using Neuro- Wavelet Modelling Technique. *Ocean Eng.* **2015**, *93*, 74–83.
8. Dixit and Londhe. Prediction of Extreme Wave Heights using Neuro Wavelet Technique. *Appl. Ocean Res.* **2016**, *58*, 241–252.
9. Dixit, P.; Dmanane, P.; Kolhe, P.; Londhe. L.; Kulkarni, P. In *Estimation and Prediction of Precipitation using M5 Model Tree for Koyana Reservoir Catchment, Maharashtra, India, Conference*, National Conference on Water Resources and Flood Management with special reference to Flood Modelling, 2021.
10. Subramanya, K. *Engineering Hydrology*, 3rd ed.; Tata McGraw-Hill Publishing: New Delhi, 2008.
11. Maier, H. R.; Dandy, G, C. Neural Networks for Prediction and Forecasting of Water Resources Variable. *Environ. Model. Softw.* **2000**, *15*, 101–124.
12. Jain, P.; Deo, M. C. *Neural Networks in Ocean Engineering*.
13. Quinlan, J. R. In *Learning with Continuous Classes*, Proc. AI'92 (Fifth Australian Joint Conf. on Artificial Intelligence), Adams, A., Sterling, L., Eds.; World Scientific: Singapore, 1992, pp 343–348.
14. Sihag, P.; Karimi, M. S.; Angelaki, A. Random Forest, M5P and Regression Analysis to Estimate the Feld Unsaturated Hydraulic Conductivity. *Appl. Water Sci.* **2019**, *9*, 129.
15. Sattari, M.; Pal, M.; Apaydin, H.; Ozturk, F. M5 Model Tree Application in Daily River Flow Forecasting in Sohu Stream. *Turkey Water Resour.* **2013**, *40*, 233–242.
16. Londhe, S. N.; Dixit, P. R. Forecasting Stream Flow by using Model Trees. *Int. J. Earth Sci. Eng.* **2011**, *04*, (No 06 SPL), 282–285.
17. Londhe, S.; Charhate, S. Comparison of Data-Driven Modelling Techniques for River Flow Forecasting. *Hydrol. Sci. J.* **2010**, *55* (7), 1163–1174.
18. Rajasekaran, S.; Gayatri, S.; Lee, T. L. Supoort Vector Regression Methodology Fpr Storm Surge Predictions; 2008.
19. Solomatine, D. P.; Yunpeng, X. M5 Model Trees and Neural Networks: Application to Flood Forecasting in the Upper Reach of the Huai River in China, *J. Hydrol. Eng.* @ ASCE/Nov/Dec 2004.
20. Solomatine, D. P.; Dulal, K. Model Tree as an Alternative to Neural Network in Rainfall-Runoff Modeling, *Hydrol. Sci. J.* **2003**, *48* (3), 399–411.
21. Solomatine, P.; Maskey, M.; Durga Lal Shrestha, D. L. Instance-based Learning Compared to other Data-Driven Methods in Hydrological Forecasting, *Hydrol. Process* **2008**, *22*, 275–287.

22. Teegavarapu, R. S. V.; Chandramouli, V. Improved Weighting Methods, Deterministic and Stochastic Data-Driven Models for Estimation of Missing Precipitation Records. *J. Hydrol.* **2005,** *312,* 191–206.
23. Wu, C. L.; Li, Y. S. River Stage Prediction based on a Distributed Support Vector Regression by Proposed a Novel Distributed SVR Model to Carry out the Forecast using Data Collected from Water Level Series. 2008.
24. Website. https://indiawris.gov.in
25. Website. https://cwc.gov.in

Index

A

Absolute Percentage Error, 156
Additive manufacturing (AM), 77
  3D-printing manufacturers, 79, 80
  materials and methods
    design requirements, 80–81
    processes, 81–82
  results and discussion
    3D printing of hole features, 82–83
    design for, 84–86
    hole features, 83
    simple hole feature, 83–84
    STL File, 86–87
  technique, 78
Airfoil model, 117
  AEROFOIL 2—CLARK Y, 121, 122
  AEROFOIL 1—NACA 4412, 121
  Bernoulli's principle, 118
  CFD simulation, 119, 120
  manufacturing, 127–128
  NACA 4412 and CLARK Y, 122–123
  parameters of wings, 118
  prototype building, 126–127
  solidworks simulation, 118
  terminology, 118
  wing structure, 124–125
  winglet, 125–126
Al7075-B4C MMC, 61
Analytical formulation
  displacement amplitude, 221
  free-surface displacement, 222
  natural frequencies, 221
ANSYS FLUENT, 220
Ant lion algorithm
  design of experiments (DoE)
    response surface method (RSM), 43
  methodology, 41
    cathodic, 42
    characterization, 41
    cutting forces, 42–43
    spectroscopic analysis, 42
  modeling mathamatically
    quadratic capacity, 44
    regression equation, 45
    response surface methodology (RSM), 43
  optimization
    building trap, 50
    creating prey, 51
    elitism, 51–52
    inspiration, 47–49
    random walks, 49
    rebuilding pit, 51
    results, 52
    sliding ants, 50
    trapping in, 50
  results and discussion
    cutting force, 45–46
    TiAlN/TiSiN, 45
    stainless steel (SS), 40
    passivation film, 41
Artificial neural network (ANN), 260
Average response spectrum for analysis
  COSMOS earthquake data, 200–211
  numerical model
    SAP 2000 software, 213, 214
    SAP software, 212, 213
  response spectrum, 211, 212
  results and discussion
    base shear, 214–215
    displacement of floors, 215–216
  SEISMO-signal software, 211

B

Buckling analysis
  analytically estimated, 169–170
  linear buckling analysis, 170
  non-linear buckling analysis, 170–172
Buckling-Restrained-Braced Frames (BRBFs), 181
  earthquake records, 183
  results and discussion
    base shear, 189

hysteresis loop, 192–193
story displacement, 190
story drift, 191–192
steel buildings, modeling
nonlinear time-history analysis, 185–188
structural system, 184
Building trap, 50

C

Cabinet drier (CD), 245
Central composite design (CCD) matrix, 65
Complex modulus model, 137
Computational fluid dynamics
analysis
ANSYS FLUENT, 222
Pressure-Implicit with Splitting of Operators (PISO), 223
VOF method, 223
Constrained layer-damped (CLD), 144
COSMOS earthquake data, 200–211
Creating prey, 51
Crop
disease detection, 150
yield prediction
algorithm, 149–150
Cutting force, 45–46
Cutting forces, 42–43

D

3D printing
of hole features, 82–83
manufacturers, 79, 80
Design of experiments (DoE)
response surface method (RSM), 43
Design requirements, 80–81
Direct frequency response, 141–142
Direct solar dryer, 247
Double-pass solar drier (DPSD), 245
Drying chamber, 246

E

Electric discharge machining (EDM)
Al7075-B4C MMC, 61
EN19 alloy steel, 62
experimental details
NC Z-axis control, 62

performance measures, 63, 64
Response Surface Methodology (RSM) design, 64
Gray's relational analysis, 58
multicriteria decision-making (MCDM), 60
regression model, 60
results and discussion
central composite design (CCD) matrix, 65
material removal rate, 65, 67–69
mathematical models, 64
surface roughness (RA), 71–74
tool wear rate (TWR), 69–71
spindle speed, 59
TOPSIS and COPRAS algorithms, 59
TOPSIS method, 61
EN19 alloy steel, 57, 62
Energy-Dissipating Devices (EDDs), 181
Epoxidized soyabean oil (ESBO), 25
experimental methods
characterization, 28–29
materials, 28
sample preparation, 28
results and discussion
crystallization and melting properties, 30–33
interfacial and distribution characteristics, 29–30
mechanical properties, 34
x-ray diffraction (XRD), 26

F

Finite element analysis (FEM)
design parameters, 102–103
geometrical modeling, 101–102
Flood detection
artificial neural network (ANN), 260
data-driven techniques, 261
methodology and model development, 264–265
results and discussion
histogram for, 266–267
hydrographs, 267–269
model tree, 269
models, 265–266
scatter plots, 267–269
study area and data
stations, 261

Index 293

tools used, 262
 artificial neural networks (ANNS), 263
 M5 model tree (MT), 263
 support vector regression (SVR), 263–264
Fly ash
 experimental work
  alkaline activators, 231
  and GGBS, 232
  mix design, 232
 geopolymer, 228
 literature review, 228
  longer relieving occasions, 229
  testing, 230
 methodology, 231
 objective, 230–231
 results and discussion
  crushed and natural sand, 236
  mechanical properties, 232–234
  microstructural properties, 235
Free-surface displacement, 222
Frequency responses for equivalent stress (von Mises), 110

G

Grape samples, 247
Gray's relational analysis, 58

H

Harmonic response for equivalent strain (von Mises), 110–111
Harmonic response for equivalent stress (von Mises), 108
Hydrostatic pressure, 173
 critical buckling load, 177
 FEA and GNA analysis, 176
 internal, effect, 174–175
 plot, 176
Hysteresis loop, 192–193

I

InceptionV3 model, 159, 160
Indirect forced convection solar dryer (IFCSD), 246
Indirect type solar dryer (ITSD), 245

K

Kelvin-Voigot model, 138–139
K-fold method, 158

L

Lift-to-drag ratios, 117

M

M5 model tree (MT), 263
Machine learning (ML), 148
 crop disease detection, 150
 crop yield prediction
  algorithm, 149–150
  dataset, 151
  methodology, 150
  modeling, 154
   Absolute Percentage Error, 156
   crop and season, 156
   disease detection, 160
   InceptionV3 model, 159, 160
   K-fold method, 158
   prediction accuracy, 157
   Random Forest Regression, 158
  results and discussion
   visualization techniques, 151–154
Material removal rate, 65, 67–69
Mathematical models, 64
Maximum ellipse axis, 135
Maxwell model, 138
Modal strain energy method, 142–143
Model analysis, 104–106
Model tree, 269
Moisture content (MC)
 removed from
  grapes, 255
  onions, 255
  turmeric, 255
Multicriteria decision-making (MCDM), 60
Multi-story reinforced cement concrete
 circular geometry analysis
  model description, 200–201
 rectangular geometry analysis
  model description, 197
  problem statement, 196
 results and discussion
  Story Displacement EQX, 202–204, 205
 square geometry analysis
  model description, 198–200

N

Nanoclay-reinforced jute-polyester

experimental methods
  laminates, 7–9
  materials, 6–7
  specimens, 9
flexural tests, 10–11
mechanical testing, 9
results and discussion
  flexural stress–strain curves, 19–20
  scanning electron microscope (SEM), 11–13
  tensile test results, 13–17
steel wire mesh (SWM), 6
tensile specimens, 10
NC Z-axis control, 62
No-load test, 248, 251, 252
Numerical analysis, 143
  constrained layer-damped (CLD), 144
  NASTRAN, 144
Numerical model
  SAP 2000 software, 213, 214
  SAP software, 212, 213
Numerical models
  complex Eigen value, 140–141
  direct frequency response, 141–142
  modal strain energy method, 142–143

O

Onions, 247–248
Optimization
  building trap, 50
  creating prey, 51
  elitism, 51–52
  inspiration, 47–49
  random walks, 49
  rebuilding pit, 51
  results, 52
  sliding ants, 50
  trapping in, 50

P

Phase lag, 136
Pranhita sub-basin, 275
Pressure-Implicit with Splitting of Operators (PISO), 223

Q

Quadratic capacity, 44

R

Random Forest Regression, 158
Random walks, 49
Raw turmeric, 248
Rebuilding pit, 51
Regression equation, 45
Regression model, 60
Response spectrum, 211, 212
Response surface methodology (RSM), 43
  design, 64

S

Seaforam, 26
  polypropylene and POE, 27
SEISMO-signal software, 211
Shear transformation, 135
Sliding ants, 50
Sloshing
  analytical formulation
    displacement amplitude, 221
    free-surface displacement, 222
    natural frequencies, 221
  ANSYS FLUENT, 220
  computational fluid dynamics analysis
    ANSYS FLUENT, 222
    Pressure-Implicit with Splitting of Operators (PISO), 223
    VOF method, 223
  results, 224
Solar collector, 246
Solar collector orientation test, 248
Solar dryer
  cabinet drier (CD), 245
  double-pass solar drier (DPSD), 245
  experimental methods and materials
    direct solar dryer, 247
    drying chamber, 246
    grape samples, 247
    onions, 247–248
    raw turmeric, 248
    solar collector, 246
  indirect forced convection solar dryer (IFCSD), 246
  indirect type solar dryer (ITSD), 245
  results and discussion
    moisture content (MC), 253
    no-load test, 248, 251, 252
    solar collector orientation test, 248

Index

Spectroscopic analysis, 42
Square geometry analysis
   model description, 198–200
Stainless steel (SS), 40
   passivation film, 41
Standard linear model, 139–140
Steel buildings
   modeling
      nonlinear time-history analysis, 185–188
      structural system, 184
Steel Moment-Resisting Frame (SMRF), 181
Steel wire mesh (SWM), 6
Story Displacement EQX, 202–204
Story displacement, 190
Story drift, 191–192
Strain–time history, 136
Stress–strain relation, 135
Support Vector Machine (SVM), 149
Support vector regression (SVR), 263–264
Surface roughness (RA), 71–74

T

Tensile specimens, 10
Thin-walled structures
   analysis model, 169
   buckling analysis
      analytically estimated, 169–170
      linear buckling analysis, 170
      non-linear buckling analysis, 170–172
   cylindrical shells, 168
   hydrostatic pressure, 173
      critical buckling load, 177
      FEA and GNA analysis, 176
      internal, effect, 174–175
      plot, 176
Three data-driven techniques
   methodology and development
      models, 280–281
   results and discussion
      histogram, 282–283
      and hydrographs, 283–284
      models, testing, 281–282
      scatter plots, 283–284
   study area and details, 276
   techniques used, 276
      artificial neural networks (ANNS), 279
      M5 model trees (MTS), 277–278
      pruning, 278

      smoothing, 278
      support vector regression (SVR), 278–279
Tool wear rate (TWR), 69–71
Total deformation, 107
Trapping, 50

U

Ultrasonic horn
   design of, 98–100
Ultrasonic vibration assisted turning
   ANSYS®
      finite element analysis (FEA), 95
   element type
      boundary conditions, 104
      MESH, generation of, 103
   FEM investigation, 96
   finite element analysis (FEM)
      design parameters, 102–103
      geometrical modeling, 101–102
   harmonic analysis, 97
   modal analysis, 97
   results and discussion
      frequency responses for equivalent stress (von Mises), 110
      harmonic analysis, 106–107
      harmonic response for equivalent strain (von Mises), 110–111
      harmonic response for equivalent stress (von Mises), 108
      model analysis, 104–106
      total deformation, 107
   ultrasonic horn
      design of, 98–100
Unmanned aerial vehicle
   airfoil model, 117
      AEROFOIL 2—CLARK Y, 121, 122
      AEROFOIL 1—NACA 4412, 121
      Bernoulli's principle, 118
      CFD simulation, 119, 120
      manufacturing, 127–128
      NACA 4412 and CLARK Y, 122–123
      parameters of wings, 118
      prototype building, 126–127
      solidworks simulation, 118
      terminology, 118
      wing structure, 124–125
      winglet, 125–126
   CAD models, 117

electronic selection
  electronic connections, 129
  MICRO-UAV, 128–129
final prototype, 130
lift-to-drag ratios, 117

V

Viscoelastic material
  analytical models
    classical models, 137
    Kelvin-Voigot model, 138–139
    Maxwell model, 138
    standard linear model, 139–140
  frequency domain behavior
    complex modulus model, 137
    phase lag, 136
    strain–time history, 136
  hysteresis behavior
    maximum ellipse axis, 135
    shear transformation, 135
    stress–strain relation, 135
  numerical analysis, 143
    constrained layer-damped (CLD), 144
    NASTRAN, 144
  numerical models
    complex Eigen value, 140–141
    direct frequency response, 141–142
    modal strain energy method, 142–143

W

Water Resource Information System (WRIS Platform), 286
Water stage graph, 284–285

X

X-ray diffraction (XRD), 26